The Moulton Bicycle

ALSO BY BRUCE D. EPPERSON
AND FROM MCFARLAND

*Roads Through the Everglades: The Building
of the Ingraham Highway, the Tamiami Trail
and Conners Highway, 1914–1931* (2016)

*Bicycles in American Highway Planning:
The Critical Years of Policy-Making, 1969–1991* (2015)

*Peddling Bicycles to America:
The Rise of an Industry* (2010)

The Moulton Bicycle

*A History of the
Innovative Compact Design*

BRUCE D. EPPERSON

McFarland & Company, Inc., Publishers
Jefferson, North Carolina

LIBRARY OF CONGRESS CATALOGUING-IN-PUBLICATION DATA

Names: Epperson, Bruce D., 1957– author.
Title: The Moulton bicycle : a history of the innovative
compact design / Bruce D. Epperson.
Description: Jefferson, North Carolina : McFarland & Company, Inc.,
Publishers, 2018 | Includes bibliographical references and index.
Identifiers: LCCN 2018018394 | ISBN 9781476673257 (softcover :
acid free paper) ∞
Subjects: LCSH: Moulton bicycles—History. | Moulton, Alexander. |
Banham, Reyner.
Classification: LCC TL437.5.M68 E67 2018 | DDC 629.227/2—dc23
LC record available at https://lccn.loc.gov/2018018394

BRITISH LIBRARY CATALOGUING DATA ARE AVAILABLE

ISBN (print) 978-1-4766-7325-7
ISBN (ebook) 978-1-4766-3240-7

© 2018 Bruce D. Epperson. All rights reserved

*No part of this book may be reproduced or transmitted in any form
or by any means, electronic or mechanical, including photocopying
or recording, or by any information storage and retrieval system,
without permission in writing from the publisher.*

Front cover image is of Alex Moulton in front of The Hall,
at the time of the Raleigh buy-out, 1967, photographer unknown
(The Moulton Bicycle Co Ltd)

Printed in the United States of America

*McFarland & Company, Inc., Publishers
Box 611, Jefferson, North Carolina 28640*
www.mcfarlandpub.com

Table of Contents

Introduction 1

1. A Bicycle Standard of Living 15
2. But Today We Collect Ads 29
3. Cycling in a New Key 44
4. Who Killed Roger Rabbit's Moulton? 73
5. Really, What Makes a Bike? 89
6. History Repeats Itself, Once More 102
7. Alternative Wheels 121
8. Vanished into the Clouds 140
9. Yesterday's Tomorrow Is Not Today 158
10. Clip-On; Plug-In; Burn-Out 173

Chapter Notes 189
Bibliography 209
Index 219

Introduction

Banham invented the immediate future.—Robert Maxwell (1967)[1]

Promises of the future are invariably jumbled reflections of the present. To name an exhibit *This Is Tomorrow,* as we did in 1956, is either ironic or intentionally inexact.... Yesterday's tomorrow is not today.
—Lawrence Alloway (1990)[2]

I

This is, superficially at least, the story of two men and a bicycle. In November 1962 an inventor and reluctant industrialist named Alex Moulton introduced at London's Earl's Court Cycle Show a radical new small-wheeled, dual-suspension bicycle he had been developing for five years. A few months later his fellow Englishman Peter Reyner Banham, an editor at *Architectural Review,* wrote about it in the general-interest magazine *New Statesman.* It was one of the first articles to appear outside of the cycle trades or business press. In the following months, Banham followed its progress in several other magazines, including the *Architects' Journal, Living Arts,* and *Design.* Partially as a result of this coverage, the Moulton quickly became a high-end consumer novelty despite its price, about twenty percent higher than similarly equipped conventional bicycles. Eventually, about 100,000 were produced before Raleigh, Britain's largest cyclemaker, bought out the Moulton Bicycle Company. Afterwards, Raleigh made another 55,000 or so before production ended in 1974.

But this is not a superficial "fanzine" story, not by any means. Banham had received a doctorate in art history in 1959 from the prestigious Courtauld Institute of Art under the famed architectural historian Nikolaus Pevsner. Banham's doctorial thesis, published in 1960 under the title *Theory and Design in the First Machine Age,* remained in print for over thirty years. Pevsner also helped him get the *Architectural Review* job. But independently of his work as a critic and historian, Banham was busy carving out, along with a group of young artists, architects and designers, an entirely new intellectual discipline that would later be called "design studies" or "material culture" studies. It left buildings to the architects and museum art to the establishment critics and looked instead to the basic "stuff" of everyday life: tools, appliances, entertainment, infrastructure, packaging and advertising. Impressed by such middlebrow, general-interest magazines as *Popular Mechanics* and *Astounding Science Fiction*, Banham wanted to make design studies more accessible than either art or architectural criticism, so he split off his writings on material studies from his more scholarly architectural material, putting it in short articles and regular columns he prepared for mainstream magazines like *New Statesman*, *The Listener* and *The Times Literary Supplement* that were still typically a notch or two above the mass-circulation periodicals of the era.

Over the course of a forty-year career, he published over 750 articles. This in addition to fourteen books that he wrote, co-wrote or edited. And newly discovered articles, mostly pseudonymous newspaper reviews of theatre and art exhibits from the 1940s, are still being uncovered.[3] His friend, the urban planner and historian Peter Hall, himself the author of two seminal histories, observed that "all his writing had the immediacy, the vitality, the correctness of the best journalism." He once asked, "if other academics had come to research and teaching that way, would not academic writing be immeasurably better than it is, on average, and would not the academy be an infinitely richer and more interesting place?" But after pondering his own question, decided: "but perhaps not; the world produces very few Banhams."[4]

He had grown up in Norwich, which he automatically equated with bicycling ("anybody who knows Norwich knows that the cyclist is still the king of the road there"), and after the war he regularly used to cycle around his hometown, then in London, at least until his banged-up old roadster was stolen in the early 1950s. He was so impressed

Reyner Banham preparing photos for one of his books at the University of California at Santa Cruz in 1980. Banham was professor of architectural history at UC-SC from 1980 to 1988 (photograph by Carol A. Foote, Special Collections, University Library, University of California Santa Cruz, MS0259-Neg-bk-5808507.tiff).

with the Moultons he saw at the Earl's Court show that he soon bought one. Emily King, who at that time was a curator at the Victoria and Albert Museum, recalls Banham riding back-and-forth to the museum in South Kensington from his office in Gower Street in 1972 as the two worked on a proposed industrial design exhibit. In his writings, the Moulton bicycle started to take on a life of its own as a metaphor for the displacement of a staid, traditional, pre-war British working class by a new generation of young, upwardly mobile, trend-seeking consumers. One article, "The Atavism of the Short-Distance Mini Cyclist," is still considered an avatar of the emerging '60s pop culture, and it continues to be reprinted, some twenty-five years after Banham's untimely death.[5]

On the other hand, Alex Moulton was born into affluence, educated at Marlborough College, then Cambridge. An engineer all his life, he was a researcher and inventor, not an industrialist, and while his bicycle was still in the development phase, he tried to convince Raleigh to manufacture and market it. After an initial flurry of interest, they it turned it down, largely because of their own precarious financial condition. The domestic market for bicycles in the UK had shrunk from just over a million units in 1950 to about half that by 1961. Raleigh was in a somewhat better position than the others because it had swallowed up so many of its rivals, but its production for the domestic market was still essentially flat, increasing from 380,000 to only 410,000 during the same eleven years. But it had greatly expanded its Nottingham factory at the same time it acquired its main competitors in the mistaken belief that Britain's postwar economic expansion would bring dramatically improved bicycle sales. That excess capacity was now sitting dark, silent and empty. "Riding bicycles," Banham once said, was one of "those long term, permanent working class things ... only, of course, the working class don't ride bicycles any more."[6]

So after being turned away by Raleigh, Moulton built his own small factory on the grounds of the family estate at Bradford-on-Avon and started selling bicycles, often through high street department stores. To Raleigh's chagrin, while the working class may have given up their old black three-speeds in favor of Morris Minors, Ford Cortinas, and Austin 1100s, they did take to the Moulton in the same way they took to transistor radios, portable record players, and other newly affordable "lifestyle" accessories. By 1965, forty thousand had been sold, so many that Moulton had to outsource production to Ludlow and Fisher (later Fisher-Bendix, a maker of large household appliances), near Liverpool. Dumbfounded, Raleigh hired a coterie of young, college-educated marketing specialists, mostly from the tobacco industry, and poured a small fortune into research to find out what was going on.[7]

At the time, this kind of data-driven research was almost unheard of in England; after all, the last of the postwar restrictions on such basic commodities as coal, milk, eggs, and sugar had only been lifted in 1954. What Raleigh's marketing people discovered, to their amazement, was women. They found the Moulton attractive. Its lack of a top tube and step-through frame was far more comfortable and practical than the diamond-frame roadster, and they liked shopping for it in department stores instead of cycle shops. Bicycle shops were dark, dirty places where they were patronized and leered at. "The Depression spawned a whole generation of cycle retailers who were good at repairing, but not good at selling," complained one Raleigh sales executive. Buying a Moulton in a high street department store was like buying any other household fashion accessory: a Braun mixer, Olivetti portable typewriter or Eames chair. So Raleigh made women their new target audience.[8]

Raleigh's response was to introduce its own fashionable small-wheeled bicycle, the RSW 16, which it trumpeted in an expensive introductory marketing campaign. The only problem was that while it looked a lot like a Moulton, it didn't work like one, and while it did make a splash, and Raleigh considered it a success, its shelf life was only a few years. Its real advantage was that from an engineering perspective, it was something like the incredible plastic bicycle. It could be stretched this way and pulled that way and twisted around like a pretzel to come up with almost anything the marketing people needed.

At one point the American market was awash with something called the Stingray, a dragster-imitating bicycle that sold like hotcakes to teenagers. Raleigh twisted the RSW 16 around to come up with a swoopy thing called the Chopper. Cycling purists hated it, deriding it as a toy. Banham, by now working in the States, loved it, at the same time fully agreeing with its critics. It *was* a toy. But more importantly, it succeeded because it was a kid's toy that copied the *adult* toys kids saw in action around them every day: hot rods, chopped motorcycles, dune buggies, racing karts.[9]

One of Raleigh's competitors introduced a mid-size utility bicycle in the mid–1960s that was boring as dishwater, but sold steadily. Raleigh again worked over the RSW 16 to come up with a bicycle called the Twenty, after its 20-inch wheels. Boring it may have been, but the Twenty was also a marvelously practical bicycle, well designed and built, and Raleigh eventually sold hundreds of thousands of them in the United Kingdom, North America, and the antipodes. Like the Thermos bottle, the Twenty so dominated its field that the brand name of its most popular variant became the generic name for all 20-inch utility bikes: the Shopper.[10]

Forty years later, when the third generation of Alex Moulton's bicycles had inspired other firms to develop innovative, well-built, high-performance (and quite expensive) small wheeled bicycles, amateur shade-tree bicycle mechanics lacking the $2000 to $10,000 needed to buy one turned to the lowly, ubiquitous Raleigh Twenty to provide their "alternative wheels." Radically altered and fitted, sometimes at great effort, with lightweight racing bike components, these hopped-up Twentys (note the spelling) could approach the performance of the new generation compact bicycles at a fraction of their cost.

Years after his passing, two of Banham's most cherished visions, technological innovation as the destroyer of tired, obsolete social conventions; and flexible, adaptive, affordable technology as the great social equalizer, had merged into one. Back in 1963, the original Moulton had, in his words, freed the cyclist "of the cloth-cap and racing pigeon culture of which cycling is (was?) an integral part," and thirty years later the Raleigh Twenty had leveled the playing field, allowing one to affordably shoulder their way into the high-tech "funny bikes" game Alex Moulton had single-handedly created—without having to cough up the requisite four- or five-figure check, provided that you were willing to suffer a little grease under the nails.[11]

So while this may superficially be the story of two men and a bicycle, it is about a great many other things as well. It is about technology, and a time in British history when it was acceptable, even fashionable, to think about technology optimistically, affectionately. "Technology was an obsession in an almost romantic way," recalls Banham's son Ben, "for him it was about optimism, the future and immense possibilities, the astonishing capabilities of man." Banham's colleague Peter Hall recalled that "he really believed, as many thinking people of his generation did, but few of their children seem to do, that technology was a liberating and beneficial force."[12]

Alex Moulton felt the same way: "there was a dynamicism in the creative world about that time, everything was very restricted during the war—we were very conscious of having won the war. We felt we were able to make new things ... so we had a confidence in the manufacturing and engineering world that we could do things."[13]

Moreover, this is a story about the emergence of popular culture, and about the social and economic changes that made this new culture possible. One of these changes was the technological dynamism just referred to. Often forgotten is that a lot of the technology that worked its way into the consumer markets of the 1960s was a byproduct of cold war defense spending and the nuclear arms race. Solid state electronics, specialty plastics and composite materials, digital computing, graphical user interfaces and the internet all originated from military research.

"He was never naïve about it and a lot of other people were," recalls Banham's wife, Mary. "He started as an aero-engineer, so he knew." Banham himself recalled, during the much more skeptical 1970s, that "We were not just simple technological optimists. Technology was horrors as well as wonders. We had seen too many napalmed bodies being drug out of foxholes in Korea. It was too close to Hiroshima. The promise of technology was not only gratification and pleasure, but destruction and mutilation. We were part of the first generation that had to try to adapt to the atom bomb. But for all that, technology was still, for us, the promise of a better future."[14]

A second change was in the economic sector. The late 1950s and '60s was a time of tremendous change in Britain's material condition. Wartime scarcities had continued almost unabated until 1949, with some rationing and price controls extending up to 1954. But after the Korean War, the resources that had been needed for defense obligations and national recovery were freed to create rising personal incomes. In 1950 less than five percent of British households had a television set. In 1955, that figure was 40 percent, and by 1961, 79 percent.[15] "Pop was the product of the era of affluence and consumerism experienced by western industrial society in the 1950s and 1960s," notes Banham's student Nigel Whiteley. "An incontestable fact of the 1950s is that private affluence substantially increased for the vast majority ... everyone was encouraged to buy and be a good consumer ... [such] changes were bound to have an effect on tastes."[16]

Richard Hamilton, who participated with Banham in the loose-knit "Independent Group" at the Institute for Contemporary Arts in London during the 1950s, defined "Pop" as "popular; transient; expendable; low cost; mass produced; young; witty; sexy; gimmicky; glamorous; and Big Business." The success of the Moulton bicycle was largely based on tapping into just this image. Movie stars such as Peter Sellers, Julie Christie and Goldie Hawn were photographed with their Moultons; stage and television star Eleanor Bron actually toured around Europe on hers.[17] Minister of Parliament Quintin Hogg was frequently photographed riding past the Houses of Parliament in a Savile Row suit. In America, the author J. D. Salinger bought two so he and his wife could take their kids around New York's Central Park during its car-free weekends. *The* iconic

Moulton photo from the period shows one of the early production machines parked in Carnaby Street, surrounded by three *Vogue* models in tweed outfits.

Similarly, Raleigh put the RSW into the hands of every photogenic celebrity who would take one. "To help place our bicycles in the leisure-society picture we need the assistance of the trend-setters," explained Raleigh's domestic marketing director, Peter Seales, "these are the people who can set the lead with the RSW 16, and ten of thousands of people will be willing to follow."[18]

But creating such a "design shift" for a new, high-tech bicycle was a highly problematic endeavor. The problem was that the typical bicycle consumer over the last twenty-five years had been a utility rider, Banham's "cloth cap and racing pigeon" man, or as one unnamed Raleigh executive described them more precisely (if not somewhat cynically), "C2-D-E males of 40 plus."[19]

The Independent Group had looked to the United States, already in the throes its consumer revolution, for inspiration. But America had no bicycle culture to offer—if anything, it had long ago snuffed out what little remained of its pre-war adult sport cycling clubs. As British households became wealthier through the 1960s, they emulated the Americans and skipped over high-tech bicycles entirely and went straight to automobiles. Banham held out the hope that a new class of "urban radical cyclists," affluent, young, hip and socially conscious, would make up the Moulton cycling vanguard. His hopes were dashed. Then, with the great American bike boom of 1969–73, he held out renewed hope of a pop-culture-inspired transoceanic design shift. Again he was disappointed, as the bicycle of the moment became a slight re-working of the vintage European road racing machine.

His high-technology "alternative wheels" vision was sharply at odds with the simple-is-better ethos of the emerging environmental movement, placing him, for the first time, outside the vanguard. He complained about "the failure of the seemingly radical bicycle to link up with political radicalism," but resigned himself to the fact that "a movement which is increasingly wary of advanced technology—if not downright Luddite—is bound to be suspicious of bikes that use square tubes or space-age alloys."[20]

This brings us to the last theme to be explored in the following pages, the risks of making bold predictions, of getting it wrong, and the attendant dangers of unexpected consequences. In 1967, Raleigh bought out the Moulton Bicycle Company. Three years later, Raleigh introduced the Raleigh-Moulton Mk.III, which Banham called "probably the best (certainly the best made) Moulton ever." However, Raleigh offered it in only a single variant, using low-grade carbon steel tubing and an off-the shelf component group. It was like designing a new sports car, then giving it the tires, wheels and suspension from a standard econobox. It was also poorly marketed. It sold only a few thousand units and in 1974 Raleigh discontinued it. Alex Moulton and Raleigh soon parted ways.[21]

In 1979, speaking before the Royal Society for the Encouragement of the Arts, Manufacturers and Commerce, Alex Moulton recalled being told by a Raleigh executive that the Moulton bicycle's enduring significance was that "it opened up the floodgates of new design possibilities for bicycles." Moulton agreed, but complained that some of these "new design possibilities" had not "always been to my taste," a thinly veiled reference to the RSW series, which he believed had been intended from the start to kill off his small operation. "Alas for prophesy, as they say," rued Banham about this time, "'Moulton' never quite made it into the pantheon of legendary technological names." He was premature. Indeed, fellow architecture critic Robert Maxwell mused: "one wonders if Banham takes seriously the proposition of refuatability," a polite way of asking if Banham really cared if any of his bold Whoosh-Zoom-Wham-Bang predictions ever hit their mark, or if he was just interested in their rhetorical impact.[22]

For a year or so after they parted ways, Moulton and Raleigh engaged in a low-grade skirmish over naming rights and the ownership of patents. It was eventually settled, but as a result Alex Moulton determined that the best way to avoid further problems was to circumvent the traditional bicycle industry entirely. He would do this by pursuing a strategy of uncompromisingly high design and quality. He would keep all the manufacturing in-house at Bradford-on-Avon and would limit output to whatever the small shop could comfortably handle. Moreover, he would straightforwardly ask whatever price was necessary to economically support his standards of design and construction. In 1983, he introduced a space-frame, small wheeled, full-suspension bicycle, the AM, that many consider the best bicycle ever built. Ever. Now in its third generation, about 200 a year are fabricated at the Moulton estate, almost all to custom order, with prices running as high as $17,000. Since the recent death of Alex Moulton, the firm has been acquired by another specialty bicycle maker, Pashley Bicycles, which makes less expensive standardized variants in larger numbers at its factory in Stratford-on-Avon. The premier and custom bicycles continue to be built at Bradford-on-Avon.

The spaceframe Moultons have spawned a new market for innovative new designs that range all the way from break-apart full-sized bikes that otherwise look and function like a regular cycle, to complex, multi-jointed, micro-wheeled folders that can fit into a small suitcase or golf bag. The one thing they all have in common is that they are expensive. Not as expensive as an AM, but $1000 to $3000 is typical.

Banham had been intensely interested in hot-rod culture even before he moved to the United States in 1976. "Peter and other particularly working-class boys had taken a great interest in American magazines," recalled Mary Banham. "They acquired a great respect for American know-how, and it was mostly technical … it went into Peter's interest in dream cars of the '50s." His contribution to the Independent Group's lecture series in the early 1950s had been "Vehicles of Desire," a study of Detroit tail-fin, bumper-bomb styling. The point wasn't to ask if a Cadillac's tailfins were in good or bad taste; it was that car styling, like toaster styling or chair styling, or the styling or any other consumer good, had to be evaluated on its own merits, not using the old, established gold standard of architecture.[23]

As the insolent auto design of the '50s and early '60s gave way to the smaller, plainer, more international look after the oil embargoes of 1973 and 1979, auto technology drifted out of Banham's oeuvre, only to later re-emerge through hot-rod culture. It came up in a roundabout way, to illustrate an emerging trend combining technology and architecture. Back in 1958 he had discussed the idea of "clip-on" technology, where a basic platform (in this case, sci-fi robots) can be quickly and easily changed by pulling off one set of tools and clipping on another. By the mid–1960s he had merged that concept into the "plug-in" architecture appearing in an underground architecture magazine called *Archigram*. "Plug-in or clip-on, it's the same magpie world of keen artifacts, knock-out visuals and dazzling brainwaves assembled into structures whose primary aim seems to be to defy gravity," he explained.[24]

It was easier (and less incredible) to illustrate the idea of "clip-on" or "plug-in" modularity in the case of racing cars than buildings (or robot assassins, for that matter). Noting that even the most sophisticated formula one racing teams were starting to buy an increasing proportion of their cars, such as brakes, gearboxes and even entire engines from suppliers, he pointed out that American hot-rodders had been doing this for decades. They gradually built up reliable models such as the Model T or Chevrolet Bel-Air V8 (or later, the Volkswagen Beetle) starting from stock through to street-legal boulevardiers to full-out competition vehicles using standardized parts found in thick mail-order catalogs published by J. C. Whitney and Warshawsky's. He believed this incremental "clip-on" culture would eventually make the pure, designed-from-the-ground-up racing car obsolete. "We aren't all endowed with absolute originality," he noted, and buying your speed incrementally, from a catalog, "enables you to concentrate on your areas of talent and get the rest done by experts. You can make up as much of an original lifestyle as you want and conform where you feel the need."[25]

In the face of the generation of precision-designed, precision built, very expensive compact and folding bicycles, a group of what were essentially bicycle hot-rodders started to convert cheap, readily available existing low-tech utility-bikes into performance machines by retrofitting them with alloy parts meant for racing bikes. The platform they chose for their chopping and channeling and converting was the lowly, ubiquitous Raleigh Twenty, the low-visibility variant of the RSW 16 reluctantly introduced by Raleigh in 1967–68 to counter the success of a competing product, the Dawes Kingpin.[26]

Introduced in 1964 as a bland utility bike, a roadster for the new apartment-dwelling postwar generation, the Kingpin sold steadily until Raleigh could no longer ignore it. Although they weren't excited about it, with their prior RSW experience, the Twenty emerged as the best of the 20-inch wheeled Shoppers. The shade-tree mechanics took decades-old Twentys, substituted racing-bike handlebars, stems, and seats; even brakes, cranks and derailleur gears. Sometimes they could be swapped out as-is, at other times they had to be heavily modified.

What qualified them for Banham's "clip-on" categorization was their combination of price and quality: by the mid–1980s Japanese firms had covered the market with high-quality, relatively inexpensive racing bike components, some of them direct knock-offs of classic premium Italian and French equipment. Butchering a pricey Campagnolo crank in an attempt to fit it into the Twenty's proprietary-threaded bottom bracket would be a catastrophe; doing the same with its virtually identical SR (Sakae Ringyo) copy, merely an inconvenience.

So Moulton did end up making it into the pantheon on legendary technical names, but not in the way Banham thought he would. Instead, his became incised on the same wall as the famous, old-world, high-art mechanical engineers, joining men such as Lancia, Daimler, Maybach, Rolls and Delarge. On the other hand, the bicycle that the original Moulton indirectly gave rise to, the unloved, ugly duckling Raleigh Twenty, proved to be the compact bicycle for the masses, and like the Model T or the original Beetle, a cheap, rugged, reliable platform that young (almost always) men with time and dedication could turn into something that could challenge, even if it couldn't beat, the best of the high-art engineers.

II

The only thing Alex Moulton and Reyner Banham had in common, other than Moulton's bicycle, is aeroengines. Moulton's engineering studies at King's College, Cambridge, were interrupted by the start of World War II, and he began work in the test shop of the engine division at Bristol Aeroplane. Similarly, Reyner Banham left Norwich in 1939 to enter an engineering management course jointly run by Bristol Technical College and the same Bristol Aeroplane firm. He too, worked in the engine division. "I never really thought of doing anything else but engineering even when I was at school … all the Banhams before me were technology men," he later told an interviewer. Neither man ever mentioned the other when recounting their experiences at Bristol, so it is unlikely they ever met. After the war started, Banham moved into the factory full-time. About the same time, Moulton was appointed personal assistant to Roy Fedden, Bristol's head of engineering. After the war, Moulton worked briefly with Fedden on an unsuccessful automobile project, then moved back to Bradford-on-Avon to join the family business, the George Spencer, Moulton & Co. rubber works.[27]

Moulton began work on the bicycle in 1958 after developing, with the British automobile designer Alec Issigonis, a suspension system for the Morris and Austin Mini. It had no moving parts; opposing rubber cones seated in metal cups provided all the springing and damping the car needed. The technology was adapted from Moulton's Flexitor springs, developed for automotive trailers, then railroad rolling stock. Although it appeared simple, the Flexitor, which twisted rubber like a torsion bar, depended on a rubber-to-metal bond so solid that that it could be strained in sheer up to several thousand pounds per square inch. The Flexitor had taken years to perfect. The bicycle, in turn, used a coil spring and a rubber bumper buried in the head tube for the front suspension, and a rubber-to-rubber bumper in back. This dual suspension system made possible the bike's essential element: its wheels were only sixteen inches in diameter. This, in turn allowed the top tube to be eliminated and replaced with a "step through" frame. The front and rear cargo racks were also so much lower that, when fully loaded, the bike's center of gravity was nearly a foot closer to the ground than for a conventional diamond-framed machine.[28]

One of the great mysteries is why Banham and the Moulton bicycle have become so intertwined. A literature search indicates that he wrote about mini-cycles in general, or the Moulton specifically, exactly seven times between 1963 and 1982.[29] One of these is the essay, "The Atavism of the Short-Distance Mini-Cyclist" from 1964, a wide-ranging critique of pop culture, or to be more precise, a biting critique of its abandonment by the avant-garde of the period. It was accompanied by *the* iconic photograph of Banham cycling in Queen Anne's Gate near the Architectural Press offices.[30] A colleague, Gillian Naylor, in a 1997 memorial lecture, describes it:

> Here he is in the 1960s a hero of modern life–1960s modern life of course, not 1970s "on yer bike" modern life. And man and machine are in motion: he's in command of what is now a suitable sub-

ject for a design historical case study, the Moulton bike. He's wearing what seems to be a dress suit, not the ubiquitous duffle coat. He hasn't got bicycle clips, but he's wearing a cloth cap ... a toffish cloth cap as far as I can see; his shoes are shining and his beard is patriarchal. And he's obviously in control there, and making a complex and male-oriented statement about design, class, mobility and modern life.[31]

But according to Banham's own recollection, his cycling exploits were a lot less heroic than Naylor gave him credit for: "The most illuminating comment on this came from that south of the Thames social critic R. R. Langford, who said to me when he heard the news, 'is that your bike outside I take it?' And I said 'yes.' He looked at me and said, 'cycling, that's a bit atavistic, isn't it?' The point went home because I have had one or two accusations of atavism recently ... but this is regarded as both atavism and (apparently) showing off. But the thing remains true, the working class is where I come from."[32]

And this is what makes Banham's identification with the Moulton so special. Banham's 1960s cultural criticism was grounded in his 1930s upbringing in provincial Norwich. He was not raised on a cultural diet of the British Museum and the British Tate, but of American comic books, Saturday morning cinema, and, later, Buicks and Cadillacs. The culture of consumption, optimism, and future-thinking became his intellectual bread-and-butter. Over this was laid the traditional academic modernism of the Courtauld Institute, the *Architectural Review*, and his mentor at both, Nikolaus Pevsner. As Penny Sparke notes:

armed with two seemingly conflicting value systems informing the twentieth century's material culture, from the 1950s onwards Banham's self-appointed role—especially through his membership of the Independent Group (IG) was to resolve the tension between those two systems and, in the words of his fellow IG member, John McCale, to learn to love 'both Bach and the Beatles.[33]

Alex Moulton riding an F-frame Moulton Bicycle, Bradford-on-Avon, c. 1964 (courtesy Alex Moulton Bicycle Co., Ltd.).

Unlike Banham, Alex Moulton grew up surrounded by wealth. He was born in The Hall, built in 1599 for John Hall and purchased by the Moulton family in 1848, along with the adjacent textile mill, which was converted to the production of one of the marvels of the era, vulcanized rubber. "We children lived in the still-Victorian household of our grandparents," he later recalled. By the time he entered prep school at Marlborough College "I had no doubt that I wanted to be a designing engineer." But he later recalled that his most enduring lesson was learned at Bristol Aeroplane. "Bristol was famous for its radial air-cooled aero-engines—the Pegasus, Mercury, Hercules, and Centaurus," he later explained to an interviewer. Nevertheless, the differences between Bristol's designs and those of the other famous manufacturers, Rolls-Royce, Pratt & Whitney, Curtiss-Wright, and BMW "was idiosyncratic," that is, incremental. In one fell swoop, between 1944 and 1949, the development of the turbojet rendered all this marvelous technology obsolete. The revolutionary nature of this change—a "jump change" in Alex Moulton's lexicon—made an enormous impression on the young engineer.

"What is the nature of events that causes a whole species [of technology] to become doomed to extinction?" Moulton mused. "Leaving aside change for fashion's sake, the thrust behind an invention can be traced to one man's driving interest in realizing his imagined jump of improvement."[34] For Moulton, such wide-ranging technological leaps were not cultural or economic phenomenon, they were personality-driven, emanating from the imagination of a single mind.[35]

Thus, the stacking of contrast upon contrast. Moulton the man of privilege, of private school and university, and Banham, the self-professed "bob-ender" ("in the days when a bob-ender meant a certain class of persons"[36]); Moulton, the heroic individualist, and Banham, if not a socialist, then at least a progressive; and most of all, the evolution of Moulton's bicycle itself, originally Banham's symbol of a radical new consumerism, seemingly beaten by the system, only to reemerge as a new form of Parthenon-fronted Platonic engineering ideal, a two wheeled Rolls or Bugatti, the antithesis of everything Banham stood for, replaced by hacksawed Raleigh Twentys, the former products of the evil empire. And lastly, the empire itself now rendered obsolete, literally non-existent, its factory torn down to make way for a housing development, its name owned by a foreign venture capitalist, its bicycles shipped in containers from China.

III

A book such as this that tells its story by scanning a broad horizon of technical and cultural change by necessity relies on the cooperation and active assistance of professionals and academic specialists from a number of different fields to provide accuracy and depth. Nobody can simultaneously be an expert in bicycle history, cycle technology, architecture, town and transport planning, and British economic history. Foremost among those who have taken an active interest in this project, and have given generously of their time, is Tony Hadland of Oxfordshire, the leading historian of the Moulton Bicycle Company, and, more generally, the British bicycle industry.

After I published a small paper describing the connection between Reyner Banham, the Moulton bicycle, and the later Raleigh Twenty "hot-rod" culture, I shelved the idea of expanding the project because it overlapped so much with Tony's earlier work on the history of Sturmey-Archer and the Moulton bicycle, then after 2000, his equally detailed work on Raleigh Industries. However, it was Tony himself who urged me to continue. He provided me with insights into the personality and methods of Alex Moulton and eventually made the necessary introductions to John Macnaughtan and Dan

Farrell of the Alex Moulton Bicycles, Ltd., that permitted a visit the Moulton factory and The Hall estate in Bradford-on-Avon. (Tony, being an old and familiar friend of the firm, was allowed to act as my Beatrice in exploring much of the facility.) I was also allowed to tour The Hall and see the famous Oak Room, Alex Moulton's office and studio.

It was apparent that while John Macnaughtan was terribly pressed for time, he still gave me several hours to answer questions, point out important prototypes (in storage since Alex Moulton's death) and talk about his time as a senior executive with Sturmey-Archer. Similarly, Dan Farrell answered my technical questions and was able to talk about Alex Moulton the man, having worked for him many years. Dan was also generous enough to give me a copy of Alex Moulton's *The Moulton Formulae and Methods*, a semi-privately published adaptation of his own handwritten deskbook of engineering equations, together with graphical explanations of how to apply them. To me it is, aside from one of the original bicycles, the most direct connection one can have to the mind of Alex Moulton.

A participant racing a modified Raleigh Twenty in the "Tin Can 10" club time trial in the English Midlands, 2002. "Tin Can" is a euphemism for a hub gear; and the Tin Can 10 was traditionally limited to bicycles with 3-, 4- or 5-speeds (photograph by Tony Hadland).

During this same trip to England, I was also shown the same level of hospitality at Brompton Bicycle Ltd.'s factory at Greenfield, west of London, by Nick Charlier, Brompton's former public relations and communications executive. What I found most interesting at Brompton was the people on the shop floor more than any exotic machinery or high-tech processes, and I think you will agree this comes out in the photos taken in Brompton's new factory that appear in Chapter 8. It is proof of what can be possible when a clean, bright factory, modern industrial engineering and production ergonomics intersect with a product that the employees really believe in.

In regards to the history of Raleigh Twenty "hot-rodding" I would like to thank John S. Allen and the late Sheldon Brown, whose work continues posthumously on in his website Sheldonbrown.com, maintained by his friends, including Mr. Allen. Mr. Brown's website was also an important source for Raleigh in America retail catalogs from the late 1960s and early 1970s, which are much different from the United Kingdom

Banham cycling his Moulton on Carterets Street in Westminster, just around the corner from his office at the Architectural Press, 1963 (Royal Institute of British Architects, RIBA 8487).

retail catalogs. I also thank Sheldon Brown's wife, Dr. Harriet Fell, who was kind enough to grant permission to reproduce images from her late husband's collection. Mr. Allen was also an important source of information his in his own right, having first written about his converted Twenty in the magazine *Bicycling* in 1981. Alas, that first Twenty, pictured in Mr. Allen's 1981 book *The Complete Book of Bicycle Commuting* now rests in that great bike rack in the sky, after having seen some thirty-plus years of use and one or two misadventures.

The Veteran-Cycle Club of Britain is making a heroic effort to collect, catalog, and make available what may end up being several hundred oral history audio- and videotapes, as well as historic commercial recordings, such as British Pathè movie trailers of the Earls Court Cycle show, business shorts, and noteworthy cyclesport events. Tony Hadland migrated the many interviews he and John Pinkerton collected over the years to this site. I used several of these in the preparation of this book, including those with

David Duffield, Jack Lauterwasser, John Macnaughtan, Vic Nicholson, and John Woodburn.

The paper was initially presented at the 22nd International Cycle History Conference, May 2011 in Paris. The hospitality of our hosts, the Centre D'Histoire Des Techniques et d l'Environnment, was, and is, much appreciated. In particular, I thank Nicholas Oddy, who championed unsuccessfully for the initial publication of an article taken from an early draft manuscript of this book. (It was published two years later it in a new journal, *Mobilities*.) Other participants at the ICHC and T2M (the Society for Transport, Traffic and Mobility) conferences who have helped by providing support, timely critiques or information include Peter Cox, Martin Emanuel, Evan Friss, Zack Furness, James Longhurst, Robert McCullough, Ruth Oldenziel, and Manuel Stoffers.

As always, the most thanks to Nora, who somehow manages to get this most reluctant tourist back and forth across the ocean in one piece, on the right train, to the inevitably needed chemist's shop, and generally keeps the academic show on the road.

1

A Bicycle Standard of Living

"Whereas most historians work to dislodge past events from the dust which has already immobilized and which might efface them, Banham seems to prefer events which are still in motion, on which the dust has not settled."
—Robert Maxwell (1981)[1]

"He would have trampled on his grandmother to snuggle up to a passing trend."
—Jonathan Meades (2014)[2]

I

"The working class is where I come from," Banham often pointed point out. "I don't know that I'm particularly proud of this working class, cloth-cap bit in my background, but I think it gives me the right to speak on certain subjects." His friend Robert Maxwell[3] once noted this readiness to turn such ostensibly modest origins into a personal iconography of "a Banham who, in spite of being an academic, does not wish to be associated in any way with academicism." Otherwise, he did tend to be reticent about the details of his family life.[4]

Peter Reyner Banham (in private life, he was always Peter, in print almost always Reyner) was born in Norwich, about 75 miles northeast of London, in 1922. Other than a passing remark that "all the Banhams before me were technology men," he never discussed his parents. His father, Percy Banham, born in 1899, was manager of the Norwich Gasworks, as was his grandfather Charles. Banham's mother, Violet, born in 1896, was a housewife. Percy and Violet were married in Hampstead in 1921, a year before Peter was born. Violet's maiden name was Reyner, which is how Banham got his middle name. He had one brother, and according to 1930's census records, a sister, although his wife, Mary, never spoke of her.[5]

They lived in what Banham once described as a pre-war "spec-builder-suburban garden" home. This was 25 Larkman Lane, a 1930s detached home on a leafy road then well west of the city. The "spec-builder" comment was a clear, if discreet, message that they did *not* live in the Earlham council houses, which he later called "some of the most abysmal local authority housing put up in Britain between the wars." In 1939, he recalled watching as "the field where we had dug caves and defense-works, potted larks with air-guns and horsed around generally was rapidly covered with houses full of slum-cleared families whose younger offspring answered to such calls as 'Come yew haer, gal Gloria, do else I'll lump ya one!'"[6]

Hardly the sympathetic voice of a fellow proletarian, it suggests out that he grew up at least one or two classes above "gal Gloria." "His father was works manager for the Norwich Gas Works," recalled Mary. "He got a scholarship to the Cathederal School [King Edward IV Grammar School] which was the grammar school there, very traditional, Latin and Greek based, which he never regretted. His brother went to the big boy's grammar school out in town, but he went to the Cathederal School. And then they had this scholarship from Bristol Aircraft for boys in public schools, and at that time it was—all cathederal schools were—considered public schools." John Hewish, a contemporary of Banham's at Bristol Aeroplane, explained that the award actually came from the Society of British Aircraft Constructors, and that it gave the recipient his choice of firms to study with. In Banham's case this was a joint program run by the Bristol Technical College and the Bristol Aeroplane Company."[7]

In any event, whether his family was working-class or lower-middle class, or even solidly middle-class, it is really irrelevant to his point, which is that he and his mates were "all brought up in the Pop belt somewhere. American films and magazines were the only live culture we knew ... we returned to Pop in the early fifties like Behans going to Dublin."[8]

For a young man, typical, not affluent, but neither poor, nor delinquent or openly rebellious, growing up in Norwich "the culture, the live culture in which we were involved, was American pulps, things like *Mechanix Illustrated* and the comic books (we were all great Betty Boop fans), and the penny pictures on Saturday mornings."[9] Banham's wife Mary, recalled that the phenomenon later known as "pop" germinated in the British working classes a decade or more *before* the war through American technical, popular science, and science fiction magazines:

> People like Richard Hamilton and Peter and others, particularly working class boys, had taken a great interest in American magazines, American ads, all the way back through the 1930s, their growing-up period. It had been, I don't know if you could call it an influence, but they acquired a great respect for American know-how. And that was mostly technical. That's what the Americans were good at, at that time. And it went on into Peter's interest in dream cars of the '50s, and a lot into Richard Hamilton's paintings. So it was an ongoing thing, it wasn't a sudden interest.[10]

As we shall see, Alex Moulton, who was very definitely not from the working class, would also recall the influence these heavily illustrated technology magazines had on his decision to become an engineer.

Mary Banham noted that "all his childhood he made model aeroplanes, that's why he was so keen on aeroplanes—his whole family was, his father had been in the air force during the First World War, and his brother was too, and they were dead keen on aeroplanes, anything with engines in it."[11]

Banham moved to Bristol in 1939. The students at Bristol's mechanical engineering program alternated between the classroom and the shop floor. However, when World War II broke out in September, the instruction program was discontinued and the students were moved to the factory floor and test-sheds full-time as fitters (i.e. skilled technicians).¹² He applied to the RAF but was rejected for poor eyesight. Besides, he was needed for reserve duty at Bristol. Wartime work for the overstretched technicians was, in some ways, no less arduous than conscription, and finally took its toll: "well, you can't go on doing 24-hour shifts one after another forever and just before the end of the war I was invalided out." He moved back to Norwich.¹³

During his time at Bristol, he had begun reading seriously about architecture. He recalled waiting to catch a bus, only to miss it because he was so engrossed in Nikolaus Pevsner's just-published *Outline of European Architecture*. "I can still see the back of that blasted bus as it pulled away, graven in my mind's eye as a marker for the moment when I became an architectural historian."¹⁴

Back home, he made ends meet through a series of odd jobs, including spending "six memorable months" as the master of a "remand house," a residential shelter for juveniles. It was there that he really got his education about the "D-E" bottom end of the English class structure. He recalled one of his charges telling him how he had stood up in court "to give his mother the farewell she deserved: 'Yew only want me outa the house so yew can goo off with that bleedn' Yank.'" According to Banham, it was true. That was how a lot of kids got kicked out. Many estate households acquired "transatlantic hangers-on" who assumed the roles of the men who were away, missing or dead.¹⁵

Banham's theorized that was why Britain's postwar consumer society first emerged—a full decade early—at the *bottom* end of British society. It was the result of what he called "PX affluence": cast-off T-shirts, packs of Luckies, comic books, sacks of sugar. This beneficence "fell at random and with maximum psychological and social disturbance on an uprooted working class for whom rationing was just a continuation of undernourishment by other means."¹⁶

II

Two weeks before the German surrender, Mollie Panter-Downes, writing in her "Letter from London" column for the *New Yorker*, cautioned that "some of the responsibilities of the future are already in sight. The Government warned months ago that the feeding and clothing of the people in the liberated countries would pose a strain so severe that it would have to be shared by all Allied stomachs and backs. That strain, it seems, has begun." Rice and powered milk, which had never been rationed during the war, suddenly disappeared from store shelves. "Many people who are not at all scaremongers think that this may be just the beginning, for this densely populated island, of a post-war period which could make the war years seem, in retrospect, almost comfortable."¹⁷ She was right. Six months later, she informed America that "the factories, which people hoped would be changing over to the production of goods for the shabby, short-of-everything home consumers, are instead to produce goods for export." She predicted that "the Government will have to face up to the job of convincing the country that controls and hardships are as necessarily a part of a bankrupt peace as they were of a desperate war."¹⁸

John Lehmann, editor of the magazine *Penguin New Writing*, recalled that:

Night raid on London by German bombers, December 1, 1940. The dome of St. Paul's Cathedral is lit by the fires from burning buildings surrounding it. Virtually the entire square mile to the east and north of the church, including Aldersgate and the Barbican, was wiped out (International News Service photograph/Library of Congress).

> The adrenaline [of the war] was no longer being pumped into our veins. We endured with misery and loathing the continual fuel cuts, the rooms public and private in which we shivered in our exhausted overcoats, while the snow blizzards swept through the country again and yet again. Were there to be no fruits of victory? The rationing cards and coupons that still had to be presented for almost everything from eggs to minute pieces of scraggly Argentine meat, from petrol to bed-linen and "economy" suits, seemed far more squalid and unjust than during the war.[19]

After the ravages of the 1930s depression, followed by the war, followed by this torpid peace, the concept of a consumer society was inconceivable to most Brits. In 1947, only 68,700 households had a television set, with wait lists averaging a year or longer. The *Daily Express* derisively called the first post-war Earl's Court Motor Show in October 1948 "The Biggest 'Please-do-not-Touch' Exhibition of All Time." Thirty-two British manufacturers exhibited more than fifty different models. All except the most expensive had wait lists of at least a year; for the new, small, low-priced Morris Minor, introduced at the show, the list grew from zero to thirty months in five days.[20]

There were many causes for the crisis: an overvalued pound; a brutally harsh winter of 1946–47 that shut industry down for three weeks in February 1947 for lack of coal; unrealistically high peacetime defense budgets (£6 million in 1938, £207 million in 1947); a refusal to acknowledge the loss of captive imperial markets; obsolete domestic infrastructure (especially in energy and transport); intransigent unions; rampant black marketeering; a frozen social structure. But above all, debt. The war had cost Britain a quarter of her national wealth, about 7.3 billion pounds. Less than half of this was accounted for by the destruction of physical assets such as housing and factories; £4.2 billion had gone to pay for supplies purchased from abroad, mostly from America, mostly before 1940. Lend-lease had ended that drain, but debt to other nations still increased from £0.5 billion to £3.4 billion between 1939 and 1945.[21]

The war had thrust Britain's industrial system onto center stage, and it had not fared well under the harsh glare of the public spotlight. A 1942 investigation by the *Times* indicated that arms output was 40 percent below the maximum feasible. The public opinion research organization Mass Observation reported that "All is not well in our our war-production at present. Something is seriously wrong somewhere. Different people point the accusing finger at different wrong points.... The amount of pointing, especially in all types of printed matter, adds up to a veritable forest of inked fingers."[22]

On the other hand, the war economy *had* irrevocably started to flatten out Britain's notoriously skewed wage and salary structure. The war had pulled three million men out of factories and into uniform. Despite an influx of 1.7 million women, the total workforce—at a time output increased by 40 percent—shrank from 18 to 16.7 million. The estimated real "unemployment rate" was a *negative* seven percent. As a result, wages—especially at the bottom—lifted upwards, closing in with those of semi-skilled and skilled operatives. In 1935 a manual laborer in an aero-engine factory or an auto plant made 70 percent of the wages of a skilled fitter or body-maker; in 1949 he made between 85 and 86 percent. Moreover, administrative, technical and clerical employees increased from 13.5 percent of the nation's workforce in 1935 to 18.6 percent 1948.[23]

Even so, the disparity in postwar Britain between blue-collar workers, no matter how skilled, and the professions would have been startling to an American: the average English professional in 1949 earned as much as one skilled operative, one semi-skilled worker, and one common laborer put together.[24] In addition, this flattening was not uniform across the board. The gap didn't close for women and boys, for example. In the bicycle industry, for boys and youths—who earned approximately *one third* of adult men—the gap was actually greater in 1947 then in 1938.[25] While the numbers of administrative, technical and clerical employees nationwide had grown 6.1 percent during 1935–48, the number of common laborers had decreased by only 5.9 percent during a longer period, 1931–51. A sizeable number of the middle class had not climbed up from below, but fallen down from above: professionals and proprietors.[26]

Even when they held their station, the professions after the war were heavily impacted by the new income tax scheme needed to pay off debt and rebuild the nation. It levied a standard rate of 45 percent on a much broader cross-section of the population: by 1949 twelve million individuals were paying into Inland Revenue as opposed to four million in 1938.[27] Overall, the changes in the wartime British economy could be less accurately described as "a rising tide that lifts all boats," than a bell-shaped curve being drawn in towards the middle. Historian Alan Marwick comments that "heavy taxation, rationing, a positive nutrition policy and higher wartime wages had produced a leveling of standards—up as well as down."[28]

As Banham recalled, this "leveling out by leveling down" was a disconcerting social change for the middle class. "The immediate postwar period was a baffling and deeply frustrating time for most British citizens. There was greater scarcity and deprivation than during the war itself. While some areas had been hit hard, the nation's infrastructure and housing were, as a whole, intact. Yet, the nation was hit by a succession of economic blows that left it in a state of seemingly permanent austerity."[29]

The lower-middle and middle classes felt these blows far worse than the working classes, who were used to doing without, and who had ingratiated themselves more skillfully into the shadow economy created by the American "occupation." It was very difficult for the establishment to realize that Britain now faced a new and different postwar world. Its government leaders still looked in the mirror and saw *Great* Britain. For example, they tried to maintain the pound as a reserve currency for the commonwealth, but no longer had the dominating gross domestic product to discipline their former colonies. The Bank of England, the reserve bank for the commonwealth, found itself saddled with responsibility without sovereignty, a situation much like the European Central Bank finds itself in today with the euro.

On September 18, 1949, the government suddenly devalued the pound from $4.03 to $2.80. Treasury minister Edwin Plowdin's recollection of his meeting with Foreign Secretary Ernest Bevin and Chancellor Stafford Cripps indicates just how lost these Whitehall dinosaurs were in the world of modern monetary economics:

> There were two rates put forward, $2.80 and $3.00, and I think the majority of us felt that $2.80 was the right rate.... Stafford was there and his view was that $3.00 was the right rate and we argued for the lower rate. Ernie then turned to me and said "What effect will this have on the price of the standard loaf of bread?" Fortunately, thinking he would ask this, I'd sent a cable to the Treasury asking. It was a penny. We put that forward and he said, "Oh all right, but I hope we can have a whiter loaf. It makes me belch, this stuff." So it wasn't the $2.80 argument that was decided, it was the price of bread that decided it.[30]

On the other hand, Oxford fellow J. M. Mogey pointed out that the Labour Party leaders were making their decisions in a vacuum largely because they had no data to work with. Social science was then at a stage equivalent to the heroic days of geographic exploration a century earlier: "Some bring back mere traveller's tales; others crude estimates of a distance obtained by counting paces or by trundling a recording mechanism attached to a bicycle wheel; yet others set out with a plane table, compass, and clinometer and bring back a fair map. Our explorations of social space are like the efforts of the man with a bicycle wheel; we have many observations, few measurements, and our instruments are not very precise."[31]

This was not limited to government and academia. Industry, too, was largely run by a combination of tradition, inertia, and class privilege. In 1949, the *Daily Mirror* pointed to a recent report which found that output in the American steel industry was "anything from half as much again to nearly double the rate as ours." But as historian Nick Tiratsoo notes, "What is clear about these years is that neither side [management nor unions] was willing when it came to it, to follow the American gospel of productivity, with its pervasive emphasis on new methods and new techniques of doing things."[32]

Hartley Barclay, editor of *Mill and Factory*, wrote that on average, production in American factories by 1952 used roughly twice as much machine horsepower as did industry in Britain. That gave American workers a purchasing power twice that of their English counterparts, triple that of Germany, and almost four times as much as Italians. Americans owned almost three hundred million more shoes, fifteen million more telephones and twenty-one million more radios than the British. Modern Americans enjoyed an "automobile standard of living," Barclay stated, while Europeans, once the masters of the world, were resigning themselves "to a bicycle standard of living."[33]

The advantage that Britain had enjoyed immediately after the war because of its relatively undamaged physical condition and the dire circumstances of many of the other former combatants, which delivered up captive export markets, had evaporated by the early 1950s. "Every year since the war we had been warned by one Minister after another about the growing threat to our overseas markets from Germany and Japan," warned the *Iron and Coal Trades Review*, "but so far the markets have been able to absorb the growing output from all of them. There are now signs that the whole picture is changing. In one trade after another the sellers' market has gone, or is about to go. This year [1952] may well see the first head-on clashes between the world's large exporting nations."[34]

The only thing this boy was able to save from the ruins of his blitzed home was his stuffed toy. London, 1945 (photograph probably by Toni Frissell/Library of Congress).

Aside from foodstuffs (especially meat, sugar, eggs and, briefly, bread) and clothing textiles, the biggest need was for housing. It was difficult to get a handle on the extent of destroyed housing in the chaotic days after the German surrender, because "destroyed" was as much an economic as an engineering determination. As many as a third of the "damaged, uninhabitable but repairable" homes (Class "C(b)") needed work costing over £250, but a large proportion of the homes were worth £450 or less, making them marginal candidates for repair. Yet, it was probably faster to repair than to raze and rebuild them. Nationally, the Ministry of Health counted 175,000 Class C(b) units, 115,000 in London alone.

In April 1945, the Ministry counted 26,000 homes destroyed or damaged beyond repair (Class "A" and Class "B") nationally. The County of London tallied 69,000 Class A and B buildings (both residential and commercial), including 18,000 dwelling units.

The five outlying cities of Coventry, Hull, Plymouth, Portsmouth and Southhampton reported 21,500 Class A and B homes. That was an average of seven percent of their 1938 housing stock. At the low end, Hull lost five percent. Plymouth and Southhampton topped out at nine percent.[35]

But as Banham noted, the war damage wasn't generalized. It was confined largely to parts of port cities and manufacturing centers. The systemic problem was the housing stock that survived the bombing: it was already old and dilapidated when the war started. In 1950, out of 12.4 million total dwelling units, 1.9 million had three rooms or less, 4.8 million had no bathtub, 2.8 million had no private, indoor toilet. Thirty-eight percent were built before 1891; 20 percent before 1851. The official government estimate was that the nation needed 700,000 new housing units in 1948. Most independent experts put the figure at between 1.4 and 2.0 million. As a comparison, Britain (not including Ireland) had built a total of four million dwelling units between 1919 and 1939, of which 1.5 were council homes.[36]

The London County Council and the surrounding boroughs built only 1,302 dwelling units as of January 1947. Private developers built another 2,686. But during the first two post-war years the slow pace was primarily due to shortages of material and the prioritization of manpower to repair work. In London, 140 million bricks were recycled from destroyed and razed buildings. At the peak of the repair program, 33,000 workers were imported into London to repair the Class C(b) homes; they were released in late 1945. By January 1946, 70,000 Class C(b) units had been returned to service nationally; a year later the aggregate total was 137,000.[37]

By 1952, Coventry, Hull, Plymouth, Portsmouth and Southhampton had built a total of 33,500 new units, not counting temporary pre-fabs. Yet they still had a combined wait list for 61,700 council homes. Between 1952 and 1959, the five cities had added another 43,000 units, but the wait list was only slightly smaller. In London, new construction from April 1945 to January 1948 on both municipal contracts and for private developers was 23,600 dwelling units, with an additional 68,000 Class C(b) units back in service. In 1960 the Housing Ministry was forced to admit that the two million estimate had been right: despite the fact that public authorities and private developers had built a million homes during the fifties (and demolished 300,000), the backlog was still 850,000 dwellings.[38]

The question became, what to build in their place? The modernist architects, inspired by the work of Le Corbusier and his Unité d'Habitation in Marseilles, wanted to build 20-story slab blocks of apartments. Others, principally J. M. Richards, editor of *Architectural Review*, rejected the Corbu solution as inhumane, urging Scandinavian planning models that mixed townhouses, mid-rise flats of around four stories, and a few apartment towers for young people, the elderly and childless couples.

Elizabeth Denby, a well-known housing consultant, admitted that "the form of high-rise flat is unacceptable to many English families," and that it was a design "in which architects delight ... even though I still have to find one who lives in such a block himself!" *Architectural Review's* Ian Nairn, who probably would have become just as famous a critic as Banham had his career and life not been cut short by illness (his last book published at 38), wrote that "people are being driven from the centre, not by congestion but by the wrong sort of development.... Elizabeth Denby has plenty of unpublished evidence to show that what working-class families really want was the type of building they had before—a house and garden, cosily planned and near their work."[39]

City administrators, by and large, wanted the replacement housing to equal the density of what had been there before because they did not want to lose the tax revenue resulting from de-densification and dispersal out to the suburbs or New Towns. Mollie Panter-Downs observed the same thing when the London County Council unveiled their postwar housing plan in October 1943: "Both East and West Enders have now got a chance to say what they think of the plans for repairing London's battered face after the war. Parliamentarians and ordinary citizens have been staunchly united in disliking the thought of being anything but Londoners; judging by the comments on the County Council's plans, most people have no wish to included among the half-million souls who, according to the plan, will move away."[40]

On the other hand, the city officials didn't want the slab blocks or towers, either. When it was pointed out that with the space needed for modern parks, civic centers and roadways, keeping the population at pre-war densities meant accepting high-rises, they either grudgingly acquiesced or did what local governments do best—avoided making any decision at all.

Very often, the compromise was an uninspired middle course: three- or four-story flat-roofed walk-up buildings with a single entry, minimal interior hallways (or balcony access to each front door), 6 to 12 flats per building; recognizable in much of the (western) world as council housing. The poster child for bungled council development was Kirkby, six miles northeast of the Liverpool Docks. "On the dreary flat, wet plain of South-West Lancashire, it repeats many of the less pleasing features of similar developments elsewhere," wrote Liverpool University sociologist John Barron Mays, "an atmosphere of organized anonymity prevails along its length and breath; a new, raw hardly lived-in place, unsoftened by time and unrelieved by local color."[41]

It was an overspill area, established in 1950, but unlike the famous "New Towns" around London (Welwyn Garden City, Hemel Hempstead, Stevenage; eight altogether) it was never formally designated as such. It was supposed to be primarily a new-tech business park with surrounding housing, but it was 1960 before the Development Corporation got around to preparing the business park. By then the population had ballooned from 2,000 to 52,000, but none of it could be built on the biggest contiguous parcel, because it was reserved for the business park. So the housing ended up being shoehorned in various nooks and crannies around a vast mudhole that was the any-day-now-soon-now business park.

With limited money and land, what resulted was a much higher proportion of narrow, three-story townhouses and endlessly repeated six- and eight-flat walkup buildings than was the case in any of the official New Towns. The outcome was so bleak and gritty that BBC-TV based a police series, *Z Cars*, there. Although it was ostensibly located in an everyplace-noplace subtopia called "Newtown," everyone recognized it instantly as Kirkby. (The American crime show producer Jack Webb transported *Z Cars* to Los Angeles, turning it into the long running TV series *Adam-12*.)[42]

Even worse than the bad housing design and site planning was Kirkby's utter lack of support services, such as shops, pharmacies, doctor's offices, banks, government offices, and community centers. Unfortunately, this was true even for the formal New

A British prefabricated housing unit on display in Washington, D.C., 1945 (photograph by J. S. Lakey, U.S. Farm Security Administration/Library of Congress).

Towns and estates. "It is significant," notes historian Nick Tiratsoo," that despite wartime promises, few of the new estates in the blitzed cities were provided with any community facilities, and ended up, therefore, looking like their much criticized 1930s predecessors."[43]

Some local politicians, hearing an almost constant barrage from their constituents in the center city redevelopment areas that they would rather have ground-level homes than high-rises, began to clamor that the delays were being caused by the necessarily longer lead-time for the large-scale construction of apartment buildings as compared to rehabilitating existing terrace homes or replacing them with rowhouses. But when a four-woman team from the survey organization Mass Observation surveyed a thousand housewives in working-class neighborhoods across the country as to their housing preferences, they discovered no consistent response outside "a high degree of acceptance of conditions as they are." Moreover, they found "that when people were asked whether they like the neighborhood or not [as opposed to their old house or flat], less than one person in a hundred mentioned any form of activity that involved co-operation with their fellow citizens ... people are passive minded, letting things be done to them, hardly thinking of what they could get done, if they would co-operate with their neighbors."[44]

The medical journal *Lancet* studied the condition known as "suburban neurosis," supposedly rampant during the 1930s in the pre-war council estates where former slum-dwellers had been decanted to. Now, in the New Towns, separated from old friends and families in the blitzed inner-city slums, suburban neurosis was allegedly on the rise again. The *Lancet* editors concluded that "suburban neurosis" was primarily "transitional neurosis." That is, the very real stress incurred during and immediately after relocation tended to exacerbate preexisting psychological problems, primarily among young women already undergoing the stress of recent marriage and parenthood. The problem appeared worse than it was, because couples with young children were heavily over-represented in the New Towns. Moreover, the women were often unemployed, an indicator of homebounded-ness. As they settled in, their "neurosis" improved as the shock of the new went away, new routines and relations were established, and mobility increased. It appeared that one of the most important factors was the installation of sidewalks. Without sidewalks, mud ruined expensive shoes and made prams useless.[45]

The husband-and-wife architects Alison and Peter Smithson, who would come to know Banham well through the Institute of Contemporary Arts, like most of the International-School influenced architects of the period, believed that their modernist designs provided the solution to such social ennui, and waived away the traditionalists that insisted on neighborhood reconstruction based on of the familiar garden-fronted rowhouse. "You might argue that the back garden and front pocket handkerchief necessary to look out on. But what fills the windows of your day rooms is the houses opposite and the back behind.... How many gardens in your street are gardened for other reasons than that of keeping up appearances? ... The argument that suburbs are what everyone wants is invalid. We are not a medieval community that actually directs its individual houses to taste. Folk-build is dead in England."[46]

III

Mary Mullett and Peter Banham met in Norwich shortly after the war. Mary had grown up in the suburbs northeast of London, near Dagenham. Her father was a draftsman and her mother, after the children had grown, went to work for the civil service. In 1944 Mary graduated from London's Central School of Art and Design, then attended the Educational Institute for a year. The Institute was across the street from the Courtauld Institute of Art, and she used its library while she earned her teaching certificate.

In 1945, she was hired to teach art at a convent school in Norwich. "It was a Catholic school, but it was only 30 percent Catholic," she later recalled. "They [the students] were not suppressed in any way.... The sisters were wonderful teachers. Very dedicated." She lived in a cottage in the center of town next door to the Maddermarket Theatre, started by Nugent Monck in 1921 as a repertory theater. "It was sort of a magnet for all the young people in town." Peter Banham was hired as temporary stage manager soon after his return from Bristol until the regular stage manager, who was overseas and hadn't yet been demobilized, could return. "He had come back from the war and was looking around, not really knowing what to do with himself, really, because he had gone straight from school off to the Bristol Company.... I didn't really start to take much of an interest in him until we began to interact at the theatre."[47]

This was the start of a life-long pattern: Banham, the man who disliked and claimed to avoid administrative tasks whenever possible, hired to run something or another because of his talent for management. On the other hand, he loved to write and to lecture. He probably started off contributing theatre reviews to the local evening newspaper, the *Eastern Evening News,* sometime in 1945. It is known for sure that he started submitting art exhibit reviews for the *Evening News* in 1946 under the byline "PBR," then a few months later for the larger-circulation *Eastern Daily Press* using the name "Reyner-Banham."[48]

He was also hired by the Worker's Education Association (WEA or EA) to give lectures. "He was doing this series on local history in a town called Wymondham, twelve miles outside Norfolk, where all his family—both sides of his family—had originated. There was still a lot of his family living there," Mary recalled. "We used to cycle in the winter through the fens in the moonlight. It was very romantic. But cold." Banham would stand up to give his talk "and here are all these relatives, not the women, just the men—sitting back, arms crossed, 'tell us what you know, boy,' and he was scared to death." After it was over, "we would go around to meet the aunts."[49]

There are differing accounts of how Banham got into the Courtauld Institute of Art. According to Banham, he traveled to London to hear Nikolaus Pevsner lecture. Pevsner was on the Courtauld's faculty, and Banham wanted to introduce himself to the already-famous historian of the modern architecture movement. He applied for admission, and was rejected. "Norwich is a great graveyard of promising careers," he recalled later, "where people who might just have made the big time with an effort relax and go small-time instead…. A man with a modicum of money or talent behind him and an entertaining wife beside him can have a very nice life in Norwich in a sickening sort of way." He was determined not to let that happen. He applied again in 1949, and was accepted.[50]

But according Mary:

> We went to the Castle Museum to take some extramural classes run by Cambridge on art history. Unlike the EA courses, the extramural classes required essays…. He wrote such wonderful essays for this class that the woman who ran the class, named Helen Lowenthal, she was from the Victoria and Albert, she was a wonderful teacher, but she was so impressed with his essays that she said, "what are you doing hanging around this town doing all sorts of daft jobs? I can get you into the Courtauld Institute." And he didn't have a clue what the Courtauld Institute was, so he opened his mouth and said "Huh?" So I kicked him under the table, hard, because of course I knew what the Courtauld Institute was, I was in awe of it…. She talked to Anthony Blunt, and he was in. That was in 1949.[51]

Actually, the stories don't conflict much. Helen Lowenthal established the Victoria and Albert's educational outreach program about 1950, and earlier a similar program for Cambridge. She would have been influential enough to get someone's foot in the door of Anthony Blunt's school. On the other hand, it's hard to believe that Banham had no idea what the Courtauld was, although he may not have appreciated the importance of Lowenthal's offer to the extent Mary did. Mary said that "we didn't find out until very late" that Banham was in, so it is likely that he was not immediately accepted and was on a reserve list and received the call-up only after a late cancellation, so his memory of a first-round rejection is understandable.

Peter and Mary married in 1946. "We lived together, and we got married because his family, especially his mother, worked for all sorts of voluntary organizations and such, which wouldn't have approved of us living together … it was actually fairly common then, especially during the war. It surprised me when things like white weddings and the like came back right after the war." Mary made clear to an interviewer a half-century later that they did *not* have a white wedding, although the two families did pool their ration coupons so they could have a proper wedding cake.⁵²

His exhibit and book reviews began regularly appearing in the bi-weekly *Art News and Reviews* in March 1950, while he was still a student. Pevsner, who sat on the editorial board of *Architectural Review*, urged senior editor J.M. Richards to hire him. Richards took him on as a part-time "literary editor" that fall. The typical path for a Courtauld graduate was to start at a provincial gallery for a couple of years, then start working one's way back to London, according to Mary. But she believed Pevsner and Anthony Blunt helped Banham get the *Architectural Review* job because he badly wanted to stay in London, was already an experienced journalist, but above all, "was not a gentleman and said what he thought." Journalism it may have been, but it was not a job to turn your nose up at. "The influence of the *Architectural Review* on contemporary British architecture after Nikolaus Pevsner joined the editorial staff in 1941 can hardly be overemphasized," recalled architecture critic and journalist Anthony Vidler.⁵³

From 1952 to 1964, Banham worked in the offices of the Architectural Press, publisher of *Architectural Record* and *Architects' Journal*, located at 9-13 Queen Anne's Gate, just beyond the famous statute of Queen Anne. Carteret Street turns to the right just in front of the building in the background covered in scaffolding (photograph by the author).

While Pensver was his academic advisor, Banham always considered the period between 1952 and 1959, when he was promoted to *Architectural Review's* assistant senior editor, as his true apprenticeship, with Richards as his mentor. One commentator noted that "just as Banham wrote of the work of the Californian architects Greene & Greene, whom he much admired, 'it looks casual, but is fundamentally formal,' so too is the artifice of Banham's apparently relaxed writing style to conceal a rigorous and demanding mind at work." The regular paycheck didn't hurt either—his trusty roadster bicycle, Boanerges von Heinz II (aka "The Death Rattle") had been stolen from the rack in front of the Hammersmith Library a few months before he graduated and, lacking a driver's license, he was having to make do with public transit and London's notoriously expensive cabs.⁵⁴

Pevsner had also nominated him as a doctoral candidate at the Courtauld Institute.

Their relationship was summed up by Mary: "Peter loved Nikolas dearly and disagreed with him profoundly." Pevsner had expected that Banham's Ph.D. dissertation would be a follow-up to his own *Pioneers of the Modern Movement*, an account of architectural genius passed through the generations in a linear, orderly progression. It left off at the start of the career of Walter Gropius and his Bauhaus contemporaries. But Banham was not overly impressed with Gropius, and most decidedly did not believe in the concept of a linear, orderly evolution of architectural expression guided by timeless purities of form. Instead, he held to an alternative view that accepted (indeed, idealized) temporality and expendability of form.[55]

Pevsner would normally have rejected such logic out of hand, but Banham had the good sense not to throw it in his face, but simply presented his case studies, focusing on some of Pevsner's overlooked players, most notably the futurists F. T. Marinetti and Antonio Sant'Elia, and let them stand on their own merits. Also, as Mary noted, Pevsner "and all those of his generation, because they were refugees, they were such generous scholars who believed everything should be shared, and were not automatically afraid of any kind of opposition, as seems to be so fashionable at the moment." Six years in the making and published commercially in 1960, it remained in print for over thirty years.[56]

Architectural critic Shantel Blakel believes that Banham's appreciation of the Moulton bicycle was a direct outgrowth of the two themes first explored in *Theory and Design*, futurism, and technology as a beneficial force:

> Banham's appreciation of the F-frame [Moulton] was a logical extension of his writing from 1960 on the theme of an imminent second machine age, the successor to the era described in his *Theory and Design in the First Machine Age* ... [it] had been a way of pulling the Italian futurists out from the shadows of architecture's emerging modernist canon, and was also a retort to writers and commentators like George Orwell, who had depicted technology only as a means of control, or C. P. Snow, who was writing about a deepening divide between lay society and a technocratic elite. The simplicity and innocence of the bicycle, in this sense, became a way of selling technology as something incontrovertibly democratic and good, and not just a fast and efficient way of getting around town.[57]

2

But Today We Collect Ads

> Yes, I've heard of that New Brutalism. I understand it was sociological as much as architectural, a kind of institutionalized insensitivity to peoples' needs.
> —Headmaster J. Little, Hunstanton School, Norfolk (1984)[1]

I

In 1952, the same year Banham graduated and started at *Architectural Review*, he was introduced to a lecture and study group at London's Institute of Contemporary Arts (ICA) that, for lack of a better name to use in the ICA's monthly minutes, Dorothy Morland, the Institute's assistant director, called "The Young Independent Group." They quickly dropped the "young," and have gone done in history as the Independent Group, or IG. Many were gallery assistants, interns, and other junior staffers. "I was not a founding member, myself," Banham later admitted. "I was looking for a rolling bandwagon, and I think I found the right bandwagon. But the Independent Group was hardly a rocket ship to fame. Most of the world had never heard of it until it was safely dead."[2]

Banham once called the IG'ers "a rough lot," but that was typical Banham reform home bravado: of the original members, only artist Eduardo Paolozzi, the son of Italian immigrants whose Edinburgh ice cream kiosk had been destroyed by vigilantes in 1939 had a real claim to working-class hardship. Asked about Banham's quip, artist Richard Hamilton replied, "I would say we were a mixed lot rather than a rough lot. Because of the war, we came into a similar kind of experience, but from different experiences ... the architects were from a more traditional background, where the painters were more of the roughnecks. Paolozzi for sure. But not Henderson, whose mother worked for Peggy Guggenheim, running her gallery. It was incredibly mixed."[3]

Nigel Henderson claimed that for him, the genteel, high-art modernism in which he had been raised in the Guggenheim gallery turned to dust during the war: "houses chopped by bombs while ladies were still sitting in the lavatory, the rest of the house gone but the wallpaper and the fire still burning in the grate. Who can hold a candle to that kind of real-life Surrealism?"[4]

Henderson was also friends with the husband-and-wife architects, Alison and Peter Smithson. Both had grown up near Newcastle, she just across the Tyne in South Shields, he a little south of there in Stockton-on-Tees. He was about half-way through the architecture program at Durham University when he was drafted into the Royal Engineers. He ended up, late in the war, in Burma. She, born Alison Gill, was still in high school when evacuated to a relative's home in Edinburgh. She returned after the war, finished high school, and also entered architecture school at Durham University, where they met. He graduated before her and spent two years at the Royal Academy in London. When she graduated in 1949, they married and started as technical assistants at the schools division of the London County Council.⁵

They were a serious, purpose-driven duo. Both were only children. Christopher Woodward, who worked for the firm of A+P Smithson as a junior architect in the 1960s, described their respective personalities. "Peter thought, taught and lectured, took photographs, and made drawings. Mainly thinking and writing, Alison supported and drove him in these. In a reversal of gender sterotypes, he personified intuition, she the will. She drew very little." Both confessed to being easily bored and frustrated. "We are both variously conscious of wasted time," Alison told a writer. "I hate crowds and we never get into situations we can't get through quickly. For example, we'd never shop in a supermarket on Saturday."⁶

In the early '50s, when she had time on her hands, Alison played with fashion design. "She was a Mary Quant before Mary Quant even thought she was Mary Quant," said the artist Magda Cordell. Mary Banham recalls Alison showing up at an event in one of her creations: "she knitted herself a kind of green gnome suit, like a baby grow, with a hood, and she looked absolutely extraordinary in it." Mary believed "she had a touch of genius, there's no doubt about it." While "I was terribly fond of her," she also admitted that "she was terribly difficult to get on with." She wasn't the last to hold this opinion, to the later detriment of the couple's career. But in 1950, while still working for the schools division,

Left to right: Peter Smithson, Eduardo Paolozzi, Alison Smithson and Nigel Henderson in Limerston Street near the Smithsons' flat. This is an alternate to the photo used for the Group 6 poster for the *This Is Tomorrow* exhibit at Whitechapel Gallery, August 1956 (photographer unknown [probably Nigel Henderson], © Nigel Henderson Estate/Tate, London. 2017; MB2071).

the Smithsons floored everybody by winning the competition for a secondary modern school at Hunstanton, on the coast in Norfolk. He was 26, she was 21.[7]

"Somebody who was a real heroine of this period was Dorothy Morland," recalled Peter Smithson shortly before his death in March 2003. "She acted as a bridge between our generation and the previous generation.... The institution was intended, like the Museum of Modern Art, to be propaganda for that type of art, and for Picasso. It was exactly like MoMA, rich people interested in art. But Dorothy was not quite like that. She was the wife of a surgeon, I think, in the London University hospital.... She opened the door of the ICA through her friendship with Eduardo Paolozzi. Eduardo's wife worked in the gallery. She was a seamstress."[8]

Mary Banham agrees: "The powers to be, like Herbert Read and Roland Penrose, who were actually on the board, wanted to get rid of us, because they knew we were questioning their own fundamental principals, and of course they had been the young Turks back in the 1930s ... but it was Dorothy Morland, who was on the board, who said, 'you lot must remember what it was to be a difficult youngster—let's give them a chance.'"[9]

Morland recalls that the ICA's gallery director, Richard Lannoy, approached her "to say that there was some dissatisfaction amongst a group of younger members who felt they weren't getting an opportunity to exchange views." Lannoy and magazine art editor Toni del Renzio sent out invitations "to about fifty people" for a lecture by Paolozzi in April 1952.[10] Paolozzi barely spoke; the "lecture" was a rapid-fire presentation of color images taken from American glamour magazines, using an overhead opaque projector called an epidiascope. A couple of years later, Alison Smithson wrote an explanation (in verse) of why the IG found Paolozzi's work so important:

> Why certain folk art objects, historical styles or industrial artifacts and methods become important at a particular moment cannot easily be explained.
> Gropius wrote a book on grain silos.
> Le Corbusier one on aeroplanes,
> And Charlotte Perriand brought a new object to the office every morning;
> But today we collect ads.[11]

"It was exceeding primitive, recalled Peter Smithson, "his text was grotty. But this is the funny thing to describe. In that context, at that time, it was hilarious. It was funny."[12] Banham laughed along. But he was also impressed with Paolozzi's epidiascope magic. For him, it was an accelerated course—shock therapy, if you will—in the future of what England's increasingly consumerist culture would look like:

> We goggled at the graphics and the colour-work in adverts for appliances that were almost inconceivable in power-short Britain, and food ads so luscious you wanted to eat them. Remember we had spent our teenage years surviving the horrors and deprivations of a six-year war. For us, the fruits of peace had to be tangible, preferably edible. Those ads may look yucky now, to the overfed eyes of today, but to us they looked like Paradise Regained—or at least a paper promise of it.[13]

Mary Banham recalled that "the Smithsons had a kitchen in Limerston Street that was entirely papered in ads for food out of American magazines. It was a kind of pornography of food, because we were so much rationed in those days."[14] In Westminster, Churchill, confused and exasperated by rows and columns of weights and measures in the newly downsized ration table, asked the Minister of Food to show him what an individual's ration actually looked like. "Not a bad meal, not a bad meal," Churchill murmured contentedly when the exhibit was stacked on a tabletop in front of him. "But these," the minister replied, "are not rations for a meal or for a day. They are rations for a week."[15]

Banham was actually not among the fifty invited to Paolozzi's show, but had gate-crashed his way into the lecture. Afterwards he talked it up around town. "Banham was very vociferous, rather lecturing, really, about Eduardo's show," recalled Henderson. "I think it was because the visual wasn't introduced and argued in a linear way but shoveled, shriveling in the white-hot maw of this epidiascope." Morland recalls that when Paolozzi left the glued collages under the light too long, a couple of them almost caught fire. "They looked like what they were," Banham recalled. "The very beginnings of the process by which mass media imagery and pop iconography get transferred into pop art, but we couldn't know that, because pop art didn't exist yet."[16]

Four months later, Morland and Lannoy asked Banham to take over as secretary of the Independent Group. More than one IG member later commented on the irony of a group formed to emphasize the "non-linear" didn't really take off until someone comfortable with bureaucracy and administration was asked to take a leadership role. Although he did not organize or curate the three so-called "IG" exhibitions ("Parallel

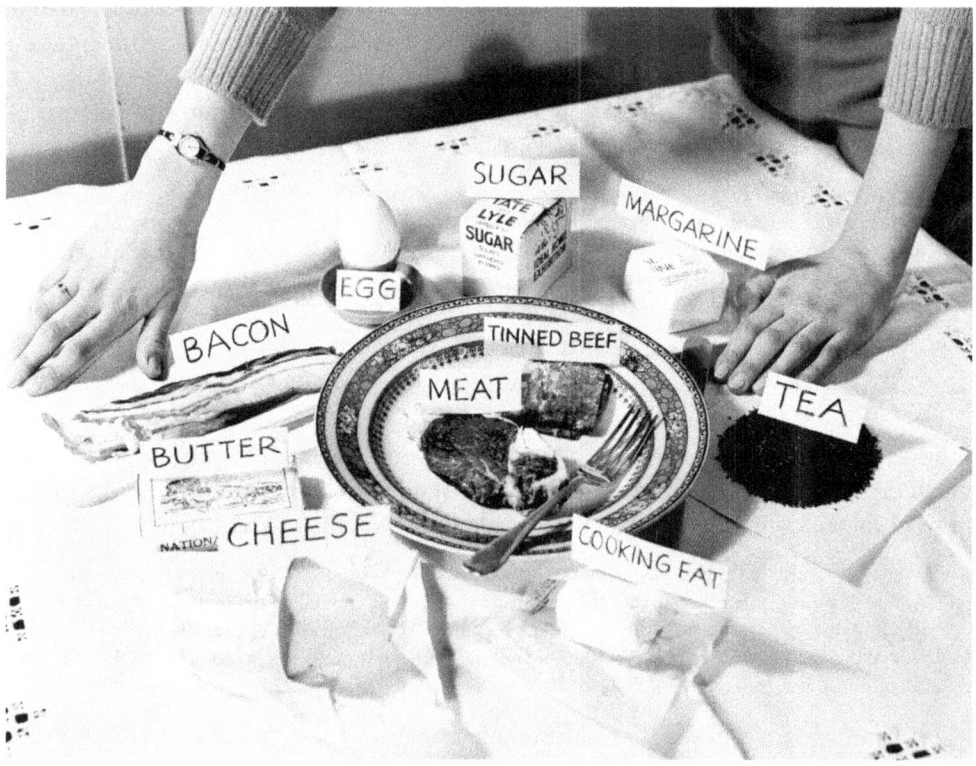

One week's rationed foodstuffs for an adult, Britain, 1951. Some staples, such as bread and (for children) milk were not rationed. Other substitutes, such as olive oil, canned milk and tinned ham, were also unrationed, but expensive (National Food Survey Records, (Department for Environmental Food and Rural Affairs, Neg. No. 28513208, November 2, 1951).

of Life and Art," 1953; *"Man, Machine and Motion,"* 1955; "This Is Tomorrow," 1956), he did organize a series of nine lectures over the winter of 1953–54, and later became the group's unofficial archivist, along with John McHale. He was forced by the demands of his Ph.D. dissertation and family illness to relinquish his formal IG duties in 1954, but continued to be active, showing up often and occasionally giving talks.

But architect Colin St. John ("Sandy") Wilson believes that it was Banham's ability, mostly through his continuing editorship in *Architectural Review* and his increasing number of articles in other journals, to propagandize for the ideas of his fellow IG'ers that made him so important: "The undeniable fact is that Banham was chief provocateur, spokesman, and historian of our goings-on, and in his capacity as active journalist as well as scholar he was able to publicize them in a regular way. His *Theory and Design in the First Machine Age* was written for *us*—the dedication says so—and much of it was hot off the anvil."[17]

The Smithson's school at Hunstanton was completed in 1954. Their unflinchingly, almost violently Spartan design, inspired by Mies van der Rohe's Illinois Institute of Technology Metals and Materials Building, created a sensation. Banham reviewed it, also in *Architectural Review*, the first time simply under its name, "Hunstanton School at Norfolk," later under the banner of its architectural theme, a stylistic name invented [maybe] by Alison Smithson: "The New Brutalism."[18]

"Parallel of Life and Art" contained no art objects, but it did have photos of original works by Paolozzi and other IG artists, as well a photograph of Jackson Pollack in the process of creating one of his drip paintings. Most of the photos were raw, technical, functional: x-rays, materials undergoing destructive testing, tribal masks, tumors, microscopy. Because there was insufficient floor space for free-standing display panels, the blow-ups were suspended from the ceiling, with viewers walking under them. It ran for five weeks. It had 443 visitors. A German art magazine unwittingly

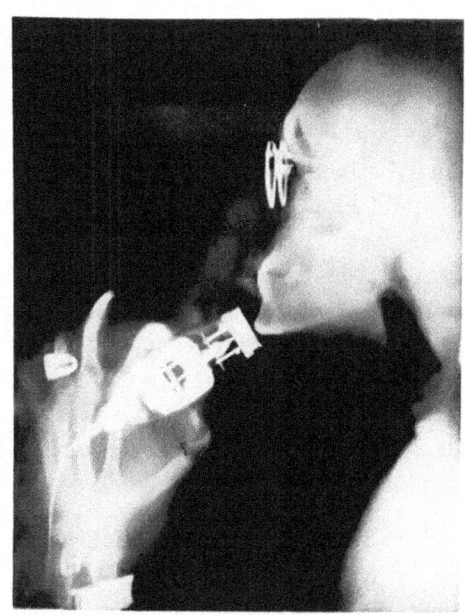

Cover of the catalog for the exhibit *Parallel of Life and Art*, ICA Gallery, London, September–October 1953 (author's collection).

printed a photo of the exhibit upside down and nobody caught it for weeks. Afterwards, some of the blow-ups were moved to the Architectural Association in Bedford Square. At a stormy "meet the curators" panel discussion, the AA's students accused the "Parallel of Life and Art" organizers of trying to create "a cult of ugliness." Banham later wrote that he regarded "Parallel of Life and Art" as the *locus classicus* of the New Brutalism movement.[19]

In March 1953, Banham lectured to the IG on automobile design, more specifically the American, high-finned, bumper-bomb, glass-and-chrome styling he called borax.[20] It was later published in an Italian industrial design journal. He had previously written about the contrast between auto styling and architecture, particularly about what a botch architects had made of it when they turned their attention away from the permanence of buildings to the "here today, gone tomorrow" word of automobile design. But now, he was focusing on the virtues of industrial design as a thing onto itself.

"Architects, for entirely valid reasons, are habituated to think in terms of a time scale whose basic unit is about half a century, he said. "But a sixty year old car, even if in perfect mechanical condition, would nowadays cause traffic jams and invite accidents if let loose among the nimbler products of more recent decades … the subject matter of Industrial Design has changed, because the foundation stone has crumbled—there is no longer universal acceptance of Architecture as the universal analogy of design."[21]

Banham was inspired by Deborah Allen, auto reviewer for the new, low-budget, New York based magazine *Industrial Design*. She was an unlikely automotive critic. She lived in Manhattan, took the subway, and didn't think very highly of cars. "It was hard to write about them, because they were senseless," she admitted. A trained designer, she believed a lot of Detroit body styling was the result of lazy artists and rushed artwork. As a result, little more than rough sketches went to the clay shop, so the model makers could pretty much do as they pleased. On one Lincoln, she actually identified which clay-working tool the model maker had used to incise the side swoop running the length of the car from headlight to tailfin.[22]

But her thinking changed one evening when a friend took her on a ride through Connecticut in a 1955 Buick. "I saw how he lived in his car and he enjoyed it," Allen recalled. "I was so amazed that there could be some sense in this car. It was a revelation." She sat down that evening and wrote one of the most famous car reviews in history:

> The Buick's designers put the greatest weight over the front wheels, where the engine is, which is natural enough. The heavy bumper helps to pull the weight forward; the dip in the body and chrome spear express how the thrust of the front wheels is dissipated in turbulence toward the rear. Just behind the strong shoulder of the car, a sturdy post lifts up the roof, which trails off like a banner in the air. The Driver sits in the dead calm at the center of all this motion; hers is a lush situation.[23]

Banham wrote an article comparing it to the writing to Bernhard Berenson, at the time the greatest living connoisseur of Italian Renaissance art and culture. "This is the stuff of which the aesthetics of expendability will eventually be made," he gushed. He took, in his lecture slides, to copying her layout technique of cropping down car photos to individual elements like headlights or tailfins or windshields, then laying out ten or a dozen side-by-side as if they were window lintels in a first-year architecture class. Richard Hamilton created a series of abstract paintings under the name "Hers is a Lush

Situation." All clearly suggest a woman behind the wheel of a Detroit behemoth. What shifts, depending on how you view the painting, is which individual element depicts the car's body and which the woman's.[24]

However, neither Allen nor Banham were forging an entirely new path. In 1936, Nikolaus Pevsner had written *Pioneers of Modern Design*, a companion book to *Pioneers of the Modern Movement*. Historian Clive Dilnot notes that Pevsner's book was "animated by two powerfully linked ideas. First, design is of great importance and significance in the modern world. Second, precisely because of this, the form that designing takes in this emerging world is of social and ontological importance." In other words, good design was worth fighting for; it was worth getting in scrapes and arguments about; it was worth risking losing friends over. Pevsner's *Pioneers of Modern Design* was revised and re-issued in 1960, the same year as Banham's *Theory and Design in the First Machine Age*. "Banham's book marked the beginning of a period of intensive study of Modernism and its origins, which meant that the Pevsnerian program of study was at last taken seriously," Dilnot notes. Ironically, while *Theory and Design* may have contradicted many of the conclusions Pevsner drew in his *Pioneers of Modern Architecture*, Dilnot believes that it broke through the barrier of academic stagnation and allowed Pevsner's *Pioneers of Modern Design* to achieve, in the long term, equally legitimacy to *Pioneers of Modern Architecture*.[25]

Lawrence Alloway believed that the importance of the Independent Group was not so much its focus on mass media or on material culture studies, but the fact that they were happening simultaneously: "We started looking at it [mass media] at a time when the cars were bigger and the tail fins higher, the camerawork more mobile, the color more vivid, the screen wider. It was an expanding technological period, so that to the traditional appeal of the mass media, this improved technology was being added ... our liking of Pop was very much linked to this pro-technology attitude."[26]

By 1956, Banham was appearing regularly in the biweekly magazine *Architects' Journal*, which, despite the name, was marketed as much to builders, contractors, and material suppliers as architects. His column "Not Quite Architecture" was the first of his ventures into short, snappy, non-academic pieces that reached beyond a narrow intellectual audience. Sometimes they stuck to architecture (book reviews, mostly), at other times they ventured far afield, frequently into technology, whether at the Brands Hatch auto race track, the barber's chair, or the stereo shop. "Although Banham's material culture studies never found their way into any of his books," notes his doctoral student Penny Sparke, "they formed, none the less, a vital element within his career as a journalist through contributions to, among other periodicals, the *New Statesman* and *New Society*, which flourished from the late 1950s and through the 1960s and 1970s."[27]

Mary Banham agrees:

> He saw his books as being serious works for serious readers and students, his students. Often, there wasn't a book he could show to his students, so he'd write one.... His articles definitely were more for general consumption. His main aim was to popularize without talking down. That was an Independent Group thing; communicate, communicate. The future; technology; and communicate. That's what the Independent Group was all about, and they all believed in it entirely. He was a popularizer.[28]

Not everyone was thrilled by this journalistic eclecticism. "Some academics would dismiss Banham's history writing because he 'lowered' himself by writing accessible criticism," recalls Nigel Whiteley. "The academic may write about communication, but does not always practice it, whereas Banham reaped the benefits of writing journalism for an intelligent lay audience." He was also, in the words of Robert Maxwell, "a damned good writer."[29]

This does not mean Maxwell always agreed with the way Banham went about his business. "There are still some doubts to raise about his method," he complained in 1981 after Banham had left England to teach in the United States. "We, the architects, have been constantly chastised by him for our backwardness to the point where we appear to be more concerned about the past than he, whose job it is as a historian to analyze the past." In the end, Maxwell concluded that "his reputation has been so strongly linked to his concept of a history of the immediate future" that he stopped worrying much about the past—his focus instead, shifted to prediction, "his anticipation of a future increasingly focused on the possibilities of technological development."[30]

On the other hand, Adrian Forty, who shared an office with Banham at University College, London before Banham decamped for the States, disagreed. "Banham took being a historian exceedingly seriously," he argues. Banham was fully aware of the dilemma: he believed that a rapidly advancing, technologically based, consumer-driven society was changing lives much faster, and to a much greater extent, than architecture ever could, but yet he had an academic responsibility to record and explain how that technology was impacting architecture in an objective and reasoned way, even if those architectural changes were not terribly important to the lives of average people.

"His pragmatic, everyday solution was to make a distinction between what he wrote as criticism (which he treated as ephemeral) and what he wrote as history (which he regarded as permanent) ... before popular culture was acknowledged as anything other than a threat to Western civilization, Banham's recognition that one might subject ordinary, everyday products and buildings to serious historical study was radical and exciting." Penny Sparke noted that "Banham has learnt from the ad-man the technique of selling through evocation and suggestion.... There is a strong parallel with Tom Wolfe's 'New Journalism' and other exponents of the same style on both sides of the Atlantic, which manifest a similarly playful, racy style to simulate the message of Pop Culture." Or, as Banham himself once put it, "History is, of course, my academic discipline. Criticism is what I do for money."[31]

The IG sharpened Banham's thinking on the dichotomy between high and low culture. "For the Independent Group, it was essential to evaluate the workings of mass culture, to judge why one image was more effective than another, to affirm the skill of the graphic designer or marketer in creating an advertisement. Mass culture was worthy of just as much serious attention as high culture," notes Anne Massey. "*Mad* magazine as successful satire, Elvis as a powerful totem in contemporary myth and *Double Your Money* as a sophisticated, ritual drama."[32]

It is not possible to overstate the risk of personal and professional ostracism that came with such an approving, or even nonjudgmental, approach to American popular culture. Labour Party leader Hugh Gaitskell, in his private diary, ruminated over the level of insipient anti–Americanism he saw around him, dormant since World War II and re-kindled into life by the Korean War:

> The truth is we have a dilemma. We do not like to admit our relative weakness, because we should then look too much like a satellite. But if we try and live up to a military standard we cannot afford, that means economic trouble. A poor relation who is driven to live beyond his means by his rich cousins will not feel well disposed towards them.... It is easy to see how powerful anti–American prejudice can be when to this already difficult relationship is added the genuine fear felt by many people that America will land us all in a war. Moreover the war if it comes will engulf and destroy Britain and Europe while very probably leaving the territory of America physically untouched.[33]

Sometimes, there were good reasons to question the generosity of the Americans. Lincoln Kirsten, founder of the New York City Ballet, helped provide funding for the ICA to hire Anthony Kloman as public relations director in the spring of 1951. Rumors abounded that another Kirsten-backed project, the Paris-based Congress for Cultural Freedom, publisher of the magazine *Encounter*, was actually a CIA-backed effort. Later, Kloman secured $48,000 in funding from the John Hay Whitney Charitable Trust for a sculpture design competition, *The Unknown Political Prisoner*, the winning entry to be temporarily installed on a hilltop in West Berlin where it could be seen over the wall in the East. Reg Butler won the competition, but his design was never installed, as the left-wing American magazine *Ramparts* reported that both the Congress for Cultural Freedom and the Whitney Charitable Trust were pipelines for CIA funding, and that the senior staff at *Encounter* as well as Herbert Read and Anthony Kloman at ICA knew this, at least in a general way. The administrators of the ICA "had been politically naïve at that time," explained Dorothy Morland in the 1970s.[34]

According to Banham, the IG "Ceased to exist after the spring of 1955, just stopped holding meetings and faded away." Mary recalled that "Everyone soon passed on to other lives and other places." Asked what caused the break-up, she replied "because we all got jobs and didn't have time."[35] However, most of the IG members place the end of the group a few months later, in August 1956, when the "This Is Tomorrow" exhibit closed at London's Whitechapel Gallery. But everyone agrees that any particular date is arbitrary, and that everyone was becoming very busy.

The Smithsons designed a house for the Sugden family in Watford in 1955 and prepared a full-size mock-up of a pre-fab housing unit called the "House of the Future" for the *Daily Mail's* Ideal Home Exhibition in March 1956. That same year, Peter started teaching part-time at the Architectural Association. Then they won a commission for the three-building Economist Plaza in London in 1959, which many consider their best work. Richard Hamilton had been teaching in Newcastle since 1952; Lawrence Alloway was appointed assistant director of the ICA in 1953, then took a job at the Guggenheim Museum in New York in 1961. Toni del Renzio was hired as art director at *Harper's Bazaar* in 1958. John McHale was back-and-forth between London and Yale University almost continuously between 1955 and 1960.[36]

The Smithsons' House of the Future was a conceptual, self contained unit of about 750 square feet intended to be molded from plastic composites, designed to be readily available, and flexible in application. It could be used as a single-family detached home, attached into small groups of garden apartments, or stacked into mid-rise flats. Planned for a design year of 1981, the House of the Future was anticipated to cost little more than a well-appointed caravan, despite being fully appointed with such labor-saving devices as a "radar range" (not yet available for home use), trash compactor (ditto), dishwasher (only in America), and multiplex entertainment center. "The whole thing would be mass-produced, rather than the parts," Peter Smithson recalled. "To change anything in it," Alison said, "would be like trying to get a bigger glove compartment for your car." That is, it would be more complicated and more expensive than getting rid of the whole thing and buying another car. It had a tremendous impact on Banham, much greater than the subsequent "This Is Tomorrow" art exhibit, although the latter is much more famous.[37]

Two years later, the Smithsons continued their experiments in this direction with two units they called "Appliance Houses." These were intended for industrial mass production using current (1958) technology, mostly using manufacturing techniques for caravans and pre-hung windows and doors. However, by 1960 a growing demand for their services (especially the Economist Plaza) brought these experiments to an end.[38]

Both the House of the Future and the Appliance Houses were ingenious mixtures of traditional building techniques and Detroit-style car design. Banham believed they represented a radical break with architectural traditions. When industrial design was in its infancy "there were valid reasons for giving architects hegemony over the training of designers and the formulation of theory." Thus, industrial styling followed in the wake of the "mother art," architecture. But the Smithsons, in putting on display a housing unit that was merely a shell whose function was to mount and organize a collection of mass-produced appliances, had turned the old order on its head. As a result of their work, Banham began to see industrial design as something independent and autonomous of architecture.

"Houses are, therefore, not like consumer products because they are not portable; consumers' attitudes to them are not the same," Nigel Whiteley explains. "On the other hand a house could be like a consumer product if it was thought of as a piece of industrial design. This, Banham felt, was a bigger mental leap than might be imagined, for it required the architect to become immersed in technology."[39]

It is hard to overstate the degree to which housing created a political identity in Britain up to this time. Despite the efforts of the postwar Labor government to make council housing respectable and broad-based (even to the point of offering some council homes for sale to long-term occupants, a laughable idea before the war; who would buy one?) the type of housing and social class were inextricably linked. Madge Martin, the wife of an Oxford clergyman, visited the Smithsons' model at the Ideal Home Exhibition in 1956: "We queued up to see the incredible 'house of the future'–and decided *no*." Despite her misgivings, most new homes were being built along modern lines: more open, more informal; more flexible, if, for no other reasons than the changing relative prices of materials and the need make a home adaptable to new appliances. Brick, stone, slate and tile were become more expensive; form-poured cement and balloon-framed interior walls less expensive.

One thing builders agreed wholeheartedly with the Smithsons about: kitchens were going to need more space and electrical capacity for the white goods that were sure to come. In 1955 only 18 percent of households had a clothes washing machine, but Hoover was about to do away with the clothes wringer, replacing it with a machine with two tubs, the second for spin-dry cycle, and was already talking about a separate forced-air dryer that would eliminate the need for hanging out laundry entirely, but it might need a stepped-up 220 volt household connection.[40]

This is what Banham was talking about when he referred to "technology," and it is why he saw it as such a positive and egalitarian force. It is also what he meant by "Pop": not really low/highbrow or proletarian/bourgeoisie, but more of a continuum: absolute disposability on one end (a fast-food cup or hamburger box); absolute permanence on the other (Sandy Wilson's new British Library). He decried those who, for instance, lamented Detroit's admittedly crass commercialism. "The old, standardized and unquestioned, public-school-pink proposition that all common taste is bad, and all commercialism is evil, appear to need some revision," he wrote "The concept of good design as a form of aesthetic charity done on the laboring classes from a great height is incompatible with democracy as I see it."[41]

Banham's point was that it was absurd to expect that objects designed "for a short useful life should exhibit qualities signifying eternal validity," abstractions like "rule proportion," "pure form," or "harmony of color." But when he referred to things with a short life, he did not *always* mean ephemera. A building may stand for a millennium, but its equipment has to be replaced every forty or fifty years, and the furniture will hold up no longer than twenty or twenty-five. Nor is everything equally valuable: a mathematical formula scratched out on a sheet of graph paper used to verify the safety factor of a roof truss may wind up in the wastebasket alongside the morning's sports section, but it would be a mistake to claim that the sports page was just as significant because of its equal transience.

All are part of pop culture, but the mere fact of their ethereality does not automatically mean they are shoddy. "Contemporary popular culture, which is a function of an industrialized society, is distinguished from other "fold art" by its refusal to be shabby or second-rate in appearance, by a refusal to know its place, Banham wrote. "Yet the articles of popular culture are made, not to be treasured, but to be thrown away."[42]

Even if that "throw away" period may be ten minutes for the hamburger box, five years for a car, or thirty years for an industrial air conditioning unit. In the end, this is what the members of the Independent Group had been trying to accomplish all along. "The protagonists of 'Pop' art," explained Banham, "maintain that there is no such thing as good and bad taste, but that each identifiable group or stratum of society has its own characteristic taste and design of style.[43]

The concept of "Pop" as both a form of contemporary culture and an art form was first realized only a few months after the IG stopped meeting. If "Parallel and Life and Art" was the *locus classicus* of New Brutalism, then "This Is Tomorrow" was the *locus classicus* of 'sixties High Pop. Or at least one-twelfth of it was. Held in the summer of 1956 in the Whitechapel Gallery, "This Is Tomorrow" was broken into twelve sequential, walk-through spaces, like market stalls. Each was put together on a low budget by a team, most combining an artist, designer, and architect. About a third were IG participants.

What put "This Is Tomorrow" into the history books was Team 2's space, a sensory stimuli exhibit put together by Richard Hamilton, John McHale, and Joel Voelcker, with help from the Cordells and Hamilton's wife Terry. It featured a 16-foot high movie poster of Robbie the Robot, an only slightly smaller one of Marilyn Monroe, blaring music, an optical distortion corridor with a squishy floor that emitted strawberry or lavender odors when you walked on it—all in a 12x16 foot space. According to Banham, a lot of people came in off the street, walked through Team 2's space, peeked around the corner, saw that the rest of the sections were mostly traditional art gallery stuff, turned around and left.[44] John McHale later made an observation that seemed to slip past everyone else at the time:

> I can't speak much for the other groups, but our section did a very good social forecast of a decade hence. It was very interesting. To pull that off was really quite extraordinary, because it was done mostly on intuition and acute observation of what was going on in that period. For example, it had multiple projection, a full range of audio-visual means being used, the background music was rock on a jukebox playing the most popular numbers of the period, and that kind of ambiance was carried out throughout the exhibit.... If you took the whole exhibit and transported it ten years forward, you got London in the 1960s. Swinging London. Very interesting.[45]

II

But for much of this period, the IG'ers were carrying on in the absence of their friends Mary and Peter Banham. Soon after Peter got his first editor's job at the *Architectural Review* in September 1952, Mary got pregnant. Their daughter Debbie was born in June 1953. Ben followed in 1955. In October 1957 Mary went to the doctor with a pain in her thigh, which she thought was rheumatism, which ran in her family. Her general practitioner sent her straight to Putney Hospital. After two days of tests, the doctors determined that it was a fast-growing cancerous tumor. Her left leg was amputated at the hip.

Mary's father had just retired, although her mother was still working for the civil service. They had moved into a house in Kent "that was actually much too big for them," but that they had bought because it sat on a large lot and Mary's father planned to spend his retirement gardening. "I couldn't go back to the place we were living [in Primrose Hill], which was all stairs, so Peter and the kids went to live there [Kent], and Peter traveled up to Westminster every day, and we gave up our place in the city."

Mary was in the hospital nine weeks, then was transferred to a rehabilitation center in Burnham Beach for over two months. "Eighty men, fourteen women—mostly men injured in the coal fields. I never bought a beer the entire time I was there." Rather than get a place when Mary got out, the family decided that they would continue to live in the big house in Kent until Peter finished his dissertation in 1959.[46] It was a trying time for everyone. "My father never understood him," recalls Mary Banham of the difficult relationship between her husband and father. "My father, while a great reader, was self-educated, and he was an outdoors person.... My father did not understand why Peter couldn't go outside and help him in the garden, when Peter only had Sunday to do his Ph.D. studies, so they didn't understand one another at all."[47]

In the summer of 1959 the Banhams moved to a house in Swiss Cottage, at that time considered the north London suburbs. "I felt out on a limb," Mary admitted. "I had a prejudice against the suburbs. I'd spent eleven years trying to get out. It's different now, but when I grew up in the suburbs, everyone had 2.8 kids, everyone had the same professional views.... I thought it was the dullest place on earth." Two things changed her mind. First, they discovered that Peter Cook lived across the street. Cook, who would go on to become the spokesman for the avant-garde design collective Archigram, was just starting to make a name for himself. Just as Sandy Wilson had lived next door to the Banhams in Primrose Hill, the Cook household became a source of accompaniment and stimulation to both Banhams. "Peter Cook was wooing Peter Banham across the street to see what was happening, and it didn't happen for a little while, Peter was away or something, and when he did go and see what they were up to, he was absolutely gobsmacked."[48]

In addition "the Architectural Press did a wonderfully humane thing." They lent the Banhams the money they needed to buy a bubble car—"a Heinkel, a three cornered object ... all the children in our street in Swiss Cottage thought it was the best thing— it was their scale." Mary later recalled that with her disability, it was "what allowed the whole situation to operate, really."[49]

Banham was back in business. Although he had kept up his short "Not Quite Architecture" column in *Architects' Journal* and even added a similar column in *New Society* in mid–1958, he hadn't published a long-form piece in his own *Architectural Review* since "The New Brutalism" in December 1955. In April 1959 he published "Neoliberty: The Italian Retreat from Modern Architecture," and in February 1960, one of his best known pieces, "Stocktaking," a comparison of old traditions and new technologies on architecture. "It was a great period in London in the late '50s and early '60s," recalls Mary Banham, "you could put up antennae and pick up people with like ideas and sit over coffee in some coffee bar somewhere all night and discuss things. You can't do that anymore."[50]

III

The salesmen at the 1948 Earl's Court Show Motor Show may have bemoaned the government "export first" mandate that was keeping cars away from their customers, but as early as 1950, Britain's export volume was 50 percent higher than in 1937, and its share of world trade actually increased from 21 to 25 percent, the seemingly omnipresent Americans notwithstanding. England was slowly but surely working its way out of its economic black hole. Exports in 1951 were 174 percent of 1938; within one percent of target, and without runaway inflation and with full employment. Men killed in industrial accidents were down from about 3,080 in 1929 to 1,564 in 1950. The average manufacturing operative in 1949 worked 46.5 hours compared to 54 hours in 1939. Manual workers, who had made up 78.1 percent of the workforce in 1931, comprised only 72.2 percent in 1951. Clerical and technical workers increased from 13.5 percent in 1935 to 18.6 percent in 1948. The only static demographic was female workforce participation. It was 29.7 percent in 1931; 30.8 percent in 1951. However, the *range* of occupations they represented was far wider then in the pre-war era.[51]

In mid–1953 butchers were permitted to sell meat freely once they had provided all their regular customers with their weekly apportion. By the end of the end of the year most were reporting that ration books were unneeded, even if still officially required. You could even get what had been a ration-coupon stew bone now *gratis*, thrown in for Fido with the price of a couple of good steaks or a chop. From less than 65,000 televisions in 1947, there were 126,600 in 1949, 344,000 in 1950, 1.45 million in 1952, 2.32 million in 1953, and 3.8 million in 1954. During October 1954 alone 163,872 new home television licenses were issued. In 1952, 209,000 new dwelling units were completed; 85 percent built by the government. In 1955, the peak year for post-war housing construction, 283,000 new units were built; 63 percent by the government. In 1959, 250,000 units were added; 59 percent were built by private developers.

The mode of transport to work was starting its dramatic change, a function both of rising affluence and the population dispersal efforts of planners. With many fewer walking, a small decline in cycling, a shift from the dreaded trolley-buses (they froze up and blocked the streets every time the power failed, a frequent enough occurrence until the mid 1950s) to the dependable but smelly diesel buses, transit was still king of the road. Auto and motorcycle use was up, but there was no great surge, as there had been in the United States in the 1920s.[52]

	1930–39	*1940–49*	*1950–59*[53]
Walking	22.5	17.2	13.4
Bicycle	19.2	19.6	16.4
Bus/Trolley/Tram	23.5	29.7	25.8
Train/Underground	22.5	23.9	23.3
Car/Motorcycle/Van	11.3	8.8	19.6

Nevertheless, all through the 1950s, and especially after the end of the Korean War in 1953–54, private affluence substantially increased for the vast majority of the population. In 1951, the average weekly income for a male over 21 was £8.30 ($23.25); by 1961 that had increased to £15.35 ($43.00). Young people, unaccustomed to the idea of disposable income, had no general or identifiable taste of their own in the 1950s, but what little they did have reflected that of their families or class.[54]

Yet, as late as 1955, *Socialist Commentary's* Rita Hinton could complain that "this is still a squalid country and all too many people are still compelled to lead squalid lives. Look at the ugly towns, at the mean streets, the cramped and shoddy houses. Look at the crowded classrooms, the scarcity of teachers and the wretched playgrounds. Look at the amenities for community life offered to most of our people. Then ask how near we have come to building Jerusalem 'in England's green and pleasant land.'"[55]

But with increasing affluence earlier in life, improved global communications, and above all, the globalization of entertainment, many, if not most Britains would, by 1955, already be in disagreement with Hinton. Europe was either still pre-consumerist or anti-consumerist (i.e. socialist). England's own domestically generated culture was self-consciously anti-popular, atavistic, retrograde and insular. As Banham recalled "the emphasis and most of the content of this culture was American, because there was nothing else. Once when we couldn't get American comics we bought a copy of *Punch* by mistake and never again, do you blame us?" But this transatlantic gift was not an unmixed blessing: "it left many of us in a very peculiar position ... we had this American leaning and yet most of us are in some way Left-oriented, even protest oriented.... It gives us a curious set of divided loyalties. We dig Pop, which is acceptance-culture, capitalistic, and yet in our formal politics, if I may use the phrase, most of us belong firmly to the other side."[56]

Eventually, of course, there did develop an indigenous British Pop culture that was in many ways different, and at times more nuanced than its American progenitor. "Pop did not change the structure of society, but was itself a symptom of changes that were occurring," recalls Nigel Whiteley. "Nor did it permanently alter the basis of taste formation ... but Pop did affect taste. It made us less conservative, less sure of our taste, more tolerant, and more open-minded. It opened our eyes and taught us that design could be colorful and imaginative."[57] "When Peter and his friends exited from the ICA in the 1950s," explained Gillian Naylor, referring to the Independent Group's sometimes raucous evening meetings, "they were celebrating the Brave New World of consumption—a world of abundance, expendability, irreverence for the past, and contempt for establishment values.[58]

And so in November 1962–the year Whiteley identifies as the dividing line between British "Early Pop" and "High Pop," Reyner Banham, Senior Associate Editor of the *Architectural Review*, and as close to a freelance design and technology columnist as you were likely to find in England at the time, walked into the Earl's Court Cycle Show in London, probably looking forward more than anything to the latest Italian or German motorcycle offerings. Instead, he found a crowd thronging around an unusually large exhibit for a bicycle maker he had never heard of from a small town in Wiltshire, near Bath. He was about to come face-to-face with his own cultural revolution.

3

Cycling in a New Key

> He lives in a wild Victorianized Jacobethian manor-house just outside Bradford-on-Avon. The stables are full of peculiar pieces of machinery thumping lumps of rubber to bits; across the end of the stable-yard stands, unexpectedly, a new drawing-office block in a distinctly Frank Lloyd Wright idiom, and Moulton himself is apt to stroll into this scene with a canoe on his head, looking every inch the English bird-watcher. Appearances can be deceiving, however; some sort of nut he may be, but he is the kind of nut from whom the really creative ideas in technology come.
>
> —Reyner Banham (1963)[1]

I

Alexander Eric Moulton could never remember wanting to be anything other than an engineer. He was born on April 9, 1920, in Stratford-on-Avon to Beryl Latimer Greene and John Coney Moulton. His father was an entomologist who graduated from Oxford in 1909 and was promptly recruited by Charles Brooke, colonial administrator of Borneo, to establish a new natural history museum in Kuching. In 1912, Brooke recruited Dr. Downes Greene as his principal medical officer, to serve in Sarawak, the administrative capital of Borneo. Beryl, Dr. Green's sister, accompanied him and in Sarawak met Capt. Moulton. They were married in 1914. Beryl Moulton returned to England for the birth of her first child, John, Alex's older brother, in 1916.[2]

During World War I, Capt. Moulton was assigned to the post of Coordinator of Intelligence Services for the East in Singapore, on the Malaya Peninsula adjacent to Borneo. During the war Beryl remained with him and she gave birth to a daughter, Dione, in 1918. Now-Major Moulton resigned his commission after the war and was hired as the Director of the Raffles Museum and Library in Singapore in 1919. A year later Beryl again returned to England and Alex was born at the home of his maternal grandparents. But during another return visit home in 1926, John Moulton contracted peritonitis and, being at sea, could not be operated on. He died shortly after making landfall in England.[3]

Alex's great-great-grandfather Stephen Moulton had acquired the British rights for vulcanized rubber from the American Charles Goodyear in 1842, four years after Goodyear's breakthrough. He originally intended to broker the rights to British manufacturing interests, but was unsuccessful, so in 1846 went into business for himself, purchasing an all-but-abandoned wool mill on the former estate of clothier John Hall.

Hall had built the manor house in 1598–1601, naming it after himself. After his death, the estate passed through a couple of generations of Halls until it was left to Rachel Banton of Great Chalfield. From the Bantons it passed by marriage to the first duke of Kingston about 1710. It stayed in the Kingston family until 1802 when it was sold to one Thomas Divett, who built a mill on the grounds next to the Avon River. However, by 1842 his heirs had not kept up with the latest power-loom technologies and the move away from broadcloth, and by the mid-nineteenth century both the house and mill were in decline, the former manor house by now being used for a warehouse.[4]

By 1848 Stephen Moulton had converted the old mill to the production of vulcanized boots, capes, tarps, and other goods, and about 1850 renovated and expanded the manor house. Another Englishman, Thomas Hancock, had reverse engineered Goodyear's technique and filed a British patent two weeks in advance of Goodyear, so Moulton used a vulcanization process he had developed himself, based on hyposulfate of lead instead of Goodyear's pure sulphur. Long, plodding litigation ensued before the parties eventually settled. George Spencer, a supplier of rubber springs and buffers to the railroad industry, became Moulton's largest customer. In 1891 the two firms merged into George Spencer, Moulton & Company, with headquarters in London.[5]

The Kingston Mills lay at the foot of the hill upon which The Hall stood, the base of which formed the north bank of the Avon River. In addition to The Hall, the Moulton

Former textile mills along the Avon River, Bradford-on-Avon. Wiltshire County was once the heart of South England's broadcloth industry. The former George Spencer, Moulton, rubber factory is across the river from the tall mill building in the background (photograph by the author).

family had retained ownership of the buildings and grounds of the Kingston Mills after the 1891 merger. In 1941 these were inherited by Alex, John and Dione. The Kingston Mills needed expansion, and the only available land was a site called "the Paddock," part of The Hall grounds, but down by the river, adjacent to the existing factory. Jack Spencer, head of the firm, offered to buy the Paddock land and the Kingston Mills on behalf of the company. Their grandmother had been a relatively small shareholder in the firm itself; most of her interest came from owning the factory and the land under it. The three siblings agreed to sell. Dione accepted cash, but the brothers took stock. In doing this, they shifted from being minor shareholders who were primarily landlords to major shareholders who retained only the house and its grounds, although these were extensive. The "Abbey Mill" was built on the Paddock site.[6]

"By this time I had no doubt that I wanted to be a designing engineer," Alex later said. He believed that this resulted from a combination of an "innate curiosity" fed by reading from books and magazines on engineering with "those wonderful cutaway drawings," which he recalled "moved me enormously." That included bicycles: "the illustrations in *Cycling* had already hooked me on that wonderful machine." More than once as a boy, he had cycled the 75 miles from Bradford-on-Avon, the home of his paternal grandmother, to Stratford-on-Avon, where his mother's mother lived.[7]

His father had attended Eton, but to save money, his grandmother and mother sent him to nearby Marlborough College. While there, he converted a GN light car to steam power; a photo from the era shows the typical "hot rod": no body; fenders gone, just a seat and dashboard. Only the mass of plumbing snaking out of the engine

The Hall, Bradford-on-Avon. Built in 1610 by the clothier John Hall, it has been the home of the Moulton family since about 1848 (photograph by the author).

compartment gives a hint of the radical changes. "I achieved some distinction by being the only boy to be driven to school each term with his own lathe in the back of the car," which he set up in the woodworking shop to make parts for the ever-evolving steamer. Moulton took the mechanical science Tripos for Cambridge, passed, and was admitted. His family paid £50 for a five month apprenticeship at the Sentinel Truck Company at Shrewsbury, maker of steam trucks powered by coal, while he waited to be admitted at Cambridge.[8]

He started at 1938 at King's College, but after the end of his first year he applied to the Royal Air Force as an engineering officer and was accepted. However, there was no opening in the RAF's engineering training program until April 1940, so he applied to work temporarily at Bristol Aeroplane, where chief engineer Roy Fedden was the godfather of one of Moulton's friends. He was accepted for the interim. But a few weeks after Moulton started work, a Luftwaffe bombing raid killed several key Bristol technical staff members, including Adrian Squire, one of Fedden's assistant chief engineers and head of the Centaurus engine development program. Fedden made Moulton his personal assistant and liaison to the Centaurus program. His induction into the RAF was cancelled.[9]

This was exactly the same time that Reyner Banham was also working at Bristol and participating in an engineering management program conducted jointly with Bristol Technical College. One would have thought the two would have crossed paths, but neither man ever mentioned running into the other.[10]

Fedden, in Moulton's words, was one of the last "great engineer-autocrats," responsible for many of Bristol's famous air-cooled, radial aircraft engines; the Perseus, Mercury, Hercules and finally the 18-cylinder Centaurus. He was a temperamental man, who believed wholeheartedly in his creation, the sleeve-valve engine. The water-cooled aircraft engine designs of Rolls-Royce, Napier, Allison (U.S.) and BMW (Ger.) had their cylinders running in two straight banks, like that of a V-8 automobile. All of the equipment needed to open and shut the intake and exhaust valves of each cylinder bank ran in a straight line on the top or alongside the cylinder heads. But in a radial, such as the designs of Bristol, Pratt & Whitney (U.S.) and Curtiss-Wright (U.S.), each cylinder sprays out like the arms of a starfish, so each cylinder needs its own set of rods, lifers, rocker arms, and so forth. The amount of delicate equipment needed to open and close the valves precisely on time became enormous as radials grew to nine, then eleven and finally, with the Centaurus and the Wright R3350, eighteen cylinders, running in two spiral rows. Fedden's sleeve-valve eliminated these moving parts by using a ported sleeve that surrounded the piston cylinder and that moved up-and-down and rotated about 15 degrees, sequentially opening and closing intake and exhaust valves by first lining up, then moving over, matching ports in the side of the cylinder. However, the ports in the sleeve and the cylinder (and the cam that moved the sleeve) had to be made to incredibly exacting tolerances, and with many of the best machinists now overseas, much of the sleeve-valve engine's construction had to done by hand. Bristol's management sought a hybrid; Fedden refused to even consider it. In 1942 Bristol fired him.[11]

Many years later, Moulton said that what most impressed him about his work at Bristol was the suddenness with which all this technological prowess was swept away. The reciprocating piston aircraft engine had been virtually perfected, not just by Bristol, but by Rolls-Royce, Napier, Pratt & Whitney, Curtiss-Wright, Allison, Lycoming, BMW, Daimler-Benz, Hispano-Suizia, Mitsubishi and others. Then came the turbo-jet engine.

> This was to render a whole species of aero-engine extinct, including both families—the air-cooled radial and the inline water-cooled. Created and out of production in the course of a single lifespan! ... This is the consequence of a great invention leading the "jump" change. Leaving aside mere change for fashion's sake, the thrust behind an invention can be traced to one man's driving interest in his imagined jump of improvement—the next step in the evolutionary chain.[12]

After the war, Rolls-Royce bought out the aeroengine division of Bristol primarily to acquire its experimental gas turbine [i.e. jet] engine project, the Olympus. In different configurations, the Olympus would power several military aircraft, the Concorde supersonic transport, the HMS Ark Royal aircraft carrier and scores of electric power plants.

Moulton stayed at Bristol until late in the war when he joined Fedden's attempt to start a new car company. The Fedden auto supposedly borrowed from the design features of Ferdinand Porsche's Volkswagen, the blueprints having been copied by the British Army during the two years it had overseen the reconstruction of the VW factory in Wolfsberg and the resumption of production. It's possible, but unlikely. The Fedden did look a lot like a slightly larger, four-door VW Beetle, and it did use an air-cooled, rear-mounted engine and an independent, torsion-bar rear suspension.[13]

However, its engine was a three cylinder radial that stood upright. The VW's boxter four lay flat. That was the Fedden's Achilles' heel. The independent suspension/rear engine combination of the VW Beetle (and that of the '60s Chevrolet Corvair, an air-cooled, flat, boxter six) was always marginal, especially when pushed to the limit in a curve on an uneven road. With the engine weight in the rear, the back end wanted to pogo on a washboard road. When bumps momentarily lightened the rear end on a curve, instead of just breaking free and skidding (oversteering) as was the case for a beam rear axle, the independent suspension wanted to tuck the outside rear wheel under, causing the rear end to "skip" sideways, and when it caught again, it sometimes rolled the car. Because its engine was upright, the center of gravity of the Fedden was higher than was the Beetle's (and the Corvair's) and the prototype rolled during testing. End of funding; end of project.[14]

Moulton had the opportunity to return to Bristol but "the call of the house at Bradford-on-Avon proved too strong; it was empty then and very run down." He returned to George Spencer, Moulton & Co. where works manager Jimmy Crystal and head chemist Dr. Sam Pickles (the man who first figured out why Goodyear's vulcanizing process worked) put him through a crash course in rubber science and technology. At their urging, he returned to Cambridge in 1946 to finish his master's degree in mechanical design.

In 1948, at Alex's urging, the firm built the Centenary Building to house a research and development laboratory. He was particularly interested in the problem of bonding rubber to steel. It would seem a simple problem, but it had eluded the best minds up to that point. Getting rubber to adhere to steel so it would resist shearing or twisting (torsion) proved to be a surprisingly tricky problem. Moulton's research lab figured it out using a combination of chemicals and heat. The first application was the Flexitor, a torsional rubber spring, initially for use in the suspension of camping and cargo trailers. "But my ambition was to get it adopted for automobiles" Moulton recalled.[15]

Enter Alex Issigonis, automotive engineer. Born in Turkey, he attended Battersea Polytechnic but had to settle for a diploma instead of a BSc. degree because he was too impatient to sit and learn all the math drudgery that acts as the standard culling-out method for engineering schools. He started at Morris Motors in 1936 after two years at Humber. He gained a reputation as an expert in suspensions. Moulton later said that his aversion to calculations proved to be an advantage, because it led him to do empirical testing on chopped up "mules" out on the test track, which often proved to be a better way to get results. After the war he designed the highly successful Morris Minor, a small, four-door sedan, introduced in 1947–48. Unhappy with the width of the prototype, he had mechanics cut one in half lengthwise and move it back-and-forth until he was satisfied with the result, which proved to be four inches wider than originally planned.[16]

Issigonis and Moulton met through one of Issigonis's "test mule" projects. The Morris test lab replaced the regular suspension of a Minor with an adapted set of Moulton's Flexitor springs up front and a new product, the Rotashear, in back. Although Issigonis had left Morris by the time the test model was ready, the program went forward. It worked, according to Moulton, "splendidly." "I regarded this test as the key experiment on rubber suspension," he told an audience at the Royal Society several years later.[17]

Issigonis couldn't get along with the elderly and chronically distracted Viscount Nuffield (William Morris), who had for years promised that he would quit his sporadic meddling in management and never did, and moved to Alvis to design their new sports luxury car, the TA/350. Issigonis asked Moulton, who was by now Assistant Managing Director of George Spencer, Moulton & Co. and its new subsidiary, the Flexitor Company, to design a suspension system based on the Morris Minor test car. He developed a system of springs based on two rounded cones that faced each other point-to-point.

A cutaway Mini. First built by the British Motor Company to exhibit at motor shows, it now resides at the Science Museum, London (photograph by Nora Quinlan).

After using the Alvis prototype for a few weeks, it occurred to Moulton and Issigonis that if the hollow space within the cones could be sealed and filled with hydraulic fluid, then cross-connected, they would function as both spring and shock damper. This was the theory behind what would become the Moulton-Dunlop "Hydrolastic" suspension system. Moulton later said that "I shall always remember the impression of big car ride luxury arising from the low frequency pitch mode when we first interconnected that Alvis." Alvis eventually decided against putting the TA/350 into production, and in 1956 Issigonis moved to the newly formed British Motor Corporation (BMC).[18]

In 1952 Morris (based in Cowley, Oxford) merged with Austin (based in Longbridge, Birmingham). The result was the BMC. The move was defensive; intended to resist the American money and American methods being poured into Ford of England and Vauxhall (essentially, General Motors-UK). Austin's Leonard Lord was put in charge. He was described memorably by one colleague as "ruthless yet capable of touching generosity, frequently guilty of rudeness to the point of cruelty, yet sometimes capable of admitting and apologizing for his mistakes, [he] was both crude in speech and manner and the victim of an inferiority complex. He detested pomp and also distrusted anything approaching sophistication in the running of a business."[19]

Lord had been hired in 1929 by Viscount Nuffield to reorganize one of his many acquisitions, Wolseley Motors. He did the job so well that Nuffield eventually put him in charge of the entire Nuffield Organization, a veritable grab-bag of overlapping car companies. But the two men did not get along and in 1936 Lord left for Nuffield's main competitor, Austin. Lord vowed to get even. Now, placed in charge of the merged BMC, he got his revenge by banishing Nuffield, age 77, to his vast Oxford estate and making Morris the junior partner in the new organization.[20]

Issigonis was originally put to work on three long-range projects, a large luxury car (the XC/9001), an intermediate family car (XC/9002) and a sub-Morris Minor city car (XC/9003), but those plans went out the window in September 1956 with the Suez Crisis.

All development was shut down except for the XC/9003, which became a crash program. It started as a downsized Morris Minor with front-wheel drive (Morris had built a front-drive Minor prototype back in 1952). However, it ended up being something quite different. It was short, wide, and had only two doors. The car's short overall length—ten feet—was possible because Issigonis placed the engine crosswise with the transmission parallel and under it, not bolted onto the end, as was normally the case. To save money it sat in the engine's oil sump.

The technical wizardry largely stopped at the back wall of the engine compartment. There was no dashboard, just an instrument pod mounted to the steering column. The windows did not roll down—they slid back and forth to open and close. It was originally planned to have the Hydrolastic suspension, but to save time and money used the all-rubber system of the Alvis TA/350. It was largely engineered on the fly—there were twelve prototypes, each of them different. The decision to develop it was made in November 1956. Prototype 1 (the "Orange Box") was put together in February 1957. The following July, Leonard Lord, after being taken for a drive in Prototype 2 (the "Red Box"), and shown Prototype 3 (the "Blue Box"), approved production tooling.[21] At that point the project ceased to be XC/9003 and became the now-famous ADO 15. The first

ADO 15 (an Austin Seven) came off the Longbridge line on April 3, 1959. The first unit made under the name "Morris Mini-Minor" was produced in Cowley in early August.[22] The "design by prototype" strategy had a profound influence on Alex, who went through almost as many development units in the course of making the Moulton bicycle market-ready.

The Mini actually didn't sell very well for the first year or two. Then, as Gillian Bardsley of the British Motor Industry Heritage Trust explains:

> As the decade progressed a whole host of people who were famous for very different reasons were being seen with the car: Peter Sellers, the Beatles, Twiggy—suddenly, it was the thing to have. The car took on a life of its own and even generated its own name as its new clientele abandoned the clumsy and old-fashioned labels "Austin Seven" and "Morris Mini-Minor" and simply called it the "Mini".... Nor was it an insignificant factor that for the first time the younger generation had their own disposable income. In an era when it was still not routine to own a car, it was one of the few cheap cars which it was "cool" to own. In this context the Mini seemed strikingly modern and in tune with the times, an emblem of a free and unrestricted lifestyle and a classless icon which could be afforded by practically anyone.[23]

This phenomenon would also repeat itself with the Moulton bicycle.

In 1955 George Spencer, Moulton & Co. faced one of the most important decisions in its history. The four major corporate railroads had been nationalized into the one entity, British Rail. The resulting supply contracts were enormous, and for potential vendors, an all-or-nothing lottery. You either supplied the entire nation's needs for, say, railway buffers or you supplied nothing. British Rail had issued a request for bids to provide rubber springs for all their rolling stock. Earlier, the board of Spencer, Moulton had hired an accounting consultant, David Grenville, to apply American-style marginal

The Paddock area at The Hall. First used to house testing rigs for Mini suspensions, now the home of the Moulton spaceframe bicycle (photograph by the author).

cost accounting methods in an attempt to get a handle on the true cost of their products. Grenville prepared the British Rail bid. It won. By a large margin. By too large a margin. Grenville had submitted a bid that failed to include indirect costs. Alex Moulton, without consulting the board, fired Grenville and summarily evicted him from the grounds. The next day the board backed him up.[24]

There was only one real option. Alex invited Charles Floyd, Chairman of Avon Rubber, to The Hall. Would Avon be interested in acquiring Spencer, Moulton, along with its big new contract? Indeed. On January 1, 1956, George Spencer, Mouton and the Flexitor Co. were sold to Avon Rubber for £352,500. The building and grounds under the Kingston Mill and Abbey Mill were included. The Moulton family retained The Hall. Alex agreed to stay with Avon for one year, after which he left to start his own firm, Moulton Developments, taking many of his former R&D personnel with him.[25]

It is typically thought that with the end of the Mini's development and the start of production in 1959 that Alex Moulton's involvement with automobile suspensions ended and he turned to bicycle development full time, but this is not the case. After launching the Mini, BMC began developing the mid-size XC/9002 concept car as the ADO 16, which became the phenomenally successful Austin/Morris 1100. Equipped with the Moulton-Dunlop Hydrolastic suspension, it was launched in 1962 and actually outsold the Mini, staying in production for 14 years. Less successful was the big car, the XC/9001 (ADO 17), which became the Austin/Morris 1800, introduced in 1964, also with Hydrolastic suspension.

Part of the former George Spencer, Moulton, rubber works, with Kingston Mill Road in the foreground. The Hall is atop a hill directly behind the photographer. Moulton Developments, Ltd., Alex's first company, was founded in the newer building to the right center, the one running parallel to the Mill Road (photograph by the author).

In 1968, British Leyland took over BMC. At the time, BMC was in the process of developing the ADO 14, the Maxi. It looked awful and several glitches that should have been worked out before its 1969 launch weren't. Issigonis was replaced as Technical Director of the Austin-Morris division by Harry Webster and, with less than two years to go before retirement, he was shunted off to a basement office. This resulted in a breach in the almost twenty-year-long relationship between Moulton and Issigonis. "I related to Harry Webster, we got on fine, a perfectly sensible chief engineer," Moulton later recalled. "He [Issigonis] expected that when he went out of the mainstream, everybody else who had worked with him should go too, that was a typical attiude for him I think." Moulton was representing the interests of his own firm, Moulton Developments, and Dunlop as well, and it was typical Issigonis hubris to believe that Moulton would throw over his professional relationship with British Leyland simply because he had been replaced.[26]

Cost conscious, British Leyland wanted to axe the Hydrolastic system and go to rubber or steel springs. (They did pull the Hydrolastic suspension out of the Mini, which had been added in 1962, and put the original all-rubber system back in.) Moulton convinced them to give his less expensive Hydragas suspension, then under development, a try. It proved its worth, and went into several later designs, including the Princess (1975), Metro (1980) and Rover 100 (1987). However, the relationship with Issigonis, the evenings in Monaco over brandy and cigars after the finish of the Monte Carlo Rally in January (often won by John Cooper's Minis) or the Formula I Grand Prix in May, were over. Issigonis retired in 1972, but stayed on part-time as a consultant. By 1977 he was suffering from debilitating bouts of vertigo that required him to work from home much of the time. He became increasingly reclusive. His contract with British Leyland ended in 1987. In early 1988 Moulton made one last attempt to visit the now house-bound Issigonis at his Birmingham home. He was turned away. Alec Issigonis died of pneumonia on October 2, 1988, at age 82.[27]

II

The Suez Crisis forced Alex Moulton to start cycling. A friend from his days at Bristol Aeroplane, Geoff Warren, sold him a Hetchins, a beautiful hand-built touring bicycle with many unusual features. "He got his Hetchins, he liked it, and then, being Alex, he questioned it," says Dan Farrell, director of design at Alex Moulton Bicycles.[28] Moulton first considered a recumbent, and purchased a pre-war Grubb, a long wheelbase cycle with handlebars under the seat, much like the American Avatar 2000 of 1979. However, he found the basic layout unsatisfactory. The recumbent's hill-climbing abilities were inferior to that of a regular bicycle and the Grubb's control response was slow. Theoretically, that could be corrected by shortening the wheelbase, putting the front wheel under rider's knees instead of in front of his feet (as was true for the Avatar 2000's sibling, the Fomac Avatar), but that just proved the narrowness of the recumbent's speed/stability range—responsive designs tended to become twitchy at high speeds, while stable layouts were loggy when going slowly. It was hard to design a good "all-around" recumbent.[29]

(Moulton revisited the issue in the 1990s with his spaceframe AM series bicycles. Over the years, a few mechanically inclined owners of the original F-frame series had developed conversion kits to turn them into recumbents. Alex had a chance to look over a couple. He thought the concept had some promise. The AM was amenable to being altered into a short-wheelbase, front-handlebar recumbent, so he asked Bath University engineering students to build an AM recumbent prototype. After evaluating it, he again decided against further development.)[30]

Although Moulton was not, at this time a world-famous figure, he had certainly earned a widespread reputation within the world of automobile engineering with his innovative suspension work with Alec Issigonis and Dunlop. So why would he spend years working on a bicycle? "Because it was perceived as a finalized design," answers cycle historian Tony Hadland. "It was thought that the basic architecture was finished," Dan Farrell adds.

> He was talking once about how Issigonis had changed cars for twenty years, and he said "We could do anything. People said 'you couldn't fly to New York faster than a bullet,' but we did it. They said 'you can't improve the diamond frame bicycle,' but we did." He did. It was something about making a change on a scale like Issigonis.[31]

The decision to use small wheels appeared to be less a matter of any great inherent advantage then an accumulation of objections to the standard large wheel size. His initial concern was the placement of the top tube, which he felt made the whole machine unnecessarily awkward and dangerous. "You can't get onto it in a natural way by putting your leg through [with a top tube]," he later recalled, "and second, in any kind of accident condition or need to get off it quickly, you're trapped by it. That's the thing I thoroughly objected to."[32] (Years later, during some press event for the new all-terrain version of the spaceframe AM, he swapped bikes on one of the downhill runs with someone. After a hundred feet or so on the normal mountain bike, he got off and walked the rest of the way, to raised eyebrows and chuckles.)

However, the biggest practical advantage of the layout is that it moved the center of gravity of loads on the front and rear carriers down about ten inches. Where loads of any significant weight had to be carried in saddlebags on either side of the front or rear wheel of a regular bike, with small wheels they could be carried on top of the wheels of a Moulton on racks directly over the centerline of the bike. Moulton himself recalled that:

> The wheel size had a tremendous effect on the whole architecture of the vehicle. I sensed at once that if one could only reduce the size of the wheel, then the whole layout of the bicycle, the whole architecture, could be enormously altered and improved. I was extremely sensitive to that. I was concerned that 26 or 28 inches could be fundamentally the minimum size that could be workably used ... so I knew that everything would have to be investigated and confirmed.[33]

Alan Oakley, Raleigh's chief designer, later asserted that this explanation was somewhat disingenuous. "He thought he could do for the bicycle as had been done for the car, that is, put a rubber suspension unit between the frame and the wheel. On a large-wheeled bike this would have created a tendency to lift the frame, and thus the wheel size had to be reduced," one reporter quoted him as saying. According to Oakley, Moulton went into the project intending to build a bicycle with a dual suspension system all along. The suspension's travel distance would raise the top tube if a conventional diamond frame and wheel configuration was used. Hence, the wheel size had to be reduced to bring the top tube down. The wheels had to be much smaller if a step-through frame was desired. Dan Farrell rejects this: "the first design, the Mk. 1, did not have a suspension." Tony Hadland also points out that the first full-size prototype, built with a sheet metal monocoque, had only a front suspension.[34]

"The methods of innovation of the [Mini] and the bicycle were different, although I was working on both at the same time," Moulton later pointed out. "On the former, it was a question of hammering out the solution of each element of the system by rigorous testing in an iterative, step by step manner. On the bicycle, once having determined the validity of the key item, namely the smaller wheel with its new tyre and rim, the position of pedals, seat and handlebars were fixed, as the silhouette [position] of the rider was conventional.... It was the selection of the form for the complete bicycle which was my challenge."[35]

But when asked point-blank what came first, the idea of making a bike with small wheels or one with a suspension system, Moulton's answer indicates that Oakley may not have been entirely off the mark:

> The answer to your question is: both. Remember, I came from a rubber background, in my family ... and that led me into doing a whole bunch of car suspensions, so it was natural. I thought that any vehicle carrying a human being should have a suspension, but I also knew that with the use of high pressure tires on a small wheel, it would be mandatory. So for both reasons, you had to have a suspension.[36]

Once the basic layout had been settled on, tire availability became the next problem. Small bicycle wheels were common, but the tire/rim combinations available below 20 inches were limited and of poor quality, intended mainly for sidewalk bicycles. Moulton turned to Dunlop, with whom he was already closely working on the Hydrolastic suspension system. They prepared 14- and 16-inch versions of their popular Sprite touring tire. After testing, a 16 × 1⅜ inch tire size was selected using a 60–70 psi tire pressure and a modified, 36-spoke version of an

The first Moulton prototype, a monocoque design made of sheet aluminum, February 1959 (courtesy Alex Moulton Bicycle Co., Ltd.).

existing rim configuration known as E3J (Endrick No. 3 juvenile), not uncommon in the industry, but normally used only for kids' bikes.

(The E3J used a 349 mm bead diameter. It was referred to as "16 × 1⅜." Later, the Moulton firm introduced a narrower format tire it called "16 × 1¼" that had a 369 mm bead diameter. The intent was to allow performance Moultons to swap between wheels with wired-on and tubular tires without changing the brake blocks. When Moulton launched his spaceframe AM series in 1983, the 349 mm size was abandoned in favor of the 369 mm, which was then referred to as "17-inch.")[37] "Dunlop provided me with the necessary small tires and I was able to confirm that it was tire pressure, not tire size [i.e. diameter] that mattered," Moulton recalls.[38]

It is interesting to note that famous wheelbuilder Jack Lauterwasser, who assembled Moulton's handbuilt "S-series" F-frame models at Bradford-on-Avon in the 1960s, then built the 369 mm, 17-inch wheels for the later AM series, always believed that the Moulton would have been improved if it used slightly larger 20-inch wheels, as did happen in 1988 when Moulton produced the all-terrain version of the AM series, the AM-ATB. "The 20-inch wheel is the best improvement he's done, especially for racing," Lauterwasser said in 1997. "He should have done it years before."[39]

The decision to use both small wheels and the relatively narrow and high-pressure tire combination crossed a rubicon of sorts: it committed Moulton to using a sprung suspension system. The suspension was no mere gimmick; it was an integral part of the bike's engineering. In the 1970s, after Raleigh had purchased the Moulton bicycle company, they brought out a version of the Mini-Moulton with 14-inch wheels and no front suspension. These had to be retrofitted with a large reinforcement gusset between the head and down tubes because without the suspension this joint tended to fail.[40]

Next came the frame configuration. Moulton first used a monocoque frame stamped out of sheet aluminum. It was light, relatively inexpensive, and would provide flexible mounting options for the suspension members.[41] The first prototype had only a front suspension, a crude but easily changed set-up using a see-saw trailing link suspended in front by rubber bungee cords. However, it did not work. "I was quite wrong in my original choice of the structure," Moulton later admitted. The loads were too localized. "A bicycle demanded a 'boney' structure, not a boxy one," he rued. Even worse, the thing acted as a giant drum that amplified road rumble. "I was horrified by the noise from the resonance of the monocoque. Five and a half pounds only, very light, but horrible noise, so I decided no way."[42] Although he reverted to tubing, he chose a single, oversize monotube, essentially a cross-frame layout.

There were at least nine generations of prototypes developed after 1959. Photographs exist of the Mk. 1 through the Mk. 8B, and the production version, introduced at Earl's Court in November 1962, was either Mk. 9 or Mk. 10. (The monocque was also not numbered, and the Mk. 1 may only have been a full-sized mockup.) The Mk. 3 was the version illustrated in Moulton's November 16, 1959, UK patent application.[43] Up to the Mk. 5 of June 1960 the bicycle did not use the now-famous buried-in-the-head-tube front suspension design. Instead, it used a very long leading-arm front suspension. A front monoblade wrapped behind the front wheel. A triangle made of seatstay tubing, identical to the rear triangle then being used, projected forward horizontally from this "fork." When the bicycle was loaded, this front-facing triangle rotated

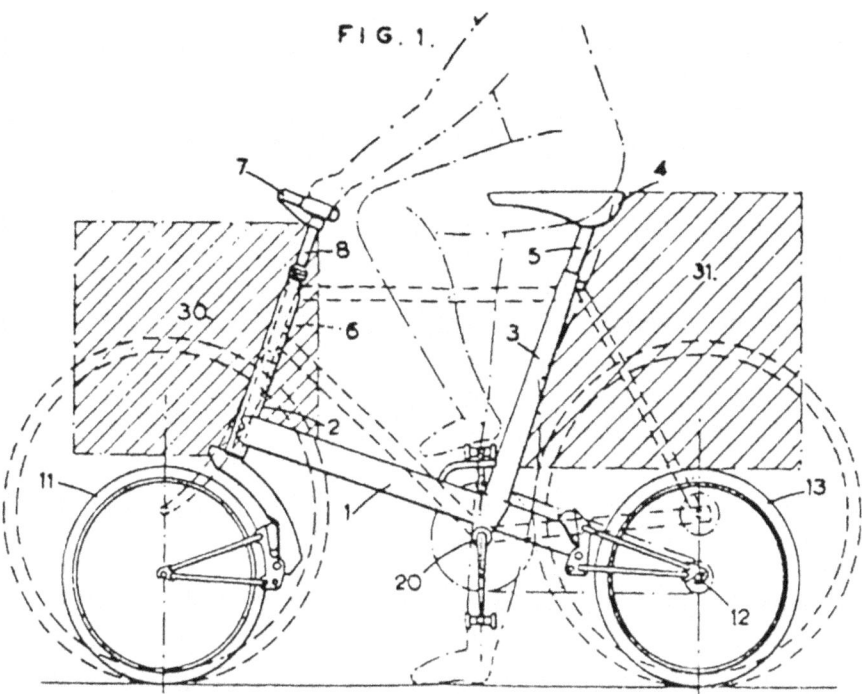

The original patent for the Moulton, 1960. Note the trailing link front suspension and the triangulated rear fork, neither of which made it into the production version (U.S. Patent No. 3083038/9).

upward, but for less than an inch, until it was stopped by a block of rubber at its base. One of the main purposes of this design, according to Moulton "was to communize the front and rear suspensions." The problem with this design was that the center of gravity of the entire front wheel and handlebar assembly wasn't in the right place, making for unstable "wheel flop" which is an important factor in a bicycle's low speed handling.[44]

The Mk. 7 of April 1961 appears to have been the first prototype to use the in-the-headtube front suspension. It consisted of a metal coil spring encapsulated within a sleeve or cylinder buried within the head tube. The metal capsule was itself mounted on a column of rubber that carried 80 percent of the bike's weight when loaded and ridden on a hypothetically ideal smooth road. The coil spring then compressed when going over a bump, and the rubber spring unit both absorbed bumping loads and damped the metal spring. Moulton later said if he were not so concerned over its odd appearance, he would have used an exterior toggle link to turn the suspension unit along with the front wheel, as is used on the front landing gear of aircraft (and which Bridgestone used in 1999 for their reproduction F-frame) instead of a splined steerer tube and serrated bushing.[45]

The Mk. 7 (or Mk. 7B) of 1961 was the first to dispense with the rear triangle in favor of a straight cantilevered rear fork.[46] The rear suspension was originally intended to be a long, narrow, horizontal triangle made out of chainstay tubing. It was about eighteen inches front-to-back and six inches high. The rear wheel had a vertical travel of an inch or so. (This is the configuration that would be used on the later Raleigh-

Moulton Mk. III, except that the triangle would be much higher on the Mk. III; an isosceles triangle.)[47]

On the Mk. 7, it was decided to dispense with the triangle and go with a simpler arrangement where the rear wheel was held in a rearward-facing cantilevered fork that had a bumper plated brazed to it that engaged a rubber block that was attached to the frame. The fork was straight, an extension of the main frame monotube. Late in the development process (sometime after the Mk. 8B of October 1961), the rear fork was again changed to make it curved. According to Moulton, this was purely an aesthetic alteration that allowed the rear suspension mechanism to be hidden under the bicycle. "With the first model, I was extremely concerned with appearance because I was doing a really brutal thing; I was imposing on the public an enormous change from the classical bicycle." However, the curved configuration placed a tremendous quality control responsibility on those fabricating the rear forks, and when consumer demand proved higher than expected and Moulton had to start farming out production, the production process used by the subcontractor, which was different from that used by the Moulton factory itself, wasn't always up to the task.[48]

Moulton later said that he had no interest in producing a folding variant, although he did include a folding feature in his patent claims. On the other hand, he did plan to produce a break-apart version and showed projector slides of what would become the Stowaway at the November 1962 Earl's Court launch. At one point in his memoirs, he implies that it was his plan to make all Moultons break-aparts once the concept was proven, but was not able to do so, probably by cost considerations.[49]

"I had no intention of making the bike myself," Moulton later explained. But after negotiating with Raleigh off-and-on between November 1959 and January 1962, the Nottingham giant decided to give the project a pass, and Moulton fell back on his contingency plan of building his own small factory on the grounds of The Hall. He gave conflicting accounts of this decision. In one interview he said: "but before I did [decide to go into production], I said to myself, am I mad? After all, if Raleigh said it was unmarketable, who was I to say differently?" But in his memoirs he wrote that "I drifted into manufacturing. I don't remember any great agonizing but just felt under the circumstances that it was natural to do."[50]

David Duffield was the newly hired marketing manager for the newly formed Moulton Bicycles Ltd. John Macnaughtan, who later roomed with Duffield when they were setting up TI-Raleigh's Australian subsidiary, described him as "the most energetic person I ever met. He'd ride a hundred miles before breakfast, then go to work for a full day." As a young man, after the war, Duffield had gone to work as a graphic artist for a couple of advertising companies, then moved to Phillips in 1952. Up to that point, Phillips had been a maker of bicycle components, but were now expanding into complete bicycles. After six months, he was moved into sales. "They had 14 field salesmen. I was the youngest, but started spending more and more time in the office working on colors and graphics and the like."[51]

In 1956 the giant Tube Investments conglomerate, better known as TI, integrated its Phillips, Hercules and Armstrong bicycle operations into a single unit called the British Cycle Company (BCC). This left the British bicycle industry with a duopoly: TI's BCC division on one hand with its three firms, and Raleigh on the other hand, which in addition to its own label offered Humber, Rudge-Whitworth, BSA, the semi-custom maker Carlton and the parts maker Sturmey-Archer. "The industry was declining tremendously, and there were just these two big groups left by 1960," Duffield recalls. He was preparing to become head of advertising for BCC when TI suddenly announced that it was acquiring Raleigh. Duffield and his colleagues were ecstatic:

> We thought, "we've got it now, we've cracked it." We were being the innovative ones, and they were still stuck in the mud. Raleigh was very conservative—they were still using Sturmey-Archer hub gears for everything when we had developed our own derailleur and so on. We thought we would have the opportunity to take over and change things because in actuality, it wasn't a merger, TI acquired Raleigh. But TI turned the management over to Raleigh, they only put six TI people into management. They offered me some job but I turned them down.[52]

Duffield left the bicycle industry and went to work in London for three years. In 1962, a friend, Roy Day, showed him a photo of the engineering drawings for a small-wheeled bicycle that Day had taken during the annual open house at The Hall at Bradford-on-Avon. Being a draftsman himself, Day recognized what looked like a fairly substantial drafting studio added to the stables at the back of the manor house. Wandering over, he looked in the window, where he saw the Moulton's engineering drawings and snapped a couple of quick shots. "He ran over and asked Alex what it was all about," Duffield later recounted. "Alex wasn't all that upset, because at this time Raleigh had said that they were not going to make the Moulton. Roy had worked for BSA Cycles (bought by Raleigh in 1957) and Alex said he was looking for people to help run his

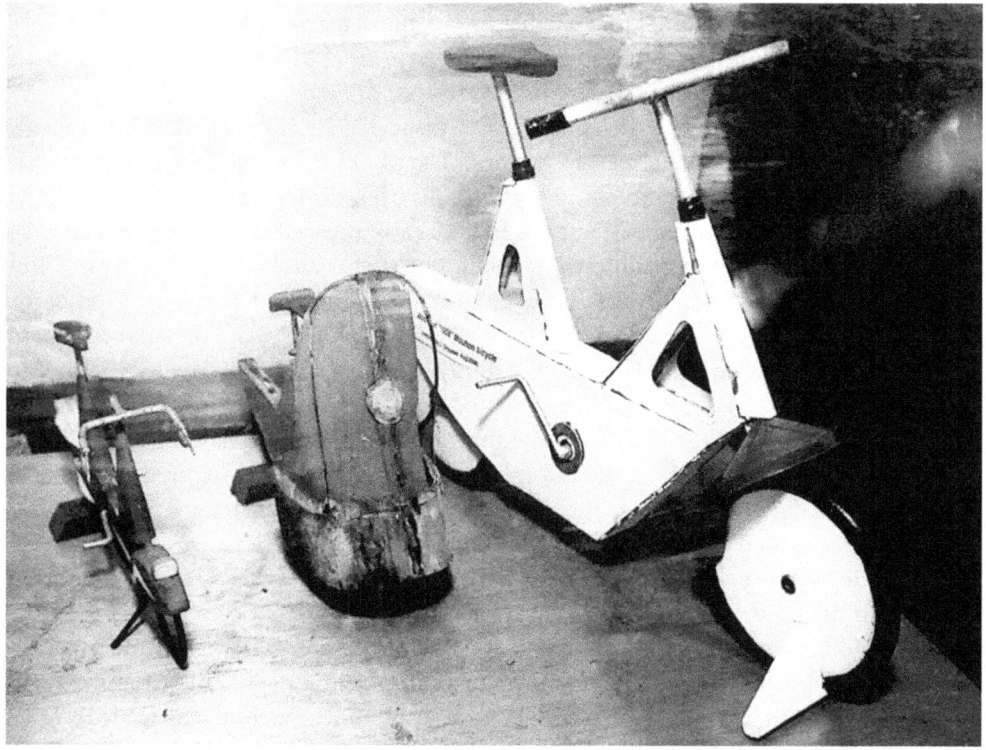

Early design models for the Moulton bicycle, 1957-1959 (photograph by the author).

operation." Day put Duffield in contact with Moulton. Duffield went to Bradford-on-Avon and test rode a prototype. "I though it was great." He became Moulton's marketing director.

Moulton ordered plans to be drawn up in mid–1962 for the factory at The Hall and by the time of the Earl's Court show in London in November, construction had already started. "I don't know if Alex asked Hercules or BCC to build it," Duffield said. "I was in a position there where I probably would have known if such a thing happened," and he never heard of any such talks. (Moulton later said he did not consider another manufacturer.) "Raleigh said the magic number that was needed was 200 per week," and that is what Moulton's new factory had been planned around.[53]

Duffield started preparing for the show while he was still in London. "I told Alex, you'll need multiple bikes there or people will think it's just a one-off thing," a show bike. His advice proved prescient. Alex had wanted one of the booth's highlights to be a special "Speed" model that racing cyclist John Woodburn would use to break the Cardiff-to-London record on November 15, just before the show. However, the attempt had to be abandoned about half-way due to bad weather, and the bike wasn't available while the team set up for a renewed attempt.[54]

Due to Duffield's foresight the Speed model wasn't missed. There were 14 bikes shown and Duffield later recalled that "we were desperately short of manpower and had to enlist friends, relations and anyone else to help us." The star attraction was a Moulton on a chest-high stand running on a set of front-and-rear bump rollers that made the suspension go up and down. "We had lots of newspaper coverage on the Friday before the show," Duffield says. There was a gentleman's agreement not to release pictures of the bike until Saturday, but the *Daily Express* cheated and carried half-page photos the evening before.[55]

Saturday was the open-to-the-public day. "We were just mobbed," Moulton recalls, "we were absolutely overwhelmed by orders." David Duffield says that is not exactly accurate. It was the goal of the company to build up a dealer network, "so we didn't really take orders," he explains. "But we had already printed up these cards to take names and numbers [to distribute to dealers], but they wanted to write checks!" But Duffield agrees that the stand was surrounded all day and that they collected cards almost as fast as they could hand them out. During the day, show president Edward Turner (accompanied by Lord Brabazon of Tara) actually came down to give the "bump rig" a look-over.[56]

Woodburn broke the Cardiff-to-London record twenty-two days after the show, on December 9, 1962, breaking the old record by 18 minutes (162 mi. in 6h:42m:29s). "I still think I'm the only person ever to break a point-to-point road record in December," Woodburn recounted some forty years later.[57]

His ride gave rise to one of the odder Moulton legends: that the Cardiff-to-London bicycle had no active suspension. Both Woodburn and Duffield claim that it had the standard Moulton setup, but according to Tony Hadland, photographs of the ride indicate that while it did have the standard mechanicals, some degree of stiffening (preloading) had been introduced.[58] Both Woodburn and a second works rider, Vic Nicholson, had been critical of the stock setup for racing in the period before the record run.

Woodburn recalled that the record attempt was done on relatively short notice. "Immediately after winning the National 25, I was riding for a shop in Finchley owned by Ted Gerrard. I was known as an independent, a semi-pro. I was approached by David Duffield to see if I would come over to Moulton's to do the record. Gerrard was okay with that. This was in 1962. We had an arrangement. I would go down on Mondays and train on the bicycle and test it and in fact do time trials." Moulton originally wanted to set a record for the standard international 25-mile (40km) distance, but later switched to the Cardiff-to-London point-to-point record, which had a reputation as a rather overlooked entry in the Road Racing Association's record book and was thus considered low hanging fruit, even though Woodburn had originally been hired based on his reputation as a short-distance time trial man.

The original plan was to use a lightweight front fairing called "the Cowl" made of doped linen stretched over a wire frame. "We originally were going to do it with the Cowl, but it was a bit like Chris Boardman's record [Graeme Obree and Chris Boardman used two different radical riding positions—since banned—to capture the one-hour record in the 1990s], people would say, 'oh it's the Cowl, not the bike or the rider,' so they decided to do away with the Cowl. Woodburn says that "I nearly killed them" because the Cowl was uncomfortable.[59]

"It wasn't easy to use, the Cowl," Woodburn recalls. "The trouble is that to get up to maximum speed you had to get your head down below the top of the Cowl, but then you couldn't see. It had a little windscreen made of clear Perspex, but it kept misting up, so I would have to pop up for a look, them drop back down. There was a woman gently cycling down the side of the road, I nearly hit her. It might have been a bit of bad publicity for Alex."[60]

About the time of the record run, the Moulton works hired Vic Nicholson. "I was invited to join Moulton because Moulton's wanted a good, local rider to ride it and do testing, and I was a local. The first day I arrived, they sat me down and said, "let's find you a job," and they made me a bicycle mechanic. The clubs were very cautious in those days about independent status, and you needed to be an employee if you weren't affiliated with a club and sponsored."[61]

Like many of Alex Moulton's associates, Woodburn found him to be capable of great warmth and charm, but focused, bored with social pleasantries, and a man who too easily lost sight of the fact that people were more than a means to an end. "I would do my training runs in the morning and then we would have these time trials with the Cowl after lunch, and frankly, it could be a bit of a nightmare, because Alex was expecting improvement every day, but the whole thing took some getting used to, the Cowl, and frankly the suspension took some getting used to … we really didn't have a lot of time." Nicholson concurs. "Alex looked at men as machines. You had to do the same thing the same way each time. One time he called over a local rowing club and told them 'look, you're not doing it right, it has to be the same way every stroke.'"[62]

Dan Farrell adds his own story. "He was a difficult man to disagree with. He could be charm itself or devastatingly rude. I walked into the Oak Room [Moulton's studio in The Hall] once at the same time as Toby the Cat. He said, 'I will see to Toby first, because he is more important.' He wasn't kidding." Toby had wandered in sometime around 2001, and he and Alex had become inseparable. "Identical personalities," observes Farrell. To facilitate his comings and goings, Alex hired a stone mason to install a cat door in the kitchen. Mind you, the walls of The Hall are a foot thick.[63]

Tony Hadland's impressions are similar: "he's very single minded. He doesn't suffer fools gladly. He can be charm itself, at other times he can be quite assertive, even quite abrasive. That's part of the personality of someone who is a thinker and a doer who actually gets things done. He's had his setbacks, but he's also had an awful lot of triumphs. A very interesting person of a type that is almost extinct, I think."[64]

Both Woodburn and Nicholson agree that the decision use the stock suspension was a mistake. Both believed that the factory needed to develop a special, stiffer racing suspension. "A racing Mini has a stiffer suspension than a regular Mini," Nicholson notes, "but Alex insisted that you could race a stock Moulton." Woodburn cites two main problems with the Moulton cycle he used: the tires and rims were inadequate and the front suspension was too soft. Unlike Nicholson, he believed "the rear suspension was okay, not a problem."[65]

Nicholson disagrees. "I found the suspension system to be a disadvantage, just a pogo stick. Every time you put pressure on the pedals, the front suspension went down. You got out of the saddle, the suspension system just went up and down. For a racing bike it was just no good at all. But Alex just wouldn't accept it ... so we devised things when he wasn't there."[66]

Together with Moulton factory mechanic Tom Crowther, Woodburn and Nicholson found that if a stack of washers was placed under the front suspension capsule, the suspension could be preloaded to about fifty percent of its travel. The washers were

An early photograph illustrating the evolution of the Moulton bicycle, taken inside the drawing studio. Moulton's Hetchins is second from left. The Mk. 2 prototype is second from right and the Mk. 5 prototype is at far right. The Paddock and Stables can be seen through the windows. Taken 1960-61 (courtesy Alex Moulton Bicycle Co., Ltd.).

held in place above the front brake mounting bolt. Nicholson did not like the way the rear suspension affected the bike's handling coming out of corners, so he preloaded the rear suspension as well by drilling a hole through the two bumper plates and the rubber in between, inserting a bolt, and tightening it until the rubber was about half compressed. Nicholson described what happened next:

> I used to have two bikes, the one I rode to work, which was stock, and the one I kept at home, which I raced with, so he wouldn't see it. But then, one day, we thought he was away, I brought it in because it needed some work, and it was down in the workshop [in the stables] being worked on, and Woodburn's bike was down there too—he was doing some testing—and his was stiffened, too. Well, John was riding down from the gate and Moulton noticed the front suspension wasn't working. "Woodburn, get off that bike, something's wrong with it." He wheels it to the workshop, where Tom Crowther was working on my bike and Reg Randall's, but Reg Randall's had a different suspension, normal, because he's getting ready to ride end-to-end [Land's End to John O'Groat's House]. So he [Moulton] pushes down on my bike. Thunk. Nothing happens. "Crowther, what's wrong with these bikes? Take it apart." Front brake off, washers spill out. He realizes what we've done, goes up in the air.[67]

(Moulton later developed a stiff suspension for the Speedsix. "His idea, of course," adds Nicholson.)

Woodburn was scheduled to try York-to-London shortly after Cardiff-London, but had bronchitis and could not make the attempt. Woodburn never again rode a Moulton in competition. Except for Reg Randall's two unsuccessful end-to-end attempts in August and September 1964 the Moulton record-breaking program went into eclipse until 1965 when Nicholson won 15 major time trials. "To say you weren't totally happy with it wasn't the best of ideas," rued Woodburn.[68] He believed that the original Moulton had the potential to be a high performance bicycle, but it needed further work.

III

Even before the 1962 Earl's Court Cycle Show had concluded, Alex Moulton phoned back to The Hall and ordered the contractor to add an extension to the under-construction factory on Holt Road. Even then it could only be expanded to 10,000 square feet. "I've never been so frustrated in my life," Moulton later recalled. "We were all set to produce 200 a week here when we launched at the Earl's Court show.... We were inundated with orders we couldn't cope with." It was starting to look like a production of a thousand a week may not be out of the question. In fact, production did not catch up with demand until November 1964.[69] According to Moulton, his fledgling firm was thrown a lifeline by Sydney Wheeler, Secretary of British Motor Corporation, who called on behalf of Leonard Lord and George Harriman, offering the services of the Fisher and Ludlow factory at Kirkby to make Moultons.[70]

The show was closed on Sunday in those days. Monday was the trade-only day. "You've had thought we had a moat of piranhas around the stand," Duffield recalls. "They saw the small wheels. 'It'll never sell.' They saw the price. 'It'll never sell.' We pointed at this huge box of cards we had collected on Saturday: 'OK, I might have one.'"[71]

Banham was apparently one of the Saturday visitors who filled out an interest card, and took delivery of his around September. His first articles appeared in *New Statesman*

The Holt Road bicycle factory being built on the grounds of The Hall in 1962. Note the racing model leaning against the unfinished wall in the foreground. This was very likely John Woodburn's Cardiff-to-London bicycle (courtesy Alex Moulton Bicycle Co., Ltd.).

(November 1) and *Architects' Journal* (November 6). About this time, he delivered the first Terry Hamilton Memorial Lecture at the Institute for Contemporary Arts entitled "The Atavism of the Short Distance Mini-Cyclist," clearly a take-off on the title of the recent (and controversial) Alan Sillitoe short story "The Loneliness of the Long-Distance Runner," a tale of mutually destructive class and generational conflict in postwar England.[72] Published in early 1964 in the magazine *Living Arts*, "Atavism of the Short Distance Mini-Cyclist" became the definitive statement linking the Moulton with the emerging high/low pop culture that had emerged from the Independent Group almost a decade before.

"It's a funny thing being back on a bike after twelve years," he said. "The reflexes don't disappear; you put your leg over it and sit on the saddle and you make off straight away, and everything you learned with such great pain, more or less at your father's knee, comes back absolutely instantaneously, its there, bred in the bone and blood. It's got to be a really weird bike, much weirder than the mini, before you come on a situation where you can't just sit on it and pedal off."[73] The Moulton might "feel loose and sloppy until you get used to it," he wrote, "simply because there is less gyroscopic inertia, [but] the response control is more delicate and immediate. It's a natural for threading through traffic, turning in narrow spaces, and craftily picking your way in general. It is, very neatly, the man-powered equivalent of the Mini-Minor."[74]

There were some changes, however. "Having one's nose out in the slip-stream is to reveal how much worse the air pollution has become." Despite this, "I find myself traversing central areas and the West End far more than I used to. Mayfair and Soho are very useful short cuts on journeys between London and London, and a bike encourages very keen shortcutmanship, because it pays off." On the other hand, "traffic sense, manners, lane discipline and all that jazz seems to have broken down almost completely. I have no complaints about the professionals ... but preserve me from the amateurs."[75]

But while it was no problem for legs acclimatized to the old black roadster to handle the new mini-cycle, it was entirely different question whether British minds could get themselves around it. The Moulton, he believed, should surely "be recognized as a minor cultural revolution ... bicycle thinking can never be the same again, and there can be no more nonsense about permanent and definitive forms ... it has put a new class of men in the saddle."[76]

Emily King, a junior curator at the Victoria and Albert Museum, recalled working with Banham on a proposal for a new exhibit, "Invention and Design." "We hoped it would stimulate a broader appreciation of design.... Banham's proposal for the show involved a study of chairs, bicycles, radios, typewriters, telephones and kitchen appliances." The museum's director, the staid and unabashedly elitist Sir John Pope-Hennessey, vetoed the idea as absurd: "I never go into my kitchen," was his only comment. (King recalled "vaguely hearing" that it was populated by a small army of immigrant servants.) King said that "I recall the astonishment still visible on Banham's face as he mounted his Moulton bicycle outside the main entrance and set off back to Gower Street. He looked like a man who had cycled to South Kensington thinking he was visiting the Victoria and Albert but had found himself instead face to face with a dinosaur."[77]

Banham had called bicycle manufacturing "one of the most somnolent industries in the world," so stuffy and unimaginative that its conservatism actually rubbed off on its customers. He complained of "the social conformism that seems to have been an integral part of the Safety subculture." It was true: the customer base for the bicycle in the United Kingdom in the late 1950s and early 1960s *was* stuffy and conservative. They *did* wear cloth caps and duffle coats. They *did* keep racing pigeons. They were, as one industry executive put it, "C2-D-E males of 40 plus."[78]

Cycle historian Tony Hadland recalls that:

> coming out of the bleak period of the 1950s when we had sweet rationing ... things were very dull, diet had been very dull, life in general had been very dull in the '50s. And then everything started taking off in the 1960s. There was that famous phrase about the "white hot technological revolution" that Harold Wilson came out with and there was all these new things, and there was sort of a belief that everything was going to get better in every way, hovercrafts, jet engines, all sorts of things were coming through, and the Moulton bicycle came through at the same time.[79]

This was a reflection of the economic realities of everyday life in Great Britain. The rise in average weekly earnings through the 1950s has already been noted; of equal importance is that this expansion extended all the way down to youth, the young men and women who had entered the work force only after the war. By 1960, the average unmarried male under age 20 had a disposable income, after food, rent, and other necessities, of £3.58 per week (roughly $12.50 in 1960 dollars, or $120 in 2016 dollars); for unmarried females under age 20, £2.70 ($9.45/$92). As one Brighton teenager recalled of 1957: "There was plenty of work. There was no trouble finding work.... Once you got to the age of sixteen, you could go into any builders or garage or anywhere and get a job."[80]

What is surprising is that this phenomenon was not seen by everybody as a reason for celebration. Sociologist John Barron Mays viewed it as a mixed blessing, at best. "In prewar days, the youth market was negligible and not worth exploiting, but since the war, teen-age incomes have, in fact, risen faster than adults," noted Mays. "It is certainly not a picture of either complete recklessness or utter depravity; it is merely one of personal indulgence, which seems to follow closely the general attitude of the time which counsels us to enjoy ourselves to the limit of our means while we can."[81]

Britain's closest observer of the new youth market, Mark Abrams, was more optimistic. There *was* such a thing as a pattern of "distinctive teenage spending for distinctive ends in a distinctive teenage world," and it could still trace its distinctiveness to the tastes and habits of the working-class. That only made sense: teens were barely independent, either still in school or new to the work force. They were only part-time or entry-level employees, so their income was still limited. Thus, while their background may have been working-class, or middle class, it made no difference: at this stage of their lives, they comprised a unique, uniform consumer group.[82]

But according to the student editors of ARK, the journal of the Royal College of Art, teens as consumers were nobody's followers, they were leaders—the avatars of the new affluent society: "The world of the teenager could well provide vital information for the new generation of professional cultural propagators. What impressed us most about the kids was the way in which they seem to understand modern styling, fashion and expendability so much better than the professionals."[83] But the most famous definition of "Pop" was taken from letter sent by artist Richard Hamilton to Alison and Peter Smithson: "Popular (designed for a mass audience); Transient (short term solution); Expendable (easily forgotten); Low Cost; Mass Produced; Young (aimed at youth); Witty; Sexy; Gimmicky; Glamorous; Big Business."[84]

Technology, played an increasing role in this. Tech was not Pop, but both Tech and Pop were contributory streams feeding the river of the British postwar consumer culture. British science correspondent John Davy noted in 1959 that England was belatedly turning to technology as a substitute for a lost empire and reduced great-power status. "As the Empire and the Navy shrink, and other nations get better at football, British national pride has tended to identify itself increasingly with science and technology. Calder Hall, the Vicount, Zeta and the hovercraft have become national idols. Each new laboratory bench is becoming the patriot's platform."[85]

Nigel Whiteley notes that "the response of the overwhelming majority of British intellectuals to the consumer society was negative." For example, socialist historian Eric Hobsbawm described the consumer-based mass media revolution as "the most appalling thing which has happened in the twentieth century."[86] What made consumerism, whether you called it Pop or not, such an emotionally charged issue was that it so much of it—certainly the technology—was seen as originating from America.

Labour Party front-bencher Hugh Gaitskell, in a piece entitled "Understanding the Electorate," wrote that:

> [M]ore and more people are beginning to turn to their own personal affairs and to concentrate on their own material advancement. No doubt it has been stimulated by the end of post-war austerity, TV, new gadgets like refrigerators and washing machines, the glossy magazines with their special appeal to women, and even the flood of new cars on the home markets. Call it, if you like, a growing Americanization of outlook. I believe it's there and it's no good moaning about it.[87]

Banham himself described Pop more broadly and less deterministically as a form of culture that was simultaneously vernacular *and* produced by professionals in a professional way. But most importantly, it was a form that could not inherently be held to the standards of "good" or "bad" taste because "each identifiable or stratum of society [has] its own characteristic taste or style and style of design—a proposition which clearly undermines the argument on which nearly all previous writing about taste in design [has] been based."[88]

As explained by Banham's fellow Independent Group member Lawrence Alloway, their sessions at the ICA reviewing glossy American magazine ads were not about entertainment: "We felt none of the dislike of commercial culture standards," he recalled, "but accepted it as fact, discussed it in detail, and consumed it enthusiastically. One result of our discussion was to take Pop culture out of the realm of 'escapism,' 'sheer entertainment,' 'relaxation,' and treat it with the seriousness of art." As a result, they were able to learn the visual language of American commercial advertising, just as entry-level students are taught the rudimentary language of visual art. In doing so they became what Alloway called "knowing consumers": those who could not be blindly dazzled, but could read and interpret an ad's semantics.[89]

The parallels between Alex Moulton and the American inventor/futurist Buckminister "Bucky" Fuller should not be overlooked. Banham's book, *Theory and Design in the First Machine Age*, was a study of the largely failed and forgotten futurist architects of the late 18th and early 19th century, such as Fillipo Marinetti and Antonio Sant'Elia.[90] He concluded that the later modernists of the 1930s, such as Corbu, Walter Gropius and his own mentor Nikolaus Pevsner had deliberately sought to erase the works of the futurists, what Banham called "the zone of silence."[91]

The futurists had been celebratory, romantic, forward-looking; the modernists, who had come to "own" architecture and design were rigid, logical, functional, austere—some would say joyless. Approaching the end, as the futurists faded and the modernists took command, *Theory and Design* grew ever more pessimistic, seemingly destined for a gloomy conclusion. But at the end, Bucky Fuller, with his egg-like Dymaxion cars, hamburger-shaped Wichita House and geodesic domes burst through as a reincarnated futurist. He was, quite literally, the book's only unalloyed hero. "Fuller for Banham personified an American ideal, the Yankee inventor who scorns traditional solutions, and whose single-minded perseverance is finally rewarded by recognition," says Adrian Forty. "It is a type that reappears frequently in Banham's writing; for instance Ernest Ransom, inventor of the concrete-frame daylight factory, Konrad Wachsmann, designer of open space-frame roofs, or the anonymous inventor of the domestic window-mounted air-conditioning unit."[92]

Mary Banham recalls that the Banhams became close to Fuller in the 1960s because the Banhams had two small children, as did John McHale, and Fuller loved children, and would sometimes drop by when he was in London on some pretext simply to play with the kids. (The death of his oldest daughter in the 1920s had led Fuller to start questioning his staid, middle-class life as a building-products executive.)

In 1980, the Banhams, now living in the United States, moved to Santa Cruz, where he had been hired by the University of California. A week after arriving, Fuller showed up to give a pre-arranged lecture. "He spent the first ten minutes telling this bemused audience, who didn't even know that Peter had arrived, how lucky they were to have him, and the reason they were lucky is because he was like a child and he had this attitude of wonder towards the world, everything was fascinating to him, even the tiniest thing, and you know, he was talking not only about Peter, but about himself. He was that kind of person ... he was a genius."[93]

Although many considered it a pretention (none of his friends, no matter how much they loved him, would ever describe Fuller as an outwardly humble or reticent man), he did honestly believe that he was not an exceptional individual:

> I was so very average.... I knew when when I started in 1927 that I could not jump very high and I could not swim very fast and I hadn't earned the best marks in the class, and I was very average and inasmuch as I was interested in what the average individual could do, I was a very good case for [self] experimentation.[94]

His friend, the pianist and composer John Cage, told an interviewer that "I would like to make it clear, as Bucky Fuller does, in his talks, where he says 'I'm just an average human being'—and then make it clear that anybody can do marvelous things."[95] Like Fuller, Moulton was an "inventor who scorns traditional solutions," the difference being that Moulton saw himself not as a "Yankee," but a product of the English midlands. In interviews, he invariably turned to the "dynanicism" and "sense of confidence" that was engendered by the Allied victory in the Second World War, and how that caused the Midlands to "burst forth with things that were pent-up," including design. While Reyner Banham and Alex Moulton may have disagreed over details, such as the value of the British Design Council (Banham was skeptical, Moulton enthusiastic) both agreed that the Moulton Bicycle was the exemplar of "one man going it alone," irrespective of whether you call it "clever Yankeeism" or "Midlands dynanicism."

Ironically, Banham's embrace of the Moulton was grounded in logic that was almost exactly the inversion of his prior embrace of Pop. As he said, British cycling had been a "long-term permanent working class" thing. But clearly bicycling was no longer a part of working-class culture. Inexpensive automobiles and the rise of household incomes had, by the early 1960s, begun to make Britain a mass-motoring society. On the other hand, he thought the Moulton would bring on "a minor cultural revolution" by replacing the "proletarian cyclist" with "a breed of cyclists who are middle-class urban executive radicals." Banham believed that with the Moulton a new type of dissemination would happen: a design shift *up* the class scale. What was once a working-class activity would now become an activity of a *new type* of consumer culled from the middle- and upper classes: the thinking classes.[96]

This is a very important point. Although he clearly foresaw the Moulton being purchased by people for recreational use, or even more extravagantly, as a type of overpriced household adornment, like a pasta maker or an espresso steamer, his idea of an "urban executive radical" clearly embraced the use of the Moulton for basic transportation. "It is," he claimed, "very nearly, the man-powered equivalent of the Mini." The reason was purely practical: London traffic didn't move anymore. (This was just a few months after the book *Traffic in Towns*, better known as "The Buchanan Report," was issued, and to the government's surprise became a best-seller.) "I have cropped twenty minutes off the journey between home and office compared to public transport," Banham reported to the *Architects' Journal* a few weeks after getting his Moulton, "the London I use has shrunk to the effective dimensions of a cathedral city."[97]

IV

"The first two years were just bedlam," recalls David Duffield. Although legend says that the company was overwhelmed from day one, Duffield confirms that "Dealers reluctantly gave orders." In fact, Moulton *had* been inundated by the information cards provided to the members of the public, but not by actual orders from retailers. The process of building up both orders and a dealer network to supply orders and service bicycles was slow, hard work. Duffield sent out cardboard stands for demonstrator bikes to sit on in stores so they wouldn't get lost amid the taller traditional bikes. They sat in the boxes. "We had to go out and do the window displays ourselves." British cycle shops of the era wouldn't dream of letting potential customers take a bike out for a test ride, "but if you don't ride it, you can't get an idea of what's it's about." Duffield's men created a stand that shoppers could use to test ride a Moulton and its suspension right in the store. "The dealers had no idea how to set them up and use them.... We actually went on television. We had a lovely little commercial that actually won an award, we did it for very little money."[98]

Production started in March 1963 at Bradford-on-Avon and July 1963 at Kirkby. Kirkby produced 8,515 bicycles in 1963, an average of about 350 units per week. Figures for Bradford-on-Avon are not available, but were probably between 6,000 and 7,000 for the year, or between 150 and 175 per week.[99] The Bradford-on-Avon factory was built for a capacity of 200 bicycles per week, but Alex Moulton later wrote that at times it could achieve 300. Kirkby had a standing capacity of 1,000 per week. Orders outstripped deliveries until the end of 1964. In 1964, Kirkby produced 31,185 bicycles, an average of 600 per week. Again, hard data is not available for Bradford-on-Avon. Output in 1964 is believed to be around 9,000, about 175 per week.[100]

The fact that production was less than the maximum capacities of both facilities can be largely attributed to various production glitches. A batch of 11,000 bad front forks was received from a supplier in early 1964 and had to be thrown out. Despite their best efforts to track down the bad units, a few bikes did make it out of Kirkby with defective forks.[101]

Vic Nicholson, the semi-pro time trialist hired to work in the factory, recounted that Alex Moulton was a poor factory manager who simply could not resist tweaking his invention in an eternal quest for perfection, even on the production line:

> He [Moulton] could be funny. He had a seat in the factory and would watch the bikes come down the production line, and if he saw something he didn't like, a washer or a screw or a fitting, he would say "stop the production line," and then re-order the item, and then start it again when the accessory came in. I felt sorry for the guy who used to do the buying, because it was "okay, we've got so many bicycles to do," and order to that quantity of specification, only to have Moulton change it, so half the accessories ended up not being used.[102]

To say nothing of the down time involved while the replacement part was ordered and shipped.

The biggest problem, however, was the rear forks. "With the first model, I was extremely concerned with appearance," Moulton explains, "I was imposing on the public an enormous change from the classical bicycle. So in order not to offend the public, I made the front and back forks nicely curved, and kept the suspension nicely hidden, But very soon, reality set in—the rear forks bent."[103]

This is not quite accurate. According to American cycling author and mechanic Sheldon Brown, there were some problems with torsional bending, but Tony Hadland says the 1964–65 crisis involved cracking along the bottom of the forks, adjacent to the bumper plate, about five inches back from the pivot bolt. This was the highest stress area of the rear fork, as the bumper plate acted as both the chainstay bridge and the point where the vertical movement of the rear wheel was transferred to the suspension system. Looked at from above, a detached Moulton rear fork was an H-shaped component, with the bumper plate acting as the cross-piece. The pivot bolt, just behind the crank, crossed the "H" at the front and the rear wheel axle crossed it at the rear.

Hadland believes the problem developed when Fisher and Ludlow switched from brazing the seams of the chainstays, which were not made from tubing, but were fabricated from rolled sheet steel, to Metal Inert Gas (MIG) welding. He attributes the problem to overheating, a symptom of atmospheric contamination. The cure involves changing the isolation gas compound and arc-weld electrode configuration; better training, more operator experience and more quality control checks. To repair existing units, the installation of two short, brazed-on demi-sleeves at the bottom of the chainstays at the bend and a re-brazing of the fork (a service offered by the Moulton Preservation Society) has proved very effective. As a production solution, a straight rear fork assembly was substituted for the curved design, although this raised the bottom bracket 0.75-inches (2 cm) and steepened the head tube angle about half a degree.[104]

There were two fundamental problems with Moulton's rear fork. The first was its basic design: a 15-inch-long, unsupported cantilever, connected only at the bumper plate and pivot bolt (called the damper bolt) in front and the rear wheel axle in back. The fork was not only subject to vertical forces by the weight of the rider and from the rebounding of the bumper plate—a problem augmented by the last-minute curved design—but also from twisting forces induced by pedaling. It was not hard for even a moderately strong rider to twist the rear fork of an early Moulton out of alignment.[105]

The second problem was the unexpected outsourcing of production. Fisher and Ludlow was not equipped for high-volume production using brazing, which was becoming obsolete except for specialty applications, and inert gas welding was then a new technology. Raleigh had automated most of its brazing by using a technique called ring-brazing, where a frame was pinned together with rings of braze inserted in each joint, then either baked in a conveyor-fed oven until the ring melted, or passed sequentially over stations where pre-directed gas jets brazed each joint using a carousel.[106]

The Bradford-on-Avon factory always used hand brazing in waist-high, three-dimensional rotating jigs, which, while efficient, still meant that its bicycles were, on a unit basis, quite a bit more expensive to produce than those from Kirkby. Fisher and Ludlow took over all production of Moultons in 1966, but even then, customers could order a special "S Deluxe" variant of the high-end models that were handbuilt at the Bradford-on-Avon factory. Although equipped slightly differently than a regular Deluxe,

The drawing studio that Alex Moulton built behind the Stables and Paddock for the bicycle company. Banham commented upon first seeing it that it struck him as being built in "a distinctly Frank Lloyd Wright idiom," and it does resemble the studio built at Taleisin West outside Phoenix, although Wright's studio was much bigger, and for many years, roofed only in canvas (photograph by the author).

making a direct comparison difficult, they were about 40 percent more expensive than a comparable Kirkby cycle, and it is possible they were sold at a loss. A Bradford-on-Avon built frame was probably somewhere between 35 and 50 percent more expensive than a Fisher and Ludlow-built one. Between April 1963 and July 1967, when all production shifted to Raleigh's Nottingham factory, less than a quarter of all Moulton frames were produced at The Hall.[107]

There was always a lot of mythology surrounding output figures for the Moulton. By mid–1965, the two factories together were producing, depending on who you asked, 800 units a week (Duffield's figure); 1,000 (Moulton's figure); or 1,440 (Raleigh's estimate). The Raleigh figure was way too high; Alex Moulton was deliberately using non-sequential serial numbers in order to make his output appear higher. Based on the memo sent to the British Motor Corporation board in November 1965 providing historical output figures at Kirkby, plus circumstantial evidence hinting at production at The Hall, Duffield's figure appears to be roughly correct for the 1964–65 period.[108]

That still made Moulton the second largest single-brand bicycle maker in Britain, surpassing Dawes and Viking. About 100,000 F-frame Moultons were made between March 1963 and July 1967, when Raleigh acquired Moulton Cycles; Kirkby made about 80,000; Bradford-on-Avon about 20,000; Raleigh made another 55,000 between July 1967 and mid–1970.[109]

It seemed for awhile like you couldn't turn around without running across a Moulton, usually with some celebrity on top of it. "It may have been a fantasy or a joke when it began; but before long Quintin Hogg and the *Vogue* models got up on two wheels," Banham later recalled. Quintin Hogg was a fiftyish, Bowler-wearing minister of parliament who took a shining to the Moulton. Banham called him a "media fiction," and he may have been as a politician, but he apparently really did use his Moulton, and for the same reason as Banham: it was the only way to get around London in a hurry without going broke on taxi fares. The "*Vogue* models" Banham refers to appear in the most famous of Moulton publicity shots. They were June Fry, Jill Wright and Jenny Wilson, surrounding one of the early models (with the soon-abandoned "billiard cue" paint job) in the middle of a deserted Carnaby Street on what must have been a Sunday morning, all in very hip tweed-cape outfits. The picture did indeed appear in *Vogue*, in 1964. Tour de France cyclist Tommy Simpson tried one at the Herne Hill velodrome over Easter holiday, 1963 and reportedly tried to talk his sponsor, Cycles Peugeot, into taking out a manufacturing license. Actress Eleanor Bron was spotted on hers around London.[110]

And then, in July 1965, Raleigh turned the table.

4

Who Killed Roger Rabbit's Moulton?

> Basically, Moulton invents a bike that's got suspension on, skinny wheels, and they're real twitchy and it rides really well. Raleigh then crush him, buy the name, produce a bastardized version of it with huge wheels, and the whole thing feels like riding through treacle.
> —Kath Hammer, York Cycleworks (1995)[1]

> JESSICA: What are you taking about? There's no road past Toontown.
> DOOM: Not yet, but several months ago, I had the good providence to stumble upon this plan of the City's. A plan of epic proportions. They're calling it a—a freeway!
> VALIANT: Freeway? What the hell's a freeway?
> DOOM: Eight lanes of shimmering cement running from here to Pasadena. Smooth, safe, fast. Traffic jams will be a thing of the past.
> VALIANT: So that's why you killed Acme and Maroon? For a big road? I don't get it.
> DOOM: Of course not. You lack vision. I see a place where people get on and off the freeway. On and off. Off and on. All day, all night. Soon where Toontown once stood will be gas stations, inexpensive motels, hamburger stands, tire stores, auto dealerships, and billboards—wonderful billboards reaching as far as the eye can see. My God, it'll be beautiful.
> VALIANT: Come on. Nobody's gonna drive this lousy freeway when they can go to Pasadena on the Red Car for a nickel.
> DOOM: Oh, they'll drive. They'll have to. I bought the Red Cars. I'll dismantle them. All of them.
> —*Who Framed Roger Rabbit* (1988)[2]

I

It is a historical phenomenon that has been blamed for so much misfortune, so often, in so many different guises, that it has acquired its own name: the Roger Rabbit plot. Named after the 1988 Steven Spielberg cartoon movie *Who Framed Roger Rabbit* it has also come to be known as the Great American Conspiracy Myth. According to political science professor Deborah Stone, the Roger Rabbit tale is a modernized version of the ancient Greek "paradise lost" tragedy, and a method of psychological self-protection common among the elderly and the displaced: "in the beginning, things were pretty good. But they got bad. In fact, right now they're nearly intolerable. Something must be done. Who will save us?"[3]

What makes *Who Framed Roger Rabbit,* the movie, so unique is that it was actually, if loosely, based on a set of historical facts. A congressional research report entitled *American Ground Transport* was written in 1973 by a staff attorney named Bradford Snell and subsequently popularized in a 1981 *Harper's Magazine* article. It alleged that in the 1930s, General Motors, acting through a front, National City Lines, bought up all the streetcar and interurban lines in Los Angeles and replaced them with busses, knowing that the resulting service would be so bad that everyone would be forced into cars, thus facilitating the early development of the LA freeway system.[4] Snell's claims were later shown to be wildly exaggerated (if not outright false) by Scott Bottles and others.[5]

Applied to Raleigh and Moulton, the Roger Rabbit conspiracy myth goes something like this: Raleigh brings out a bike, the RSW 16, that looks like a Moulton, but has no suspension, weighs a ton, and undercuts the Moulton's price. Raleigh advertises the living daylights out of it, setting off a price war that drains both firms, but Raleigh is big and Moulton is small, so Raleigh can absorb the damage and Moulton cannot. Moulton is crushed, and sells out to Raleigh for a pittance. Raleigh then buries the whole small-wheel concept, and, problem solved, goes on about its business of selling roadsters until its inept handling of the BMX and mountain bike booms facilitates an onslaught of Asian competitors, wiping it out in the late 1990s. End of story.

It would be easy to write off this narrative as naïve and simplistic: Roger Rabbit Goes Cycling. There is one problem, however. This is basically Alex Moulton's version of what happened. Until his death he was angry and bitter about what he believed Raleigh did to him and his firm. In 1966 he told a reporter that "but for them we would have home sales of about 60,000 a year." At that point, Moulton was selling somewhere around 30,000 units per year, Raleigh's domestic sales were about 330,000.[6] Nor did he believe that it had to do with aggressive business competition; it was personal. It was a vendetta directed against him:

A Raleigh RSW 16 Compact (photograph by the author).

The underlying cause was NIH [Not Invented Here] which was the first time I had come across it.... I well remember after one of my many meetings with them finding one of the officials [Jim Harrisson, home sales director] scrubbing the bike sideways along the passage with an expression of such intense hatred towards it. He actually wanted to break it.... I do not now believe that to preserve a living outsider's design and name is within the allowable corporate behavior pattern. To do so would be to cast doubts on their own virility. So the Moulton is extinct.[7]

Bicycle historian John Pinkerton recounted that "I once knew a dealer that proudly proclaimed that 'he had never allowed a Moulton across the threshold of his shop.'" Others point to a never-explained quality control crisis at Sturmey-Archer that by a strange coincidence happened to strike in 1963, causing it to deliver faulty FW four-speed hub units, just when Kirkby started producing Moultons in volume. Michael Woolf, of the Moulton Preservation Society, blames the quality control problems that came to plague the Kirkby-built bicycles to the animosity of the workers at the British Motor Corporation:

> During the early production of the F-frame, it took off so much that he actually built a bigger factory on the grounds of The Hall. That wasn't sufficient, so with his relationship with BMC—British Leyland now—they were producing the Mini and of course Alex had a major design interest in its suspension. So they used the factory to produce the F-frame at Kirkby, at Liverpool, and the lads there were actually supposed to be building Minis, and they built these things with great hatred, because building a push-bike just wasn't cool. So they weren't built to the right standards. The brazing was atrocious, and they became of victim of their own success, because they were being knocked out in this rather unskilled manner.[8]

David Duffield, Moulton's marketing manager, left the company even before it was sold to Raleigh in November 1967, citing essentially the same circumstances that Moulton raises:

> It became obvious that we would be smashed, just by the sheer weight of the money they were throwing at the [RSW 16] program. It was so obvious they wanted us out. Through the publicity; through the spending; through the pricing. That's when I left. It's not that I'm a coward, but it was so apparent that they had it in for the Moulton.[9]

But in many cases, the facts do not support these beliefs. First, consider Alex Moulton's claim that Raleigh was deliberately crippling his sales. Kirkby produced 8,500 bicycles between July 1963 (when production started at Kirkby) and December 1963. In 1964 it made 31,200. The RSW 16 was introduced in May 1965. An internal memo prepared by Moulton Bicycles in November 1965 projected a Kirkby production of 30,000 for 1965 based on a year-to-date output of 25,600. Thus the firm's estimated total output was about 39,000 for calendar year 1965, based on a Bradford-on-Avon production of 175 per week.[10] For 1964, Kirkby output had been 31,200, giving an estimated total firm production of about 40,500 based on the same Bradford-on-Avon output. That is not much of a negative impact (3.7 percent). Moreover, the memo stated that firm had caught up with orders in November 1964, six months *before* the RSW 16 was introduced.[11]

The memo goes on to state that based on orders received from January 1965 to Easter 1965 (when the RSW 16 was introduced) the Moulton firm believed that it should have received orders for 50,000 bicycles for the year. However, this would have been a 25 percent increase over 1964. The memo goes on to state that "at the beginning of Easter, 1965 there was a sudden down turn in orders received. This was due to the credit squeeze and the trade knowledge of the forthcoming Raleigh small wheeled bicycle."[12]

Thus, the restriction of hire-purchase (which the government used as an anti-inflationary measure) was also attributed as an adverse factor.

Even assuming that Moulton could have sold 20,000 more units per year absent the RSW 16, the loss for the Nottingham titan would have been minimal. The total market for bicycles in the UK in 1966 was 510,000 units, with Raleigh supplying about two-thirds of these. But the firm was also producing another 926,000 for export, with 132,000 shipped to the USA.[13] Thus, Moulton's 20,000 lost sales amounted to less than eight percent of Raleigh's domestic sales. Even had Moulton's entire hypothetical output of 60,000 been transferred to Raleigh, its total worldwide sales would have only increased by about five percent. In fact, Raleigh was making more profit off the 132,000 cycles it shipped to America than the 332,000 it sold on the home market, due to a decision in 1962 to withdraw from the American market in all sectors except the high-end bike shop market.[14]

Or the claim that Raleigh was deliberately selling defective hubs to Moulton, or was withholding shipments. The only part that Sturmey-Archer supplied was the rear hub, along with its associated hardware. Early Moulton Standards had an AW 3-speed hub (switching to the FW 4-speed unit in 1965), and all Moulton Deluxes had the FW. But they all used a fully sheathed shifter cable—a relative rarity in those days—and it was this unit that gave rise to the "faulty hubs" mythology and probably the "delayed shipment" rumor as well. The FW already tended to shift poorly into low gear unless kept in adjustment, and while the sheathed cable eliminated the need for external pulleys and guides, it was subject to the same kind of cable stretch that requires the rear brake cable of many sports bikes to have a fine adjustment knob. After 1965 Sturmey-Archer supplied a special oversize cable unit that took care of the problem. About this same time, Sturmey-Archer also started supplying the FW hub configuration that Moulton had wanted all along, but that was not in the Sturmey-Archer OEM catalog, one with 28 spokes and a 13-tooth sprocket to replace the stock 36-spoke/14-tooth model.[15]

Both the "delay" and the "sabotage" rumors can probably be ascribed to Moulton's need for non-standard equipment in large quantities on short notice. Had Raleigh wanted to inhibit Moulton's operations, it could have done a lot more damage by simply refusing to supply either the oversize cable unit or the modified FW hub unit by claiming it lacked the capacity or engineering ability. There are other examples of Sturmey-Archer making active efforts to capture (or recapture) the Moulton supplier market in areas where they did not normally have a product to offer, such as re-introducing a one-speed coaster brake in 1965.[16] The evidence points to Sturmey-Archer acting as an independent entity, seeking to maximize its own market share and profitability, and not as a puppet of its corporate parent, Tube Industries, sacrificing its interests for the benefit of its sister subsidiary, Raleigh.

The most notorious story, however, is the one cited by Michael Woolf: angry auto workers in Kirkby, insulted at the thought of being forced into making bicycles, either slapping Moultons together in a haze of disdainful boredom or even sabotaging them. An examination of the facts shows that this story is clearly apocryphal.

The frames were fabricated at Wolverhampton, then shipped to Kirkby for painting and assembly. (It appears that some or all of the frame fabrication was shifted to Kirkby at some point.) Neither was a Mini plant or even an automobile plant. They were appliance factories. In Kirkby's case, a brand new one. Fisher and Ludlow did manufacture car bodies. In fact, that was its primary business, since the 1920s. Taking advantage of its expertise in sheet metal forming and welding, it had diversified just before World War II into stainless steel sinks, countertops and racks, mostly for restaurants and the food service industry. It was purchased by Austin Motors (later part of British Motor Company–BMC) in 1953.[17]

In 1948, Hoover introduced the washing machine into England. By 1955, only 18 percent of households had them, for two reasons. First, they were expensive, over £70, and second, they were limited in what they could do—they couldn't spin-dry the clothes, so you still had to wring them out by using a tub and crank-operated wringer. In 1957 Hoover introduced the Hoovermatic twin-tub washer, the second tub being used for a spin-dry cycle. But you still had to transfer the wash over from the wash tub to the spin-dry tub. A year later, a former department store salesman named John Bloom started importing a twin-tub wash machine from Holland, under the brand-name Electromatic. He sold them direct, under a two-year installment payment plan, for only £55. The two leading department-store brands, Hoover and Hotpoint, refused to cut their prices, even after a direct-sale competitor to Bloom, Manchester-based A. J. Flatley, started making his own washer and announced that he would introduce a hot-air clothes drier in 1960. He was quickly followed by Bloom, who in early 1960 introduced a single-tub machine that incorporated the spin-dry cycle right into the wash tub.

Bloom and Flatley were selling direct-to-customer. Hoover and Hotpoint were selling high-end machines through department stores. Fisher and Ludlow decided to leap into the gap and produce a moderate price £55–£60 washing machine for the department store market. The American Bendix Corporation had a single-tub model ready to license, and would soon have a drier to match it. The Fisher and Ludlow division of BMC decided to physically separate the appliance-making division from the main plant in Birmingham and locate it in a new, purpose-built factory of its own in the Kirkby Business Park on the Merseyside-South Lancashire boundary, six miles from the center of Liverpool. In 1964, this division was renamed Fisher-Bendix and later sold off. Body panels for the Mini were supplied by Fisher and Ludlow, but these came from its original plant in Birmingham and a new "Development Area" facility in Lianelli, Wales.[18]

As mentioned earlier, Kirkby itself was a "spillover" area, a not particularly successful one. It was laid out to help alleviate the postwar shortage of homes in Liverpool city and diversify its employment beyond shipping and ship building. Some areas near the docks had been blitzed, others were being leveled as part of a nationwide slum clearance program. The Merseyside-South Lancashire area was a part of the government's Development Area program, which extended loans and grants to firms building new factories in targeted industries, including household appliances, located outside of the traditional manufacturing centers of Greater London, Birmingham and Coventry. "Not only do such manufacturers bring with them a useful measure of industrial diversification," noted one report, "but, being modern types, they have better prospects of expanding in the long term than the old, basic industries."[19]

Fisher and Ludlow was one of the first tenants of the business park. Few if any of its employees were former automobile assemblers; most were Liverpool council re-trained workers who formerly lived in blitzed slums so notorious that they were identified by government social workers only under the generic pseudonym "Crown Street," and worked at the nearby docks. They had no reason to resent any of the products they were producing. They may have been inexperienced, but there is no evidence to suggest that they felt any differently about bicycles than washing machines or restaurant kitchen racks.

Kirkby was no success when it came to city planning, but John Barron Mays and his team from Liverpool University found that "for the majority of ex-inner city dwellers the new estate [Kirkby] is desirable or at least adequate…. Certainly the residents interviewed did not seem to be unusually isolated." Only 29 percent found the neighbors less friendly in Kirkby than at their old home in Crown Street. Overall, Kirkby residents were not thrilled with their environment, but most held out hope for improvement, mostly via home ownership. None pined for a return to Crown Street. The Fisher and Ludlow plant was not a seething cauldron of social discontent, although it did experience very poor labor-management relations after Bendix bought the company, but that was later, after Moulton production had been moved to the Raleigh plant.[20]

The problem was a combination of the fork's curved shape, its fabrication from rolled sheet stock and Kirkby's use of then-new and to some degree still experimental MIG (metal inert gas) welding techniques in place of Bradford-on-Avon's hand brazing. The switch to MIG welding may have occurred when frame (or rear fork) fabrication moved from Wolverhampton to Kirkby. MIG welding had been developed in the 1940s to weld aluminum in the aircraft industry. It is a form of electric arc welding, but uses a sophisticated pistol-like hand controller. When the trigger is pulled, the arc ignites, welding "rod" (actually wire from a spool) is automatically fed to the electrode gap, and a stream of inert gas (initially, for aluminum, argon), starts to blow, isolating the electrode arc from the surrounding oxygen/nitrogen atmosphere.

Welding steel (especially stainless steel) was experimented with during the 1950s, but it was discovered that argon was not a suitable isolation gas because so much was needed. Carbon dioxide worked because it was cheap, but it tended to create globs of splattered welding wire, which could spot-overheat the material being welded. In the early 1960s, about the time of the Moulton rear-fork crisis, a perfected method was developed which mixed 5 to 25 percent argon with the carbon dioxide, used a slightly different (pulsing) electrode configuration, and sheathed welding wire. This permitted a skilled welder to work with even very thin steel sheet without overheating. Fisher and Ludlow did a tremendous amount of stainless steel MIG welding as part of its commercial kitchen equipment business.[21]

Alex Moulton had not anticipated the problems inherent in manufacturing some 600 bicycles a week in a facility that was not a bicycle factory and that used non-traditional bicycle fabrication methods. The Bradford-on-Avon plant had been built in anticipation of 200 units a week with two brazing jigs for frames and another two for subassemblies. Tony Hadland reports that while the main tubes had initially been manufactured using a roll-rivet-braze process, most production models used pre-rolled ERW brand tubing. The rear fork was an exception in that it required longitudinal seam sealing, either by brazing or welding.[22]

The Moulton's design was relatively inflexible from a production point of view. When actual sales proved to be much higher than expected, it was not possible to simply scale up the methods used at The Hall. It was not feasible to line up a dozen brazing jigs and have workers fabricate frames by hand; brazing was rapidly becoming an obsolete art. It would have taken months to train workers to do it, and most likely the per-unit costs would have been as high at Kirkby as at Bradford-on-Avon, that is, *at least* one-third more per frame *after* the assemblers were fully trained and experienced.

Unfortunately, little thought was given in 1962 to adapting the design to accommodate high-volume fabrication using a completely different method that relied on a relatively inexperienced workforce applying state-of-the-art mass production techniques. If Alex Moulton had one shortcoming, it was a lack of flexibility; he was not good at improvisation in the face of rapidly changing circumstances. We will shortly find a sharp contrast in the case of Andrew Ritchie's Brompton folding bicycle, where, after the inventor had pursued his design through several generations to a point of technical maturity, he turned control over to a production engineer, Will Butler-Adams. From then on, fabrication, value engineering and quality control issues were the major technical considerations.

But Woolf's explanation shows just how easy it was to fall into the trap of conspiratorial thinking. Even Tony Hadland, who has written two books on the history of the Moulton bicycles, cautions against such "Roger Rabbit" solipsism. "Raleigh are often portrayed as the villains of the piece as far as the history of the Moulton bicycle is concerned," he notes. "Many people believe that they only bought Moulton Bicycles Limited to kill the machines off." Both views, Hadland believes, "are false." Hadland, Duffield and others point to the Kirkby and supplier quality control problems, the Moulton's relatively high price, its unpopularity with many bicycle dealers, and Moulton's inability to offer factory hire-purchase (time-payment) plans as equally important reasons for the Moulton's failure to succeed past 1967.[23]

On the other hand, there is a near-unanimous consensus that had there been no Moulton, there would not have been a Raleigh RSW 16. Depending on who you ask, it was a clumsy, cobbled-together fake bicycle put on the street only to make all small-wheeled cycles look bad, or more commonly, it was a Moulton sales-killer. This is an argument that merits a closer examination. It probably is true that without the Moulton, there wouldn't have been an RSW 16. The question is one of causality. Opportunistic knock-off or calculated corporate homicide?

A closer look reveals a third explanation. In this version, the evolution of the RSW is largely independent of the Moulton. It was developed to address long-standing marketing needs and problems with Raleigh's dealer network that had nothing to do with the Moulton. Bicycle sales had been falling for twenty years, mostly because the industry's bread-and-butter, working-class young men, could now afford cheap cars. Rather than face that reality, Raleigh blamed its dealers for its woes, the mom-and-pop bicycle stores that had served generations of those "cloth cap and racing-pigeon men." It wanted to walk away from them and focus more on "high street" department stores such as Halford's and Curry's. It also wanted a new clientele: women, who were the predominant customers of the high street shops. But the high street stores demanded a different type of bicycle: a universal bike that would allow them to stock fewer units, and would look better sitting next to appliances and household furnishings. The RSW was created to be that bike, after the Moulton convinced them that it was worth the investment.

II

The cyclemaking firm of Woodhead & Angois was established on Raleigh Street, Nottingham in 1885. They were joined by William Ellis a year later. R. M. Woodhead and Paul Angois provided technical expertise while Ellis attended to business affairs. By 1887 they were producing about three cycles a week. That year, a man named Frank Bowden bought one of their "Raleigh" brand bicycles.

Bowden was born in Exeter in 1848. He studied law in London but gave it up to take a civil service position in Hong Kong. It permitted him to speculate in stocks and other investments on the side. He earned a sizable fortune. He left Hong Kong in 1878 for San Francisco, married a California heiress, traveled through France, then settled in London. Bowden bought his Raleigh to help him recover from a serious, but never fully disclosed, illness. He liked it so much that a year later he invested in the firm, soon buying out Ellis's share of the partnership. By 1895, Bowden had squeezed out both Woodhead and Angois, and ran the Raleigh Cycle Company as a family enterprise until his death in 1921 when his son Harold took charge.[24]

In 1934 the firm was converted into a public holding company with three operating divisions: the Raleigh Cycle Co.; the Humber Cycle Co. (acquired in 1932); Sturmey-Archer; and the trade names Robin Hood and Gazelle. Earnings suffered greatly during the depression of 1930–32, but never went into the red; they rebounded starting in 1933. In 1933 Raleigh was earning profits of slightly less than $1 million at pre-war exchange rates; by 1936 this was up to $2.3 million. The company was making so much money that it was distributing a quarter of its profits to shareholders. In 1938 Harold Bowden retired (at age 58), appointing George Wilson, a non-family member, as managing director.

At the conclusion of World War II there was every reason to believe that the future of the

Sir Frank Bowden, 1915 (*Raleigh: Pioneers of Cycle Manufacture for 50 Years—Jubilee Souvenir Brochure 1887–1937* [**Nottingham: Raleigh Cycle Co. Ltd., 1937**], **private collection**).

4. Who Killed Roger Rabbit's Moulton?   81

Workers cycling home from Morris Motors, Cowley, Oxford, just before the start of the war (*Cycling*, June 6, 1939/Library of Congress).

British cycle industry would be almost boundless, and that Raleigh would be in its vanguard. A 1944 government survey indicated that 27 percent of the working class cycled regularly, and as late as 1949 almost twenty percent of all employees cycled to work. These were not necessarily the working poor: the highest proportion of users was in the middle half of the income spectrum.[25] Although the post-war "dollar deficiency" had provoked a recession in much of the British economy, the opposite was true for the bicycle industry.

However, the likelihood that the U.S. lend-lease program would be suspended with the end of the European war, coupled with a shortage of British foreign assets and currency reserves, convinced the new Labour Party government that British firms would have to achieve a much higher level of exports than before the war to pay for such imported basics as food and industrial raw materials. After national defense, production for export became the government's number one priority. By stimulating British imports to America and by making it more expensive for American firms to ship goods to England, surplus dollars would flow to Britain. These could then be returned to the United States in the form of debt repayments, which had to be made in dollars or gold.[26]

In March 1945, the Ministerial Subcommittee on Industrial Problems issued a report titled *Post–War Resettlement of the Motor Industry*. It asserted that the British automobile industry was more focused on meeting the projected postwar demands of the domestic market than raising export levels, especially to hard-currency nations.

The Duke of York, the future King George VI, visits the Raleigh factory, c. 1925 (*Raleigh: Pioneers of Cycle Manufacture for 50 Years—Jubilee Souvenir Brochure 1887–1937* [Nottingham: Raleigh Cycle Co. Ltd., 1937], private collection).

Post–War Resettlement of the Motor Industry identified many reasons for the motor industry's export ineffectiveness. It was comprised of many small firms, each producing a wide range of models with small annual production runs; few firms had a strong overseas presence in North America; and because British firms tended to use a higher proportion of purchased parts and components, they often had weak spare-parts support networks. The report recommended government intervention (i.e. either partial nationalization or the preferential allocation of steel from the newly nationalized steel mills) to bring about the cooperative development of a large, powerful export model.

The British government was prepared to implement the forced export car project in January 1945, but by mid-year abandoned it because of the risk of generating high short-term unemployment during the implementation process. Similarly, the use of raw material incentives (or threats, as the case may be) was abandoned by fall, largely because one automaker, Ford–UK, had its own steel rolling mill at its Dagenham plant northeast of London and thus would have been immune to such a policy. The Motor

Vehicle Board of Trade reached a compromise whereby the automakers would agree to increase their exports in 1945 by 50 percent over 1938, then to 75 percent in 1947. For the most part, British automakers attempted to meet this goal by using small sports cars such as the MGA; the Triumph TR-2/3/4 series; Triumph Spitfire; Austin Healy; and Sunbeam Alpine.[27]

However successful this may have been as a workable compromise to both sides, it left one problem on the table: it didn't provide much immediate relief to the dollar crisis. All it really accomplished in the short run was to reduce the total quantity of cars available to the British domestic market, because it diverted resources away from the development and production of small sedans to gear up for sports cars for export. In the end, the American sales of British sports cars was moderately successful, but didn't really get rolling until the middle to late 1950s, after the crisis was over.

The British bicycle industry was selected as a stand-in and became part of the labor government's "Export or Die" program. It was a logical move. *The Economist* predicted that those firms engaged in the manufacture of light consumer goods, such as cycles, cutlery, and radio and consumer electronics would have a decided advantage over old-line heavy industries in the postwar export drive.[28]

The cycle industry appeared to be a particularly good candidate. By the last quarter of 1947 the bicycle division of Tube Investments, Ltd. (TI) was already exporting 75 percent of its output, and at the end of the year Raleigh was exporting at a 40 percent share.[29] There were only two caveats. First: both TI and Raleigh were hampered by periodic steel shortages. This was somewhat alleviated by 1948 when the bicycle industry was given priority access by the Ministry of Supply. "We can appreciate the difficulties in deciding priorities," said TI's chairman Ivan Stedeford, "but few claims could be much stronger than those of the bicycle industry. The conversion of steel into bicycles is among the highest of all British products and the bulk of the industry's total production is sold abroad."[30]

Second, the high export figures Stedeford referred to were being generated largely by shipments to commonwealth markets, especially South Africa, Nigeria and India, and not the all-important American dollar market. The problem was the Americans' thirty percent import tariff on both complete bicycles and parts. Norman Clarke was the president of the Columbia Manufacturing Company of Westfield, Massachusetts, and in 1953 was the chairman of the Bicycle Institute of America, the trade group for the makers of bicycles and parts, wholesale distributors, and retailers. He and his fellow executives were summoned to Washington by Eisenhower's economic advisor, Gabriel Hauge. "We were told that we were the 834th largest industry in America. On the other hand, the British bicycle industry was the seventh largest industry over there. We were told the American bicycle industry was going to help England earn dollars," Clarke recalled.[31]

He and the other bicycle industry executives were told that effective immediately, the tariff on imported bicycles and parts was going to be cut from thirty percent to no more than fifteen percent in any one category. For roadsters—popularly known in the United States as "English Racers"–it would be cut to seven percent. During the war only two American firms had produced bicycles, on a limited basis, mostly for use in factories and on military bases.[32] All had produced a variety of war material, including the two—Columbia and the Huffman Manufacturing Co. of Dayton—that produced "victory" bicycles. As far as Hauge was concerned, this proved that they could do many things besides make cycles. Clarke recalled that Hauge closed the meeting with a chilling summary: "We were told that we were expendable, that we had to do something else."[33]

The results were dramatic. Exports from the UK to the USA increased from 18,500 in 1947 to 534,000 in 1954 while domestic consumption decreased from 2.9 million in 1947 to 1.5 million in 1954. Total production in the UK stayed relatively flat at between 3.0 and 3.5 million from 1947 to 1954, but production at Raleigh increased from 465,000 in 1947 to 1.22 million by 1954.[34] "The [American] bicycle industry is feeling rather wobbly," reported the business magazine *Barron's*:

> What ails it is excessive competition from foreign-made wheels, notably those of Great Britain and Germany, which are threatening to outsell American models.... Imports of British, German, and more recently, French and Austrian makes have mounted swiftly, from only 16,000 in 1948–49 to more than 950,000 last year [1954–55].... It is lightweight foreign bicycles, perfected in Britain and made in small numbers by just a few U.S. firms, that account for the bulk of imports.... By and large, the evidence suggests the lightweight bikes have not simply invaded an old market; they have in fact largely created a new one.[35]

The UK bicycle industry responded to this bonanza with two strategies: consolidation and expansion. TI had acquired parts-maker Phillips back in 1919. Raleigh, a part owner of Sturmey-Archer since the turn of the century, purchased it outright in 1910, and acquired Humber Cycles in 1932. Raleigh closed down Humber's factory and moved all production to its Nottingham factory. This left five major firms: TI, which in addition to its bicycle units, supplied parts and tubing to the industry; Raleigh, who owned Sturmey-Archer and Humber; BSA Cycles; Rudge-Whitworth and Hercules, the largest bicycle maker in the UK and possibly the world.[36]

In 1946, Raleigh bought Rudge-Whitworth from its owner, the Gramophone Co., a subsidiary of EMI. EMI had only owned Rudge for eight years, and had acquired it mostly because it wanted access to bicycle shops to turn them into wintertime outlets for phonographs and records. The strategy didn't work, and EMI was eager to sell. Raleigh wanted Rudge only for the trademarks and goodwill. It closed its Hayes factory, moving everything to Nottingham. That same year, Raleigh had agreed to start supplying the high street department store Halford's with bicycles. With about 300 stores, Halford's was also a major retailer of record players and parlor radios.[37]

In 1946 TI bought mighty Hercules, and a year later expanded its Phillips parts-making unit into the manufacturing of complete bicycles. In 1956 TI reorganized its bicycle subsidiaries into a new firm, the British Cycle Company (BCC). There were now three players: Raleigh, TI-BCC and BSA Cycles. Immediately after TI created BCC, it began a major structural reorganization of the division. The aim, according to TI's Ivan Stedeford, was to concentrate production "in compact and efficient plants," asserting that TI would "spare no expense to make this possible." What he really meant was that ten percent of the workforce—1,250 employees—would be furloughed by early 1957 as the factories they had been working in were shuttered.[38]

In 1957, Raleigh acquired BSA Cycles, parent of Matchless, Sunbeam and New Hudson. Sir Bernard Docker, chairman of British Small Arms, and his wife, Lady Norah, were, like Leonard Lord, people from humble origins who had earned a vast fortune in industry, but unlike Sir Leonard, liked to show it off. Lady Norah put in an appearance at the first post-war Earl's Court Auto Show in a new gold-plated Daimler with her initials engraved on the side. In 1953 the couple was thrown out of the Royal Enclosure at the Ascot races, despite owning an entry, when they became uproariously drunk. There were persistent rumors that their lavish lifestyle was being unwittingly subsidized by BSA.

In May 1956 BSA's board stripped Sir Bernard of his chairmanship and threw him off the board. In August, he tried to muscle his way back on. However, due to the Dockers' spendthrift ways, they had taken out large loans from the Prudential using their BSA stock as collateral. Those loans contained an option clause that allowed the Prudential to convert the loans into a purchase of the collateralized stock. The Prudential (who already held a stake in the firm) exercised their option rights, and Sir Bernard no longer held any stock in his firm. To recoup their lost loans, the Prudential sold the bicycle division of BSA to Raleigh in early 1957 for £500,000.[39]

The Americans partially readjusted their tariff rate upward again (by one-third) in May 1955, and imports plummeted from 538,000 in 1955 to 266,000 in 1956, but *The Economist* noted that "Increases recorded by other European bicycle makers in the same period suggest that prices rather than the recent increase in the American import tariff are to blame ... after the American tariff was raised, British exports were halved, while the Dutch and the Germans rose." In January 1958 BCC workers struck at the firm's Smethwick plant over layoffs; 2,000 workers walked off the job for three weeks. The workers returned after the Ministry of Labour got the firm to agree to a more generous severance package for the furloughed workers, but chairman Stedeford warned that there was "no alternative but to accept the situation that a new level of trade exists." *The Economist* noted that "The news ... marks a public change in thinking for one, at least, of the two remaining leading producers of bicycles in this country." It appeared that Stedeford was throwing in the towel on the bicycle business.[40]

Raleigh, on the other hand, was sanguine. The firm had been looking into the market for a clip-on motor to convert a standard bicycle into a low-power moped, then into the construction of a purpose-built moped. But in 1955, George Wilson decided against expanding into the moped market. He told a reporter from the *Economist* that "Raleigh has achieved its present position in the word's markets by concentrating its entire efforts and resources on the manufacture of pedal cycles and variable geared hubs and hub dynamo lighting sets, and so bright is our view in the future of these products that we have no intention of departing from this policy."[41]

Raleigh did introduce a moped, but not until October 1958. It initially sold well, but within a year sales slumped in a market saturated with domestic and imported competitors. After the 1960 Raleigh/TI merger, Raleigh relied on models already being produced by former BCC factories; by importing products made by Motobecane of France and Bianchi of Italy; or by manufacturing those two firms' designs under license in Nottingham.[42]

Raleigh, which had opened a new 65,000 sq. ft. factory expansion to its Sturmey-Archer division back in 1950, broke ground in 1956 on a 435,000 sq. ft. expansion that increased the size of the Nottingham factory by about 30 percent. "Factory 3" was slated for completion in mid–1957. At the dedication ceremony, a vast curtain was hung behind the speakers' stage to hide the fact that it was empty, and would remain so for well over a year, until it was decided to relocate the Sturmey-Archer works there. The company had more capacity than it knew what to do with.[43]

The problem was three-fold. First, the overall market for bicycles had flattened. The home market was stable: 740,000 in 1953, 748,000 in 1958. The slump was in exports. Total UK shipments to the U.S. fell from 534,000 in 1954 to 263,000 in 1958. Gross UK production fell correspondingly: 3.0 million in 1953; 2.2 million in 1958.[44] Second, the company was having a problem with an excess of diversity in its product line. With all the brands it had taken on through acquisition, by 1952 Raleigh was producing 500 different frame and component combinations. When it added BSA Cycles in 1957 this may have gone as high as 900.[45]

Finally, the firm decided to make Sturmey-Archer the crown jewel of the Raleigh empire. As one business writer succinctly put it: "threatened with the permanent loss of overseas completed-bike sales, Raleigh decided to put concentrated effort behind the sale of components. Its objective was to have every manufacturer in the world use components supplied by Raleigh."[46] But this was a strategy fraught with problems. The best illustration of this was the identical situation that TI-BCC was having with its wholly owned parts-making subsidiary, Phillips.

TI-BCC's management was uncertain as to the strategic role that it wanted Phillips to play within the organization. Like Sturmey-Archer, Phillips was considered the acme of the group, but it was also its Achilles' heel. On a standard £9 ($25.00) 3-speed roadster as much as £8 in production costs went to the components. That meant there wasn't a lot of room for profit for the cyclemaker, regardless of whether he was an independent who had to buy everything on the wholesale market, or a division of one of the conglomerates, who lived at the mercy of the accountants at headquarters, setting the pseudo-price paid for components from their own parts-making division.

Within the TI-BCC group, there was a great deal of uncertainty and animosity about the role that Phillips should play when it came to intra-corporate pricing. Should it supply components at a discounted price in order to maximize the profits of the bicycle-making divisions, or should it supply parts at prices that reflected true costs? If the latter, should the bicycle divisions be free to seek outside sources if they could get a better deal?

The uncertainty became so great that by 1956 Phillips decided not to order expensive American-made machine tools because it was unable to do long-range financial planning. Its management simply could not predict what it would be allowed to charge TI's bicycle-making subsidiaries. Like Raleigh, TI-BCC had begun contracting with the high street Curry's department stores to carry the Hercules bicycle as their economy-price brand. But the Curry's contract called for shipments within seven to ten days of orders and in some cases four-month-old orders were still open. The store's buyers told TI-BCC management in no uncertain terms to "get their house in order." But without the new machine tools, they couldn't modernize or reorganize. In April 1959 BCC's chief executive, Jim Boulstridge, raised the once unthinkable possibility with his board of directors of pulling out of the American market to focus on domestic customers. In an editorial, the *Motor Cycle and Cycle Trader* commented that "It looks rather if the two big groups of cycle manufacturers have over-developed ... [but] neither at the Earl's Court show, nor at home in our shops do we witness any sense of urgency or enthusiasm."[47]

Similarly, there was a long-standing belief within the industry that Sturmey-Archer supplied components to its corporate siblings at below-cost prices. For example, Norman Clarke of the American Columbia Manufacturing Company always believed that Sturmey-Archer supplied 3-speed hubs to Raleigh for about half the price listed in its Original Equipment Manufacturer (OEM) catalog. Because of this, by 1960 Columbia had increasingly turned to the German Fichtel & Sachs and Union firms, then by 1965, to the Japanese Shimano company.[48]

However, Raleigh chairman George Wilson insisted that "Raleigh has no advantage whatsoever from association with Sturmey-Archer.... It is entirely independent from the policies of our cycle companies.... You need never fear that Sturmey-Archer will ever adopt any other policy than is absolutely fair." But some within Raleigh own management disagreed with this, believing the policy "offered competitors what amounted to a free gift." They argued that their high-quality components were being purchased by competitors and fitted to low-quality frames with nice paint jobs and put on the market for up to a pound ($2.80) less than "equivalent" Raleigh bicycles. In short, they believed that Sturmey-Archer should use something close to duopoly pricing toward other bicycle makers.[49]

John Macnaughtan, who later headed up the Sturmey-Archer division, explained that there was no clear-cut answer to the question of whether S-A sold hub gears and other components to Raleigh below the OEM list price. He acknowledges that there was an established "transfer price" for sales between the two. However, OEM units were sold to wholesale customers complete and ready to install. Units sold to Raleigh almost never were. Things like axle mounting nuts and washers, cable pulleys and guides, and rear cogs were left off and fabricated by Raleigh. Because Raleigh bought only Sturmey-Archer components, it could afford to invest in tooling for these simple parts that were used on almost every bike. Hence, the transfer price was lower than the OEM price, but most OEM customers were in no position to take advantage of the Raleigh units, as it would cost more to make or buy the missing parts than would be saved by buying the Raleigh-specification units.[50]

By the time of his address to the 1958 shareholders' meeting, Raleigh's George Wilson was no longer confident:

> It is a fact that never since the war, and in spite of the enormous unsatisfied demand of the war years, has the sale of pedicycles in this country ever equaled pre-war levels. There are many factors to which this development can be traced. Foremost among them is the improved standards of living of the population, which has given rise to an enormous increase in motorized transport.... I do not expect to see the demand for pedal cycles in this country ever again to achieve post-war levels or show any substantial increase.... I am unable to report any improvement in trade generally, nor do I see any indication that this is imminent.[51]

The problem was "solved" in April 1960 when TI bought Raleigh for £10.9 million. *The Economist* noted that "Raleigh Industries appears to have done rather well out of the price that Tube Investments is prepared to pay for the 50–55 percent of the declining British bicycle business that Raleigh still commands."[52]

Initially exhilarated at the prospect of merging Raleigh's state of the art production facilities and BCC's more modern, forward-thinking management, David Duffield, who at this time was a marketing executive for BCC, was shocked to discover that the management of the new combination would be turned over to Raleigh. The three directors who moved from BCC to Raleigh were gone within two years.[53] In truth, TI had walked away from the bicycle business, essentially giving Raleigh shareholders 1.35 million shares of TI stock worth £5.4 million and paying them £5.5 million in cash (a total of $30.5 million) to take BCC off its hands.

The origins of the odd merger go back to mid–1958. British Aluminum Co. (BACO) was run by chairman Viscount Portal of Hungerford, the Chief of Air Staff during the war, but by now elderly and increasingly infirm. The board proposed the sale of BACO to the American firm Alcoa, and retained two prestigious, old-line merchant banks, Hambros and Lazards, to put the deal together. Another American firm, Reynolds Aluminum, and TI, acting through a third financial district bank, Warburg's, started to buy up BACO shares in anticipation of a joint hostile takeover, touching off what became known as the Aluminum War. Three months later, Reynolds Aluminum, TI and Warburgs won, making BACO the British division of Reynolds Aluminum, co-owned by TI. It cost both firms a fortune; they had clearly overpaid. Many thought the merger of TI-BCC and Raleigh was intended to position TI-Raleigh for a spin-off or sale, perhaps to a foreign conglomerate.[54]

(To clarify an obviously confusing situation, it should be noted that TI-Reynolds, the makers of Reynolds 531 and 753 bicycle tubing, was not related to Reynolds Aluminum. TI-Reynolds, the famous tube maker, was founded in 1887 as the Patented Butted Tube Co. by Alfred M. Reynolds. Renamed the Reynolds Tube Company in 1923, it developed Reynolds 531 in 1935. It was acquired by Tube Industries in 1937. At the time of the Aluminum War its name was TI-Reynolds Tube Company. This was changed to TI-Reynolds in 1977 to reflect its diversification into the welding, bonding and fabrication of specialty metals, such as the chassis of space satellites.)[55]

On the other hand, Raleigh and BCC had been moving on converging paths for some time. Raleigh's George Wilson and Sir Francis de Guingand, head of TI-South Africa, were good friends, and in early 1959 Raleigh closed their relatively new factory in Vereeniging and created a merged TI-Raleigh South Africa at BCC's plant at Springs. Similarly, de Guingand had facilitated talks between Wilson and Ivan Stedeford that led to a joint TI-Raleigh Ireland. In August 1960, just 23 days before the death of Sir Harold Bowden, Wilson converted Raleigh to a holding company, clearly in anticipation of the merger.[56]

In one respect Duffield was wrong: it was not so much that TI-Raleigh kept its own management team in place to the detriment of BCC executives, it was more the case that it didn't keep anybody. Even before the merger, George Wilson had made a strategic decision to start recruiting top management talent from outside the company. Leslie Roberts was hired in early 1959 both as managing director and to be Wilson's successor. He came from Rootes Motors. D. Courtney Taylor, with a background at Standard Telephones and Imperial Tobacco, was brought in as the firm's manager of training. Charles Smith, managing director at Norton Motors, was selected to replace Harold Bowden after his death in September 1960. Peter Seales was hired as marketing manager in 1961 from Imperial Tobacco. He was astounded by what he found when he arrived: "If we had been independent in the early 1960s," he later said, "we wouldn't have survived." Out of 50 top and middle-management personnel in place immediately after the merger, 38 had been replaced by 1965. Almost all came from outside the bicycle industry.[57]

5

Really, What Makes a Bike?

> You know, Moulton went off, produced his bike and had immense success, getting it off the ground before Raleigh. But ours was the bike that changed the face of the business in the UK.
>
> —Alan Oakley (1973)[1]

I

The new TI-Raleigh quickly shuttered most of BCC's facilities. Archibold McLarty, Raleigh's finance director, later recalled that "we closed down and disposed of many factories, e.g., Hercules at Downing Street, Norman at Ashford, Sun at Little Aston, Brampton Fittings at Newton. We also closed the two saddle making companies, J. B. Brooks in Great Charles Street, Birmingham, and the Wright Saddle Company in Selly Oak and subsequently accommodated them under one roof.... This left the Cycle Division with two big factories, one in Nottingham and the other in Bridge Street [Birmingham]."[2]

By April 1961 TI-Raleigh had furloughed about 20 percent of the workforce in place at the time of the merger a year earlier, not counting the layoffs Ivan Stedeford imposed after the creation of BCC in 1958. "Decline is the only prospect that the cycle industry can now see for its products," reported *The Economist* dolefully.[3]

It was against this backdrop of uncertainty and change at Raleigh that Alex Moulton made his approach to the Nottingham giant. It was not a time to expect daring leaps into the technological unknown: a bloated, declining company undergoing large-scale layoffs; shedding a third of its workforce and capital assets; experiencing a turnover of more than half its senior- and middle-management executives, and whose most stable and profitable line of business was supplying the parts that its competitors were using against it.

Over the years, Alex Moulton's story about how George Wilson (who died in a car crash in December 1963) reacted to the new bicycle shifted. In 2008 and 2009, he said that he invited Wilson to The Hall in November 1959 to look over a prototype. Wilson arrived, and said the bicycle "was the most interesting thing he had ever seen." However, forty years before, in 1966, Moulton recounted that Wilson had actually accompanied Joe Wright, director of Dunlop, when Wright (along with their two wives) had visited The Hall, and that it was Wright who "was ecstatic" about the bicycle. This story is much more logical given that Moulton and Dunlop were already working closely together on the development of the Hydrolastic suspension system for the Austin/Morris 1100, which would eventually sell two million units. In this same 1966 story, Moulton attributed the decision not to proceed with production mostly to cost-cutting at Raleigh after the merger.[4]

While a few executives, such as Jim Harrisson, simply didn't like the thing, most agreed with Leslie Roberts, Wilson's successor, who never doubted the bicycle's abilities, but were equally convinced that it could never turn a profit. Alan Oakley believed that too much of the bicycle would have to be outsourced to specialty suppliers from outside the bicycle industry, a damming conclusion coming from a firm that prided itself on making all its own tooling in-house.[5]

Still others believed that the Moulton could never be made competitive unless it omitted Sturmey-Archer components and accessories (especially auxiliary groups, such as lighting systems), in favor of cheaper, possibly foreign parts. Raleigh's future hinged on using bicycles as marketing conveyances to bundle packages of Sturmey-Archer components and accessories to the point of sale. In short, Raleighs were becoming a means to an end; selling Sturmey-Archer components was the end. The Moulton couldn't do that well, because the basic platform cost about twenty percent more to make than a good roadster.[6]

In 1963, Seales hired two market research firms to investigate the British bicycle market. One was instructed to do a qualitative survey, the other a quantitative review. They were firewalled from each other to avoid "groupthink" responses. "It worked like a charm," Seales said. "It gave rise to the best piece of research I have ever seen. It was a Bible on consumer and trade attitudes." The report concluded that "the profile of the English cyclist which emerges characterizes him as associated with poverty, low social status, and hard times. The bicycle is seen as the poor man's useful and valuable friend. Their bicycles are usually pictured as old and worn, and their dilapidated condition does not matter to the owner."[7]

This was reflected in the dealer network for bicycles. "A dealer does not normally live or die from the proceeds of one product. He has no immediate reason to sell a Raleigh bike as any other, since the cut he is likely to take will remain roughly the same." Moreover, "The depression spawned a whole generation of cycle retailers who were good at repairing, but not good at selling. They are not sales-minded and are not merchandise minded, and in today's modern marketing world, they are an anachronism. They are not good business managers and are just hanging on. Most of them have been pushed off the high streets onto the back streets, into shabby shops."[8]

The goal was clear: get back onto high street. But the dealers couldn't be pulled along. A marketing team was sent out on a five-week road trip to push the firm's new "5-Star" exclusive dealer concept, similar to Schwinn's "Total Concept" stores of the 1950s. Raleigh would back up the brand with real ad dollars and point-of-sale support. In return, dealers had to dress up their shops and make a commitment to carry only the Raleigh brands. Seales: "We were met with defensive laughs, bitchiness." Enter Plan B: walk away from bicycle shops.[9]

In 1936, Raleigh had started selling to the high street department stores Halford's and Curry's in exchange for guaranteed minimum annual orders. Traditional dealers complained, but George Wilson privately told his board of directors that "the large number of cycles we could sell to Curry's would more than counteract any loss that would be sustained through malcontents." Sales to Curry's were initially disappointing because Hercules was able to provide them with a bicycle they could retail for £3, while Raleigh refused to retail for anything less than £5.25. But the post-war "American push" strained Hercules-BCC's ability to meet the demand for exported, private label English Racers to American mass merchandisers such as Montgomery Ward and Western Auto. Hence, many individual Curry's store managers refused to order from Hercules for floor stock because deliveries had become unreliable, even when ordered by Curry's central purchasing management.[10]

Raleigh was able to step into this breach. In 1962, Raleigh had 12,000 retail dealers, but the 550 branches of Halford's and Curry's accounted for somewhere between 21 and 25 percent of their domestic total. Raleigh decided that its future lay with two types of retailers: the high street stores, and perhaps 700 of the "5-Star" retailers.

Raleigh also had the weapon it needed to make the "5-Star" strategy work. For years, small shops had played off overdue payments to TI-BCC and Raleigh against each other, creating an *ad hoc* form of revolving credit. The game went like this: order from Raleigh. Accept the shipment, elect 90-day payment. As the bikes sold, use the proceeds to pay off the overdue bill on the *previous* BCC shipment you had accepted *before* the Raleighs. Once the BCC bill was paid off, the purchasing department hold was lifted on your account and you could again order. Order from BCC and accept the shipment. Use the new BCC bikes to pay off the now-overdue Raleigh bill. Repeat as needed.

One of the first things the new TI-Raleigh did was to merge the former BCC and Raleigh billing departments and send out unitary invoices. Overdue on your Hercules bill? No more Raleighs or BSAs. Late on your Raleigh bill? No Phillips or Hercules orders, sorry. As the credit merry-go-round ground to a halt, so did a lot of dealers. "It was a blood-letting exercise," recalls Seales. However, the strategy of leaving less than a thousand large, full-line, Raleigh-only dealerships was never achieved. "We weeded out most of the cottage industry until we had 1,500 aggressive dealers left," Seales notes.[11] Of these, about 200 to 300 ended up being the 5-Star shops.

Overall, specialty bike stores accounted for about 55 percent of domestic units sold; the 5-Stars made up 13 percent of this, leaving 42 percent for independent shops. The remainder was split about equally between high street department stores and suburban discounters. About 300 Halford's and 250 Curry's sold 25 percent of the total, and a slightly larger number of discounters sold 20 percent, mostly low-end kids' bikes.[12]

But leaning so heavily on the high street stores created its own problem:

> These outlets had been difficult to serve because of a general problem: there were so many bicycle variations needed to suit each individual customer's needs. There were men's and women's bicycles; sizes from 19½ to 23 inches; gears or no gears; lights, accessories; colors. There were 28 basic models, and in all there might be some 18 possible combinations of a single model. Mass retailers balked at carrying the inventories required to support such an operation.[13]

This was, in fact, only the tip of the iceberg. Seales estimated that in 1962 the company had, when domestic, export, and commonwealth markets for both Raleigh and BCC nameplates were included, close to 900 different model specifications. He was determined to cut that down to less than 200.[14]

Author Ray Gosling, a colleague of Alan Sillitoe in matters both literary and political (both were professed radical socialists), reporting in the Nottingham-based journal *Anarchy* during a 1964 strike at Raleigh, said he had been told by a manager that "Raleigh don't want to make bicycles. They want to do contract work for the motor industry. They want women and youths." While there were long-term profits to be made selling bicycles in third-world markets "Raleigh wants profits now. To sell over the world needs hard selling. It needs bosses to roll up their sleeves and sweat their way through Asia and spend hours with governments and accounts ... hard work, and they don't want it. Easier to slowly turn the factory over to the small stuff and employ women and children.... Little work and fat profits now, and the men—laid off."[15]

There was some truth to this. The J. B. Brooks division was quietly test-marketing a new self-retracting automobile seat- and shoulder- belt unit to automakers in anticipation of a national law mandating belts as standard equipment, but the project was killed in 1965 when Brooks couldn't match the OEM prices charged by its two competitors, Romack and Raydyot.[16]

At the end of 1964, Seales turned his marketing studies over to Alan Oakley, Raleigh's head of design. Oakley had already started working on the problem. "Around 1961–62, we had started looking for a different type of bicycle," he recalled. "The bike had to be unisex; it had to cover a range of adults and juveniles; it had to be different in image; suited to traffic conditions; and had to reflect the fact that there was more money to spend." Seales filled in two more conditions: its primary target audience was going to be women; and its market competition was not other bicycles. As Seales later explained, they were up against the "more skillful competition from other consumer items such as tape recorders, transistor radios, and record players." In short, "bicycles had become consumer goods, not bits of light engineering."[17]

No demographic in Britain had changed more than gender. In 1931 only 57 percent of women aged 21–39 were married. Now, in 1961, that figure was 81 percent. Before the war, the typical household was as often as not comprised of two aged parents and a spinster daughter. In 1961, the proportion of the overall workforce comprised of women was barely changed (31 percent in 1951; 33 percent in 1961). This was deceptive, however. The total workforce had expanded enormously, hiding the striking changes that had taken place within the categories of adult women and married women.

Among all women aged 20–64, six percent were employed in 1951; 42 percent in 1961. Nine percent of married women worked in 1951; 35 percent in 1961. By the time Seales made his survey, the "spinster household" was only half as common as the husband-wife dual wage-earner household, either with or without children. With better food, better medical care, and newer homes that didn't demand 40-plus hours of housework a week, the typical Midlands household wasn't made up of a daughter taking care of two aged parents, but two empty-nest grandparents hale and hearty enough to act as after-school babysitters until mom or dad got home from work. And with the wife jointly contributing to the family income, she got something that would have been unheard of before the war: joint control of the household budget.[18]

Michael Young, editor of *Which?* the monthly magazine of the Consumer's Association, recalled seeing how the reconstruction of London's Bethnal Green slum was successfully changing its family dynamics:

5. Really, What Makes a Bike? 93

"Not such a bloody uncomfortable place to be": Grosvenor Estate, London (532 units; 1928-30; Edwin Lutyens, architect). The Second Duke of Westminster built the development and leased it to the City of Westminster for 999 years for the price of one shilling to provide affordable housing "for the working classes." In 1989 the City attempted to break the covenant and put the flats up for sale, asserting that "the working classes" no longer existed. The Eighth Duke of Westminster sued to enforce the covenant. He prevailed in a court decision excoriating the City. The Estate is now a protected landmark, and still houses those who keep the lights on and the water running, but can't afford a rent of three thousand pounds ($4,000) a month (photograph by the author).

> It was clear that men's lives had been very much centered on their work; they kept a large proportion of the family's income for themselves and spent it separately from their wives in pubs and gambling and smoking. Partially they did it because the home was such a bloody uncomfortable place to be. What we saw was the beginning of change. The younger men, although interested in their work, were giving more interest to their homes, having something more like a partnership with their wives in building up their homes. And this was symbolized by the material goods that people bought. They had a terrific pride, an emotional investment, in these material goods.[19]

By now, there were about 40,000 Moultons out in the field, mostly in and around London, so Seales asked his consultant to survey their owners. Sixty-two percent were over 25 years old (25 percent for Raleigh). Fifty-eight percent owned a car; 13 percent owned two. Prior to seeing the Moulton 45 percent had no particular interest in owning a bicycle. In other words, they didn't buy a bike; they bought a Moulton. That got their attention in Nottingham.[20]

Raleigh had been sifting through a number of different alternatives, starting with a U-frame 20-incher Alan Oakley had patented in September 1959 that was similar to those that had generated a buzz at the Cologne bicycle show in 1958. Its outward appearance was identical to that of the inexpensive folders that were already starting to appear from France and Japan, but Oakley's version featured several nifty technical refinements that would have made it lighter and stronger, and would have made its production simpler and more economical. However, it still had the dowdy, U-frame look. It was rejected.[21]

By 1964 Oakley's design team had produced design sketches for bicycles more recognizable as RSW 16 progenitors. One had a unitary rear fender/rear rack made out of a plastic composite; another had a carrying handle atop the down tube (as did the early Moulton prototypes); still another had 20-inch wheels and non-cylindrical frame members made from either stamped sheet metal or fiberglass. A very similar variant of this bike, but with 16-inch wheels, was built using a sheet-aluminum monocoque frame. Built in collaboration with the British Aluminum Co. and Newland Research, it never went beyond the first prototype.[22]

The RSW 16 that eventually emerged from the development program looked different and modern, but its technology was actually quite traditional. It used regular, if oversized bicycle tubing arranged in a half-century-old cross-frame configuration. It had no suspension. Instead, it used wide, relatively low pressure tires. Tire displacement was a cheap, reliable way to duplicate the springing of the Moulton, but at a cost. The ride resistance over smooth surfaces was about one-quarter higher. Its steering was also much slower and heavier, which is its most apparent difference.[23]

On the other hand, it came with some neat gadgets. The front hub was a Sturmey-Archer GH6 dynohub, which powered the head- and taillights. The wiring ran inside the frame and was almost invisible. It also used a twist-grip shifter for the 3-speed hub, also with the cable buried within the frame. It would not be much of an exaggeration to call it a Sturmey-Archer SW 16, and a primary goal of the project was to convey top-of-the-line Sturmey-Archer components to the market in a single attractive package. "Really, what makes a bike?" asks John Macnaughtan, Sturmey-Archer's export director from 1976 to 1997. "You know, the frame and fork, it's just a coat hanger, and then you hang all your fancy brakes and lights and things, and Sturmey-Archer made those things ... actually, that side of the business is much bigger than the bicycle side of the business. The cleverest thing Shimano has ever done is *not* make a bicycle."[24]

That is why there were only two RSW 16 variants, the so-called Deluxe and a folding version called the RSW Compact. The primary purpose of the platform was to convey Sturmey-Archer components to the market as a group. Giving consumers a range of equipment choices would just defeat the purpose. (Seales and Oakley later admitted that the Compact was "a gimmick," intended primarily to drum up publicity. Folded, it wasn't much smaller than the basic bike. Its shifter cable could not be run within the frame, and its electrical wiring shouldn't have been either, as it tended to get caught in the joint when it was snapped into the riding position.[25])

The RSW's rear caliper brake's performance was dodgy, so after two years Raleigh replaced it with a purpose-built three-speed rear hub, the Sturmey-Archer S3B, that had a built-in small diameter hub brake that worked from a brake lever, not by backpedaling. In the dry, it worked about the same as the caliper brake, but in the wet it worked at all, which was a big help. "In the RSW 16, Raleigh was now able to offer mass retailers a single, unvarying product suitable to buyers of either sex and in all sizes: an all-purpose cycle which was fully equipped," said one analyst.[26] Note the emphasis on "mass retailers." Unlike the Moulton, which was offered in a number of different variants with various grades and qualities of components and accessories, if you bought a RSW you bought a full package of Sturmey-Archer's best.

David Duffield later recalled that the RSW 16 undersold the Moulton "by five or six quid" (£5.50 to £6.60; $15.50 to $18.50), but it is hard to make a direct, head-to-head comparison. The 1965 catalog price for a Moulton Standard was £30.45 ($85.25)

An early ad for the RSW 16 (*Cycling*, July 1965/Library of Congress).

and a Moulton Deluxe was £34.65 ($97.00). Raleigh told an American business reporter that the British list price of the RSW 16 Deluxe was also £30.45 ($85.25), but Tony Hadland reports that at the time of their 1965 launch, the prices of the RSW Deluxe and Compact were less, £26.50 ($74.25) and £27.85 ($78.00). Given that inflation was relatively high during this period, prices could change a great deal from year to year. In the USA, the catalog prices in 1967–68 were $94.95 for the Deluxe and $99.95 for the Compact.[27]

The Moulton Standard did not come equipped with lights, kickstand or rear bag. It used a Sturmey-Archer AW 3-speed hub until 1965, after which it was equipped with the FW 4-speed. The Moulton Deluxe was equipped with all these extras and the 4-speed hub. Obviously, the Moulton factory believed that the RSW 16 should be compared to the Moulton Deluxe, but it actually fell about half-way between the two in terms of the basic equipment (brakes, hub gears, cranks, pedals) and accessories (lights, carry bags) provided.[28]

To claim that the RSW was "five to six quid" cheaper than a Moulton was therefore a mostly accurate statement. In comparing a Moulton Standard to the non-folding RSW

16, what a customer got in the RSW was a bike with very good equipment and accessories; conservatively designed (outward appearances notwithstanding); very sturdy; no suspension; heavy; overall, decent basic transport for trips under 5 kilometers (3 miles). In the Moulton Standard, she got cutting-edge technology; a bicycle inherently capable of performing up to the level of all but the best traditional touring machines, but no accessories and in some cases fairly low-end components equal to those on a standard roadster; and occasionally, serious quality control issues. If she stepped up to the Moulton Deluxe, everything was superior except for the occasional quality control issues.

Why no suspension? First, cost. The suspension had to be paid for (roughly four pounds—eleven dollars) in the Moulton Standard by cutting out accessories such as a lighting system and kickstand, and by substituting some lower quality components, including Phillips steel handlebars, brakes, a low-end crank, and rubber pedals. (The Moulton Deluxe used G.B. alloy brakes and handlebar, Phillips alloy pedals, and the 4-speed hub gear.)

Second, the suspension involved many small non-steel (mostly nylon and other synthetic material) parts that had to be purchased from outside suppliers. It was unlikely that Sturmey-Archer would be interested in producing such a unit unless there was a promise of industry-wide sales. Finally, even before it launched the RSW 16, Raleigh was anticipating the introduction of smaller-wheeled RSW 14s and 11s, and separate suspension units would be necessary for each wheel size. "The difference in thinking," said Alan Oakley, "was that he had one idea and didn't see the family."[29] Over the next two years, the RSW 14 and 11 were added, and a few years later the basic design was adapted to create the roadster-ish Twenty and the dragster-ish kids' variant, the Chopper.

Although Seales and his marketing team poured a hundred thousand pounds into the RSW marketing campaign in its first twelve months, the RSW, like the Moulton, was not originally a big hit with bicycle dealers. "Changing the attitude of dealers was a hard slog as much as anything else," according to Seales. But it succeeded magnificently from day one with its intended market: high street department stores. In fact, this was part of Raleigh's plan. Seales's claim that Raleigh had only 1,500 dealers left by 1973 meant that there had been a massive winnowing out in the late 1960s and early 1970s. In 1962, Raleigh had 12,000 retailers, in 1964, it had 7,500. If the 1,500 figure is correct (and based on the figures in a 1981 British government report on monopoly concentration in the bicycle industry, it probably is), then Halford's and Curry's accounted for about a quarter of Raleigh's outlets in the 1970s. Raleigh was well on its way to its goal of selling bicycles like televisions and washing machines.[30]

When Raleigh first informed Alex Moulton in the summer of 1964 that they intended to introduce their own competing production, Moulton was not completely powerless, as he frequently implied. True, his own company was small and underfinanced, but he had a partner who was not: the British Motor Corporation. And the viability of its Fisher and Ludlow plant at Kirkby plant was important—it had been built with government assistance that required BMC to meet employment and utilization targets. Moulton asked for a meeting with Lord Plowden, the Chairman of TI, at the 1964 Earl's Court motor show. Plowden agreed. Moulton showed up with George Harriman, the head of BMC, who was, by no coincidence, one of TI's largest customers. (After all, every car off the line at Longbridge or Cowley had six feet of exhaust pipe under it.) The outcome: "It was agreed that Moulton would not use their legal remedies against Raleigh in relation to their marketing of a Moulton type small wheel bicycle and that Raleigh would give Moulton access to their exclusive dealer network (in particular Currys)."[31] The threat of "legal remedies" proved a toothless tiger, as Plowden probably knew: the RSW, lacking a suspension, infringed no patents, but Harriman was happy and Moulton got what he needed—continued access to the vital high street department stores.

TI-Raleigh was a huge, far-flung empire and by the mid–1960s, not every outpost was under the direct control of Nottingham. John Macnaughtan first met Alex Moulton in Johannesburg in 1967. It was Macnaughtan's first job with TI. TI-South Africa was something of an anomaly. Raleigh-South Africa and TI-South Africa had merged before their respective parent corporations, in the late 1950s. This was because the distribution chain of suppliers, wholesalers and retailers was so tied up with South Africa's bizarre apartheid-era social structure (which extended beyond just black-white relations) that both firms' South African subsidiaries had to be run as self-contained units, ignoring the vertical relationships back to each division's UK parent. For TI that meant something like 75 divisions reporting to one headquarters in Springs, which then reported to Birmingham. Both TI and Raleigh figured that if their South African subdivisions had to be run as autonomous entities, it would be cheaper and easier to merge them into one company. TI-Raleigh/South Africa was the result.[32]

The division director, Jack Catting, rejected the RSW 16, instead opting for the Moulton, which was manufactured under the Moulton name and also badge-licensed under a number of other marques. "Instead of this dreadful RSW 16 which was being imposed on them, Jack Catting liked the idea of the colonies not being run from Whitehall, so of course he chose the Moulton bicycle instead of the Raleigh, and took great pride in so doing," Macnaughtan recalled.[33]

Macnaughtan does not know how many South African Moultons were made, making it hard to be certain how many were manufactured overall. It is even more difficult to track down sales figures for the RSW. Tony Hadland believes that "over 100,000" RSW 16s were sold between 1965 and 1974. This is probably low. At its peak, in 1967, all three RSW variants (16, 14 and 11) were together selling between 750 and 1175 units per week. There was no doubt a significant downturn in RSW sales after 1970 when the Chopper siphoned away sales from the RSW 11 and 14 in the youth market, and the Twenty (introduced in 1968) diverted adult sales from the original RSW 16. However, if Chopper and Twenty sales are used as a guide, a reasonable estimate is aggregate lifetime sales (1965–74) of 150,000–175,000 for all three RSWs, but excluding the Chopper and Twenty.[34]

Raleigh also took a lesson from Moulton playbook's and used what Seales called "glamour themes" to "combat the social inferiority of the bicycle." What this meant in practice was pictorial ads featuring "beauty queens, stage and screen celebrities ... typical of these celebrities were Leslie Langley (Miss World); Peter Thompson (1965 British Open Golf Champion); Uffa Fox (world famous yachtsman); Graham Hill (noted racing driver); and Mary Quant (model)." These however, were standard paid studio appearances, as even the factory executives admitted.[35]

"Moulton sold his ideas in a very different way to Raleigh," commented one Raleigh anonymous senior manager (probably Seales). "He grabbed everybody who was somebody and put them on his bikes to be photographed, and it worked, up to a point." Yvonne Rix, Alan Oakley's assistant, who later became Raleigh's marketing director, later called Moulton marketing director David Duffield "the king of product placement." Unlike the Raleigh ads, most of these didn't cost Moulton anything, not even a free bicycle. Actually, many of these "placements" didn't even go that far; they were, in fact a discovery that some notable person had, on their own, walked into a store, bought a Moulton and was riding around on it. Reyner Banham was only one example. Others were the previously mentioned MP Quentin Hogg and actresses Eleanor Bron and Nichelle Nichols (Star Trek's Lt. Uhura). Bron may have been inspired by Banham; she was the partner of the architect Cedric Price, and Price was a good friend of Banham's. Bron even wrote a book about two tours she took in the 1960s through France and Holland on her Moulton.[36]

II

"There is no Pop architecture to speak of, and never will be, in any permanent sense, because buildings are too damn permanent," Banham once famously said. "The aesthetics of Pop depend on a massive initial and small sustaining power."[37] If this is so, then the RSW 16 can be considered the first Pop bicycle, certainly the first one targeted at the adult market. It was introduced in mid–1965. Its two smaller siblings, the RSW 14 and RSW 11 were introduced in 1967. A year later, in 1968, Raleigh introduced a variant with 20-inch wheels, the Twenty. In 1969, Raleigh introduced the Chopper. Based on a heavily re-worked RSW 16 frame, the Chopper was Raleigh's answer to Schwinn's multimillion-selling Stingray hi-rise kid's bike. It was introduced too late to succeed in the USA, but in 1970 it was introduced in the UK market along with a lightly reworked RSW called the RSW Mk. III and an all-new Moulton, also called the Mk. III, Raleigh having acquired the Moulton firm in 1967.

The RSW 14 and 11 were dropped in 1970 to leave a clear path for the Chopper. Of the three bikes introduced in 1970, only the Chopper was successful. In fact, it was a smash hit. John Macnaughtan recalls being told by Peter Seales at the RSW 16 introduction in 1965 that he, Seales, expected the original RSW to have a product life of five years, after which time it would go obsolete, at least in the UK.[38]

In 1974, Raleigh discontinued all the small wheelers except the Chopper and the Twenty. By now, neither survivor really fit the original definition of a "small wheeler" established by the Moulton and the RSW 16. The Twenty had come to replace the old 26-inch roadster as the standard urban utility bike, and the Chopper had established the norm for the (male) kids' market, at least until the BMX take-over of the '80s. According to Macnaughton, one of the primary reasons Raleigh introduced the Twenty was so it would still have a bicycle that could be targeted at women after the RSW 16 exhausted its product cycle. Seales had been pretty accurate in his estimate of the RSW's commercial durability: the lifecycle of Raleigh's small-wheel concept had lasted a little under nine years, and in actuality, you could argue that it had been on life support since the triple rollout in 1970, just about what he predicted.[39]

Tony Hadland believes that the success of the RSW was due to the fact that

> ...the idea of using a bike for a short-range shopping (or other utility) trip seemed quite a good one. The RSW traded on the Moulton's glam image ... and was "modern" at a time when technological progress was the "in" thing.... It was at arm's length and different from the dull old roadsters that people had to ride before they could afford cars, and before it became deeply unfashionable to smell of sweat, which cycling to and from work in ordinary clothes did not help."[40]

It is true that Banham found the RSW to be a "lamentable imitation" of the Moulton, but he admitted, "Raleigh can learn from any mistake, provided it is catastrophic enough," and he still preferred it to "that fossilized, reactionary design," the roadster.[41] The distinction between the two wasn't so much that the Moulton was "good" and the RSW "bad," or even, to a large extent that one was better than the other, it was that they were simply different.

"The problem of Industrial Design has not changed; it is still a problem of affluent democracy, where the purchasing power of the masses is in conflict with the preferences of the elite," he explained. "The old, standardized, and unquestioned, public-school-pink propositions that all common taste is bad, and that all commercialism is evil, appear to need some revision ... the concept of good design as a form of aesthetic charity done on the laboring poor from a great height is incompatible with democracy as I see it."[42]

In oral histories of Alex Moulton, David Duffield, and others associated with the original 1962 Moulton bicycle, one hears repeated references to how it was the most radical change to the bicycle "since the safety," (Banham's claim) or "since the ordinary," or even "since the boneshaker." But what precisely defined the difference between the RSW and the Moulton? As Banham explained, "we live in a throw-away economy, a culture in which the most fundamental classification of our ideas and worldly possessions is in terms of their relative expendability." It was absurd to apply the standards of architecture to consumer design: "buildings may stand for a millennium, but their mechanical equipment must be replaced in fifty years, their furniture in twenty ... and a research rocket—the apex of our technological adventure—may be burned out and wrecked in a matter of minutes.[43]

As Seales pointed out, the RSW's competition wasn't the Moulton, it was "tape recorders, transistor radios and record players." Britain was undergoing a sea-change. For almost twenty years, it had valiantly fought against deprivation, want, and material bleakness. Now, it was entering its own era of "high mass-consumption." Not as excessive or as overt as what occurred in America after the war, but just as real. Retail prices rose 63 percent between 1955 and 1969, but salaries increased by 127 percent, and average weekly earnings by 130 percent.[44]

The appearance of affluence was greatest among teenagers and young adults, not because they were becoming wealthier, but because they were growing as a demographic group faster than any other age cohort, while at the same time their per capita income was growing at a rate equal to their older peers. Many goods started to reach the saturation point. Eighty-three percent of households owned a TV in 1963. In 1962, washing machine sales were 1.2 million per year, vacuum cleaner sales 1.5 million, refrigerators, 1.1 million. Car ownership doubled between 1960 and 1970. The growing problem was not "how to meet demand?" but "how to sell all we can produce?"[45]

In his article on "Throw Away Aesthetic," Banham, to the horror of many of his readers, compared the engine compartment of a classic Bugatti to that of a 1955 Buick. In the Bugatti, nothing shows, because the wiring, the pumps, the manifolds—all the support equipment—are buried within the engine block. In the Buick, of course, they are bolted on top or to the side for easy removal, and if necessary, replacement. "If one opens up the Bugatti hood and finds that motor covered with oil, one's esthetic displeasure at seeing a work of fine art disfigured would be deepened by the difficulty of repairwork when the ailing component proves to be hidden away," he noted.

The attempt to transfer the timeless aesthetic qualities of art or architecture to what amounted to a transport appliance rendered the whole exercise futile. In the end, the Bugatti couldn't do its job as well as the plebian Buick. "The close link between the technical and esthetic qualities of the Buick ensures that both sets of qualities have the same useful life," Banham wrote. "When the product is technically outmoded, it will be so esthetically. It will not linger on, as does the Bugatti, making forlorn claims to be a perennial monument of abstract art."[46]

The Moulton was developed very much in this tradition. The original F-frame Moulton looks as modern today as it did fifty years ago, although its suspension technology is obsolete, and the inability to adjust its stiffness was a shortcoming apparent to many even in 1963. When John Woodburn and Vic Nicholson complained that it was too soft to accommodate out-of-the-saddle climbing, Alex Moulton told them his charts and graphs indicated that out-of-the-saddle stomping was inefficient and that they should change their riding styles.

Michael Woolf of the Moulton Preservation Society explained that the Moulton:

> is almost like the inception of an idea, with a dreamlike quality. It was perfect with all the mini-things, mini-skirts, mini-cars, and you can almost hear the Beatles in a certain way. I think everything was becoming slightly more rounded and more optimistic—the angular thing was becoming more rounded and approachable with a sense of optimism and joy, and the style of the time and it spoke to people who wanted to go forward with that optimism.[47]

But Peter Seales believed that was just the problem: "He [Moulton] didn't manage to get his idea over that that the important thing about the bike was that it was suspended." Instead, everyone focused on the image, the look. And Seales flatly admits that by working with Dunlop to create a 16-inch tire with the right balloon profile, that look could almost be duplicated in much simpler, more straightforward (many would say dumbed-down) way.[48] Indeed, Moulton himself complained many years later that "at no time did the thought enter my mind that they would use the reality of my concept to make something 'in the manner of.' Of course, these were the days before Non-Disclosure Agreements. Fundamentally the RSW didn't conflict with our patents (you can't patent small wheels as such) … [but] visually the first impact was that the two machines were quite similar." This suggests that Moulton himself believed that small wheels defined the essence of his bicycle.[49]

But it is clear that the RSW more straightforwardly shared a throw away industrial esthetic. It was the transistor radio of bicycles. That doesn't mean that it was built badly or shoddily; if anything, its quality control was more reliable than that of the Moulton, mostly because its design, despite its outwardly novel appearance, was really quite conventional, even conservative, with the factory's production strengths and limitations kept a foremost consideration.

In his 1947 book, *Design for Business*, J. Gordon Lippencott, explained that "There is only one reason for hiring an industrial designer, and that is to increase the sale of a product," cautioning that "no product, however well its aesthetic functions are fulfilled, may be termed a good example of industrial design unless it meets the acid test of high sales through public acceptance. Good industrial design means mass acceptance. No matter how beautiful a product may be, if it does not meet this test, the designer has failed."[50]

But design itself was a disposable thing. Like the Buick's bumper bombs and tail fins, it is intended to have "maximum initial impact and small sustaining power," and indeed, looking over an RSW 16 (or even more so, a 14 or 11), is rather like examining a late-50s or early 60s Cadillac; it strikes you as faintly ridiculous, but you have to admire the sheer audacity of the thing. Imagine—pumping a quarter of a million dollars in advertising into something that looks like that at a time when a quarter-million was real money!

It did boost Raleigh's sales for the better part of half a decade, and it served as the platform for two long-term successes, the Chopper and the Twenty, the former serving as Raleigh's bread-and-butter kids' bike in the UK and the latter displacing the roadster, holding on as Raleigh's best-selling adult utility bike until the 1980s when it was killed by a combination of cheap, foreign-made U-frame folders and the increasing use of inexpensive mountain bikes as urban utility machines.

Alan Oakley put it bluntly: "*ours* was the bike that changed the face of the business…. It was Raleigh which exploited the idea and set up a concerted marketing operation around it; a fashion industry in which kids of all ages were the consumer…. It introduced that element of product obsolescence which the industry desperately needed." As Banham once explained, in a throw-away world, his job as critic wasn't "to disdain what sells, but to help answer the now important question, What *Will* Sell?"[51]

6

History Repeats Itself, Once More

> As every schoolboy knows, the Raleigh company turned down Moulton's small-wheel design when he first took it to them. So how come Raleigh is now making all kinds of funny bikes, including the Moulton Mk. III? Well, it seems like all the best British companies, Raleigh can learn from any mistake, provided it is catastrophic enough. Blinkered by a stereotyped view of the cycling population as a load of dim proles who would buy the same product forever, and forms of racing so ritualized that technical progress was pointless, Raleigh failed to observe when the market was ready to change.
>
> —Reyner Banham (1971)[1]

I

The world of small-wheeled bicycles was not confined to England. A number of compact cycles had been exhibited at the Cologne, Frankfurt and Paris bicycle shows as early as 1951, when Japan Origami displayed their "Paris" folding bike, with 16-inch wheels. Amsterdam Bicycles showed their U-frame, 20-inch wheeled Panther in 1958 and Sauvage showed a 20-incher that looked a lot like a bigger version of the RSW 16 in 1960.[2]

Raleigh patented Alan Oakley's version of a U-frame in 1959, but the 1946 VéloSolex moped may have been the first mass-produced example of this configuration. The VéloSolex remained in production until 1955, after some 650,000 units had been produced. The U-frame is a very rational layout for an inexpensive small-wheeled cycle. It enables one tube to take the place of three in a diamond-frame bicycle, and a rear triangle separately fabricated by a supplier can simply be bolted on. The Oakley U-frame was innovative in that it provided for a separate head tube that was slipped into a coaxial sleeve pressed out of one end of the U-tube and was then swaged or brazed into place. However, it was never used. Raleigh did later market U-frame folding bicycles, but they were purchased from Asian manufacturers.[3]

In the early 1960s, Raleigh produced the Mobylette lightweight moped under license from the French firm Motobécane. The Mobylette did not use tubing for its main triangle. Instead, the frame was pressed out of sheet steel in right- and left-hand halves, which were then riveted and welded together. During the RSW 16 development process, one of the design alternatives considered was a 20-inch wheeled utility bicycle called the "Park Avenue" that used a Mobylette-type pressed steel frame. Like the Oakley U-frame, the Park Avenue was never put into production.[4]

Raleigh started to have some real competition starting in the mid–1960s when Dawes, largest of the half-dozen or so "specialty" makers who hadn't been swept up in the TI-Raleigh monopoly, introduced its KP500 Kingpin, with 20-inch wheels. It looked like a RSW 16 with bigger wheels, with a large monotube taking the place of separate top and down tubes, giving the Kingpin's main frame a slanted "H" shape. It was built using traditional techniques, brazed by hand on moving jigs that rolled from station to station on a conveyor, with each craftsman brazing a single joint. More compact than a roadster, but not that much different in feel, it became a steady seller, averaging roughly 500 units a week—for 25 years. A year or so later, a folding version was offered, followed by a break-apart variant, but their sales never amounted to more than ten percent of any year's total.[5]

Raleigh quietly introduced its Twenty in 1968. It was almost certainly developed in response to the Kingpin's success. While it looked like Dawes's bike, in a number of details it was a slightly better design, mostly to facilitate volume production. For example, if you look closely at the underside of a Twenty, you will see that the right and left chainstays and the two gusset stays (between the bottom bracket and the monotube) are actually each a single tube. A small-diameter tube is cut, bent to a shallow angle, then flattened at the forward end and the curve where it passes under the bottom bracket, and finally slotted and shaped for the rear dropout. It is then brazed onto the monotube, the underside of the bottom bracket, and the rear dropout all in one operation. Very neat, very strong.[6] However, the biggest difference was the hinge of the folding version, which was heavy, but as strong as the straight tube of the non-folder. The tube will break before the joint will. The hinge of the Kingpin had always been a problem, requiring two redesigns.

Very little is known about the origins of the Twenty, because its development and introduction were deliberately kept low-key. The fact that it was introduced under two off-

Advertisement for Dawes Kingpin, c. 1966 (private collection).

brand labels suggests that Raleigh's initial target may have been alternative outlets. Raleigh mostly used Sun and Triumph to sell promotional loss leaders to bike shops and high street stores (usually for Christmas or spring specials), or to sell Raleighs anonymously to discount stores and mail-order outlets. "I don't recall any major effort by Raleigh to link the Twenty to the RSW," notes Raleigh historian Tony Hadland. "The marketing was more low-key; the sensible bike, the rational hybrid of the traditional roadster with the minibike." On the other hand, as previously discussed, John Macnaugtan recalls that there was concern in management by 1968 about where the female clientele that Raleigh had cultivated with the RSW would go when it reached product obsolescence, and this was considered in the decision to put the Twenty on the market.[7]

Women, as an econo-demographic group, had continued to evolve during the 1960s. A survey of 229 well-paid blue collar workers and 71 non-managerial white collar workers in the city of Luton, 30 miles north of London, by Cambridge University indicated that in 15 percent of the blue collar households, the wife worked full time and in 17 percent of the households, part time. In the white collar households, 24 percent of the wives worked full time and 26 percent part time. Although the blue collar men typically earned more than the white collar men (this was intentional—the target sample was high-paid blue collar craftsmen, and lower-rank non-management white-collars), the white collar households averaged a higher income because the wives worked more often, earned higher wages, and they had fewer children. Therefore, it was still possible for the Raleigh sales department to talk about women as an emerging marketing group. While they were, as a group, in flux, they also had enough commonality to have some degree of coherence, at least from the perspective of a marketing executive.[8]

The same could not be said of men, and certainly not men who had been bicycle consumers. Although Banham may have talked about "new executive radical cyclists" or "a reemergence of the proletarian, working class cyclist," it wasn't happening. The Cambridge social scientists found that 45 percent of the blue-collar households owned a car. The only thing holding many of the others back is that they wanted a house more, and in the relatively small city of Luton they could get by without one: 57 percent owned their own home, not appreciably less than the 69 percent of white-collar homeowners. In fact, calling them "blue collar male workers" was probably a misnomer. Fewer than a third (31 percent) did not have any white collar connections at all. In other words, 69 percent either: (a) had a wife who was a white collar worker; or, (b) a father who is, was, or had been when alive a white-collar worker; or, (c) they themselves had previously been a white collar employee, but switched to blue collar work, almost always for higher pay.[9]

The whole idea of an alienated, angry class of manual workers looked shaky. That doesn't mean they weren't out there. As you went down the socioeconomic spectrum from Luton's new-tech blue-collar workers, past the highly paid, but brutal and dangerous old-line heavy industries (mining, steelwork, shipbuilding), down through low-paid, unskilled manual labor to the absolutely dependent, the stereotypes became more applicable. Julia Gunnigan spent 1959 as a principal at a secondary modern school in Pimlico. Her son, also a school administrator, wrote how she found that:

> no one would ever admit it, but people were happy where they were. They were especially happy with the level of complaint and grumbling, which seemed one of life's most important pleasures. A

few of her brightest pupils had passed the eleven-plus and had been offered a place at grammar school, but their parents had not allowed them to take it up. She raised the question with one pair, and they shuffled and looked embarrassed and eventually admitted to her—they probably wouldn't had she not been Irish—"We didn't want him to think he was better than us."[10]

As late as 1960, out of 2.5 million children in secondary modern schools, only 32,000 were 15 or older. But that was changing almost as fast as the age cohort born after 1946 turned 15, when mandatory school attendance ended. There were more schools, more and better teachers, more choice of curricula, and fewer families dependent on an additional weekly wage, no matter how pitiful. Overall, the idea of a "cloth cap man and his bicycle," of a working-class that retained traditional blue-collar insularity regardless of income or living situation, no longer really applied. With no one to sell it to, there could no longer be a "workingman's bicycle." Either a person needed a basic transport bicycle or they didn't.[11]

The Twenty also raised one final question. Peter Seales had said that the RSW 16's competition wasn't other bicycles, but transistor radios, tape recorders and record players; light consumer goods. One has to wonder whether the Twenty was an admission that this strategy was obsolete, that the battle was lost. The transistor goods had beaten the bicycle, once and for all. All that was left were kids, the utility market, and club cyclists.

Like the Thermos bottle, the Raleigh Twenty became so ingrained in the public mind with the idea of the semi-compact bike that the entire category of 20-inch wheeled adult utility bikes became known as "Shoppers," the name of the top-end Raleigh-badged version of the Twenty, with dynohub, lights, racks and dual carryalls. In the mid–1970s Raleigh was selling 140,000 Twentys a year, and they were being manufactured under license in South Africa and New Zealand.[12] Even then, the Twenty was only the *second* biggest seller in Raleigh's small-wheel lineup.

The Twenty first appeared in Raleigh's USA catalog in 1969. It may have been a hurried or improvised decision. The non-folder was given the name "Twenty," with the stowaway carrying the name "Folder." They appear in separate photos, with the non-folder shod with "British" 20-inch wheels (451 mm bead), but the Folder on American "BMX" 20-inch wheels (406 mm bead). Given that 451 mm tires were rare in the USA, the non-folder was likely a stock shot of a UK unit, never intended for export.

The following year's catalog has the two photographed together in matching colors, but the Folder sports a unique frame hinge with the handle on the right side of the monotube, not on top. This was probably a reaction to a recently released report of the President's Commission on Product Safety that had been highly critical of the bicycle industry for lax standards that permitted what it saw as such patently unsafe features as auto-style gearshift levers and other protrusions on the top tube of bicycles.[13] Apparently expecting an imminent safety rule that would prohibit such projections, Raleigh in America prepared a contingency design. However, no action was immediately forthcoming, and the original hinge mechanism was retained.[14]

By 1971 the Twenty was being sold in the UK in six variants, all the way from a bare-bones single speed Hercules at £26.50 ($63.60) to two different Raleighs, the standard Twenty, equipped just like a RSW, down to the rear carryall, twist-grip shifter and dynohub, at £38.75 ($93.00), and the even more lavish Shopper at £39.95 ($97.00). In

A 1969 Raleigh Twenty Folder (photograph by the author).

the American catalog, the Twenty and the Folder were relatively plain, with only a standard Sturmey-Archer thumb shifter (no twist-grip); no dynohub or lights, and the rear rack a modified 10-speed unit made by Pletscher. The only price printed in a Raleigh catalog was in 1976, $129 for the Folder, the non-folder having been dropped by now. Inflation was high in the 1970s; it is likely that the Folder was originally priced in 1969 at around $100-$110. That would have put the non-folder at around $90, the price point left open by the withdrawal of the RSW 16 after 1968.[15]

In 1972 the Twenty was dropped from the American catalog. The last year for the Folder was 1976. It may have been discontinued for fear of violating federal product safety rules. In early 1976, the Consumer Product Safety Commission did finally issue the long-rumored top tube protrusion ban, although a CPSC lawyer said that folding-bike handles like that on the Twenty would probably be eligible for a waiver because they were so low.[16] On the other hand, it could have been a pure business decision: Raleigh in America in 1977 was in bad shape. In 1974 Raleigh opened a factory in Enid, Oklahoma, to handle the crush of demand from the great bike boom. It opened just as the boom collapsed in 1975. By 1977, Raleigh in America was losing £1.4 million a year ($3.2 million). It transferred the Enid plant to another TI division that made cylinders for pressurized gases, and filled the 1977 catalog with Asian-made bikes imported under the "Rampar" label.[17]

II

The Harry Wilson Agency, a Southern California bicycle and bike parts distributor whose clients included Schwinn, noticed in 1962 that they were starting to move an almost-forgotten item: high-rise handlebars that, as near as anybody could recall, had been designed in the '40s to help paperboys with the front-loaded canvas bags they carried draped around their necks. Investigating the matter, Wilson's sales reps found that they were being bought almost exclusively by kids, who were also buying an elongated red-and-white seat called the Solo Polo available from another distributor named Peter Mole.[18]

Mole said he had bought them from the Persons Majestic Saddle Co., an Ohio maker of bicycle and exercise-cycle seats. The Persons sales staff told the Harry Wilson people they had made them in the '50s when it looked like a popular European sport, bicycle polo, might catch on in the States. It didn't, and Persons had unloaded the lot off on Mole at a surplus sale. They still had the molds and tooling, though. It turned out that kids were taking old 20-inch-wheel juvenile bikes (such as the Schwinn Hornet or Mead Ranger), removing the handlebars and saddles, and putting on the high-rise bars and Solo Polo saddles. A few of the more mechanically inclined kids were replacing the front forks with a proprietary spring-action "knuckler" fork Schwinn had once installed on some of its 24- and 26-inch wheeled cruiser bikes. To compensate for the longer fork, they replaced the 20-inch front wheel with a 16-incher taken from a child's bike like the Schwinn Pixie.

Gene Randel and Marion Moore, owners of a San Diego bike shop, put together some of their own "ape-hanger" bikes for Christmas 1962 using the Wilson bars and Solo Polo saddles. At about this time, Mole sold out of Solo Polos and asked the Persons factory to make him a new run of saddles. Then he went to Huffman Manufacturing's Azusa, California factory and contracted for a run of 500 custom-spec'd hi-riser bikes, which he called Penguin. These arrived in either late 1962 or, more likely, early 1963. He sold all 500 to California bike shops in a matter of weeks.

Al Fritz, Schwinn vice-president, was tipped off to what was going off by his west coast sales representative, who worked closely with their distributor, the Harry Wilson firm. He had them ship him a few of the hi-rise bars and saddles. Bob Wilson, who was running the Wilson company, promised Fritz that he would take 500 bikes if Schwinn would build them. The earliest bikes used stock Solo Polo saddles, dual 20-inch wheels and matching medium-width tires used for the Hornet. With foresight, Fritz changed the name to the more aggressive Stingray. When he saw the prototype, Frank Schwinn, Jr., thought Fritz had gone nuts, but Fritz was sure they could end up selling a reasonable number, say 10,000 units, so Schwinn OK'd the production run for the Wilson firm, which arrived about mid–1963.

Fritz's timing was right. Frank Jr., had just taken over control of the firm from his father, Frank Sr., who in turn was the son of company founder Ignatz Schwinn. Norman Clarke, president of the Columbia Bicycle Company, who knew both Franks well, once vividly recalled Senior: "Frank Schwinn, Sr. was a very sour, bitter and moody individual who always addressed the Bicycle Manufacturers' Association meetings with a long harange about parts makers, mass merchandisers, and the government, all of which he despised. He never smiled, always wore a black suit, was fervent in his belief in his company, and nobody else."[19]

It is hard to conceive that Senior would have approved something so frivolous as Fritz's Stingray, especially given its initial price of $51.95. Young Frank, on the other hand, was more easygoing, more openminded. He was also going out with Bob Wilson's daughter Nancy every time he was in California, and he made it a point to be in California as often as he could. (They would marry in 1964). So Frank didn't need all that much convincing before approving Al Fritz's test run. By the end of 1963 Schwinn had sold 45,000 Stingrays. A year later, during Christmas 1964, the factory was behind eight weeks on orders and dealers were selling rain checks so parents could at least put something under the Christmas tree for their kids.[20]

By the time the craze started to ebb after Christmas 1968, Schwinn had created at least three major variants, with almost a dozen different specific configurations. The three basic platforms were a dual 20-inch frame; a 20/16 chopper combination using a version of the 1930s sprung front fork; and a dual 24-inch version for mid-teens. Variants included a basic single-speed; a deluxe single-speed with accessories like fenders, a narrow front tire and super-fat rear slick tire; two-speed kickback; 3-speed hub gear and 5-speed derailleur. While the original "1963½" model, a basic coaster-brake dual-20 sold for $51.95, by Christmas 1964 the price range ran from $70 to $140. Schwinn may have sold as many as two million of them. The demand for the 2-speed "kickback" version was so high that Bendix created a special version just for the Stingray. On the two regular models, when the coaster brake engaged it shifted to low; a "kick-back" upshifted it. The Stingray hub worked the opposite: it cycled to high after braking and the kick-back down-shifted it into "wheelie" gear.

By a coincidence, Banham apparently got a chance to ride one of the early ones, a Penguin, in Chicago, during a 1964–65 visit to the United States on a Graham Foundation scholarship. ("I fell off it in a backyard," he admitted.) In his review, he got a couple of things wrong but an amazing amount right. He wrote that "these have long been known as Huffies after the Huffman Company, who first made them." At that time, nobody but a handful of industry insiders knew that the Azusa factory had actually been the first to build hi-risers, but on the other hand, no self-respecting kid would have called his hi-riser a Huffy (a term of deep derision, as most of Huffman's products were inexpensive, being targeted at the department store market). To this day, almost everyone believes that the craze started at Christmas 1963 with Schwinn's Stingray.[21]

Raleigh was becoming ambivalent about the U.S. market. At the time of the 1960 TI-Raleigh merger, the firm's strategy was to maintain its near-monopoly share over the British market and target big, high-population developing nations, especially Nigeria and Iran, with newly oil-rich economies. The American market was something of an oddity. Unlike the Third World target nations, the United States wanted the same type of cycles TI-Raleigh built for the British domestic market. The problem was that these imports were in the form of low-profit private label brands for mass-merchandisers such as Firestone Tire, Western Auto and Montgomery Ward. In 1963, Raleigh exported 802,000 bicycles into the U.S., but only 60,000 under its own name. Many of these were a legacy from before the TI merger, long-term commitments made by Hercules, Phillips, Triumph, and other British Cycle Corporation firms. Those factories no longer existed; the load had been shifted to Nottingham when Raleigh took over. Raleigh was experiencing quality control problems with both its private-label bikes and, more urgently, its Raleigh and Robin Hood models, including bikes shipped to its domestic and commonwealth distributors, who were voicing complaints at dealers' meetings and trade shows.[22]

Raleigh hired the consultants Booz, Allen and Hamilton to advise it on what to do about the situation. It recommended that the firm pull out of America in every market except the high-end bike store segment, and put its effort into building up a wholesale distribution and sales network to support a strong system of independent, exclusive Raleigh dealers. In essence, it recommended that Raleigh duplicate in North America what it was already doing domestically: walk away from 80 percent of its outlets.[23]

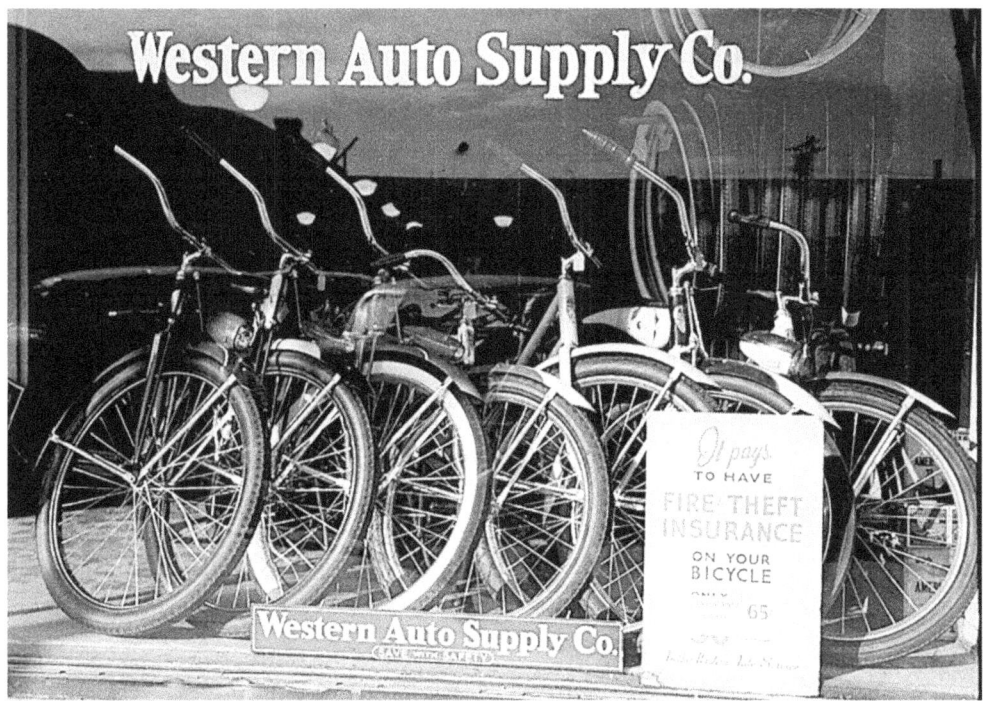

Bicycles in the window of a Western Auto store in Brawley, California, 1946. Western Auto was a nationwide chain of stores that sold tires, auto parts, large kitchen appliances, and lawn mowers (photograph by Russell Lee/Library of Congress).

Norman Clarke had bought the Columbia bicycle firm in early 1963 in a leveraged buyout when its parent, the bicycle parts maker Torrington, decided to either sell it or close it down. A few weeks later, he was approached by executives of Western Auto, who had been told "by Raleigh that they would no longer sell them English lightweight bicycles, geared bicycles, 3-speed bikes. Western Auto had catalogs out, prices out, everything ready, and Raleigh pulled out—just like that." What Clarke was witnessing was the sudden and dramatic implementation of the Booz, Allen report. Apparently, Raleigh had decided not to undertake a phased withdrawal from the North American private-label market, but something akin to shock therapy.

Western Auto initially approached Schwinn, but the elder Frank had just completed a painful 20-year process of pulling out of the same type of chain stores that Raleigh was now withdrawing from. Retrenching into dedicated wholesale distributorships and "Total Concept" exclusive bicycle shops had been expensive and controversial within the industry and engendered no small amount of bitterness among the half-dozen or so members of the Schwinn family who held stock in the closed corporation. Only a couple had anything to do with running the firm, and most of the others were dependent on dividends to support their sometimes lavish lifestyles. "Frank Schwinn turned Western Auto down flat," Clarke recalled.[24]

Western Auto next turned to Columbia. Clarke badly wanted the contract, but couldn't make the numbers work: "I told them that we can't possibly meet the price ... by using Sturmey-Archer gears. They charged us about seven dollars apiece for them."

Clarke believed that Raleigh "paid" a transfer price of about £1.33 ($3.75) each. He offered to build the bikes using Shimano 3-speed hubs. The Western Auto executives were horrified. "Okay, a week went by. Western Auto came down with company executives and said, 'Look, we need 175,000 bicycles. We'll use them if you'll stand behind them.' I said we would." The Shimano brothers had been regular visitors to the Columbia factory over the last few years, and had left samples. Clarke's young head engineer, Harold Machin, had tested many of them to destruction. He liked what he found, and reported this to his boss. Columbia was already primed to go Japanese. All they needed was for Sturmey-Archer to give them an excuse. "I hopped a plane the next day to Japan," Clarke recalls.[25]

Raleigh only exported the RSW 16 Deluxe to the U.S. for one year (1967), and the folding model, the Compact, for two (1967 and 1968). Moulton exported about 5,000 F-frames to Huffman Manufacturing's Azusa, California factory, mostly in the form of partially knocked down kits. They were sold primarily on the west coast, although there was at least one very active dealer in the New York City area. Most were apparently Stowaways.[26]

"An incredible number to those of us in the industry." In 1971, with the Great Bike Boom in full swing, lightweight bicycles were so hard to come by that Hertz Rent-a-Car experimented with a bicycle rental subsidiary. Here, District Hardware in Washington, D.C., rents a 3-speed Columbia to a customer. By 1975, the boom was over and Hertz went back to sedans and station wagons. "There were more bikes sold and never used than you could shake a stick at during those days," Columbia's president, Norman Clarke, commented many years later. (photograph by Warren K. Leffler/Library of Congress).

The smaller RSW variants were being used by Raleigh's United Kingdom dealers mostly as upscale youth bicycles.[27] The decision not to import them into the United States left a gaping hole in its North American line-up: no distinctive, deluxe youth bike. The RSW 14 and RSW 11 were selling briskly in the UK in the pre-teen market. They were certainly different from the Stingray and its plethora of imitators, and who knows, they just might have caught on in the States if some serious marketing money had been put behind them.

At first, Raleigh executives thought that the RSW 14 and 11 would continue to meet the challenge of the Stingray clones in the UK domestic market. They guessed wrong. Interestingly, Moulton was the first to try to meet the challenge. In 1965, a year before the smaller RSWs were added, Moulton licensed Lines Brothers, makers of the well known Tri-ang high-end wheeled toys to make a junior version of the Moulton for 8- to 12 year-olds. It had 14-inch wheels and a stock rear suspension, but an ersatz front suspension—the springing was provided by a bulked-up front rubber bellows, which in the adult version was merely a weatherseal for the intricate suspension system built into the head tube. The Tri-ang Junior, however, had nothing inside the bellows. On the other hand, it was £16.80 ($47.00) as compared to £30.45 ($85.00) for a Moulton Standard. (In 1966, Moulton introduced a Mini-Moulton, which was a fully functional 14-inch wheeled version of the original Moulton. It could accommodate riders up to about five-eight [1.75 m].)

To address the "Stingray problem," Raleigh first tried its own copy, called the Rodeo, which Raleigh in America introduced in 1966. It duplicated Schwinn's cantilever frame (with the bowed second set of top tubes), but it had narrow tires and a clunky Sturmey-Archer stick-shift. A year later, it added a true clone—down to the fat rear tire and Persons saddle—called the Fireball. They were also sold in the UK, with the Rodeo relabeled Mustang because Triumph already had a model with that name. They did okay, but nothing to get excited about.[28] The company was determined to change this. What happened next is the subject of considerable controversy.

According to Raleigh, Alan Oakley flew to the United States in 1967 to look over the hi-rise scene, especially on the west coast. On the flight back, he sketched out a new Stingray-killer he called the "Chopper" on the back of an envelope. As early as 1964, Schwinn had started offering variations of the basic Stingray, but until 1968 they all still used the basic Persons Solo Polo saddle. Oakley decided that the key was to focus on a new saddle design. It was to be a sprung saddle, with the springs ostentatiously wrapped around the tubes that extended down to the rear axle. To come up with the final design, he turned to Ogle Design Ltd. in Letchworth. (The ultra-deluxe Stingray Krate series, introduced in 1968, also had seat springs, but rather than external steel coils, they used rubber buffers, hidden in cylinders at the bottom of the seat stays, down by the rear axle.)[29]

On the other hand, according to just about everyone else (including the British Design Council), Peter Seales hired Ogle Design, probably the best known and most innovative industrial design firm in England, as early as 1966. Ogle's director, Dr. Tom Karen, and Raleigh designer George Ellis held about 20 meetings throughout 1966 and early 1967. Ultimately, after a very deliberate process that incorporated Ogle's design staff, Raleigh's production people, and marketing input from Raleigh in America, Ogle winnowed the alternatives down to four finalists for presentation to Alan Oakley. They all combined the same basic ingredients: a 20-inch rear wheel, small front wheel, hi-rise bars, the elaborate banana seat with the coil springs; and twin top tubes on which to mount a car-like gearshift. The gearshift design was selected to hide the fact that the Sturmey-Archer 5-speed hub needed two shift levers. The T-shaped handle was actually split down the middle and was really two independent L-shaped shifters. (Raleigh was so worried about the derailleur-equipped 5-speed Stingrays that it initiated production of a 5-speed hub at Sturmey-Archer just for the Chopper, even though the design was patented years earlier. Three-speeds got a single shift lever topped with a knob.) About the only thing both Raleigh and Tom Karen agree on is that the name Chopper should be credited to Raleigh in America vice-president Al Scalingi.[30]

The frame was modified from a RSW 16. The rear triangle was stretched to hold the 20-inch wheel, and two parallel, non-functional top tubes were added to mount the "gearshift." In fact, the frame wasn't modified quite enough: the first year's bikes were susceptible to inadvertent wheelies, and tended to develop a shimmy at high speed. A "Mk. II" version had to be developed that shortened the saddle two inches by moving the back axle forward and that lowered the frame an inch. (They are easily identifiable in photos because the seat stays of the Mk. II have a forward bend immediately above the rear dropout, where the Mk. I seat stays are straight up-and-down.)

The Chopper was launched in the USA in September 1969. This was the twilight of the Stingray era in the States, and Banham, who was in America at the time, noted that "it was a late attempt to scramble on this two-wheel bandwagon before it went over the horizon for good, but in the States, it looks very dim and ordinary among all the zoomy domestic products." Still, Raleigh managed to sell a respectable 144,000 by mid–1973 when the Consumer Product Safety Commission rules banned things like top-tube gearshifts and the external, unshielded seat springs. Raleigh chose to withdraw the model instead of redesigning it.[31]

Britain was another story. Although Raleigh did a market test in 1969, selling 500 through selected dealers in three cities for the Christmas season, it waited until 1970 for the "hard" introduction, rolling it out alongside the new Moulton Mk. III and RSW Mk. III in May. Unlike the other two, it took off with a bang, selling 375,000 in the domestic market by the end of 1972. In 1975, the year after the Moulton and all the RSWs had been cut, the Chopper still racked up sales of 100,000 units. That meant the Twenty and the Chopper accounted for almost 250,000 units out of Raleigh's total domestic sales of 595,000 bicycles. Raleigh kept the Chopper in production until the early '80s, but by the end it had been largely replaced by the Grifter, a BMX-ish bike built with traditional brazed-frame technology, road-style forks and a 3-speed hub.[32]

A 1970 Chopper. It is a Mk. II, as can be distinguished by its shorter saddle and the distinctive forward bend in the seatstays just above the rear axle. The Mk. I was overly prone to inadvertent wheelies and tended to develop a shimmy on downhills. The top-tube gearshift and exposed seat springs spelled its doom in the USA after Consumer Product Safety Commission rules were issued in 1974 (photograph by the author).

David Duffield had gone to work for Halford's in 1977. "When I joined

Halford's they did not have a reputation as being enthused about cycles ... [although] they were Raleigh's biggest customer, 200,000 bikes, and were producing another 50,000 bikes on their own." Duffield helped get the retailer re-enthused about the product at just the right time: in 1976 Skip Hess had started a firm in Simi Valley, California to make his semi-handbuilt "Mongoose" BMX racing bikes.[33]

Like Raleigh, Schwinn missed the boat. Russ Okawa, a Schwinn dealer in Canoga Park, California, recalls becoming so frustrated by Chicago's refusal to bring out a BMX bike in the 1970s for fear of cutting into its fading Stingray sales that he took matters into his own hands. He started ordering truckloads of basic Stingrays, then, in assembly line-fashion, stripping the saddles, handlebars and 28-spoke wheels off. He replaced them with BMX-style low-rise bars and mini-saddles, and 36-spoke wheels built up using high-flange hubs, aluminum rims and fat, knobby tires. The souped-up bikes cost $84 to $94, twenty to thirty dollars more than a stock Stingray. "We thought, 'that's way too much.' No one will buy these," recalls Okawa. They went like hotcakes. The factory finally got wind of what Okawa was doing because he was selling the stripped-off components to his fellow dealers at deep discounts for use as repair parts. Schwinn threatened to pull his dealership if he didn't stop.[34]

Finally, in late 1982, the factory offered a true BMX bike, the Predator. It was made in Taiwan. In August 1985, thirty-five year-old Edward Schwinn, new president of the firm, fired Al Fritz, inventor of the Stingray. Fritz had started at Schwinn five years before Ed Schwinn was born. In 1991, Ed Schwinn turned over control to a group of bankruptcy trustees. Schwinn was sold to an investment group associated with the Scott Sports Group, a maker and distributor of ski equipment, for $43.3 million. The fourteen Schwinn family members who held stock received a total of $2.5 million. (Ed collected his 3.4 percent share: $85,000.) The rest went to creditors.[35]

Unable to convince Raleigh to bring out a real MIG-welded, reinforced unicrown-fork BMX machine for Halford's, David Duffield imported private-label bikes from the USA, then from a Spanish firm, BH. At one point in the early 1980s Halford's actually sold more BMX bikes in the UK than Raleigh. "When I joined [Halford's] bicycles were 14 percent of their business. When I left it was 28 percent. BMX was the biggest transformation.... The most important part was to get the people, the work force, interested in the product." He was able to sell the organization on BMX by feeding back their bicycle profits into advertising and sponsorships at a time when they had a virtual lock on the market. "During the first twelve months, we plowed one hundred percent of our profits back into BMX. It took 18 months to take off." BMX, in turn, prepared Halford's for the transition into mountain bikes when that technology took off. Raleigh finally yielded to reality and brought out a true Nottingham-built BMX bike in 1982, six years after Mongoose started production of the originals in California. "Look where they [Raleigh] are now," Duffield told an interviewer in 2008, "twenty percent of the market."[36]

III

The Moulton Mk. III sold abysmally, even though almost everyone agreed with Banham that it was "the best (certainly the best made) Moulton ever." Between its launch in May 1970 and its withdrawal from the market in the fall of 1974, it only sold about 5,200 units. Reportedly, the factory only made one batch run.[37]

It seemed that Raleigh never had much faith in it. A prototype appeared as early as May 1969, a full year before the roll-out. British writer Peter Knottley reviewed it for the American magazine *Bicycling* after touring with it for a thousand miles. He raved about it. However, it was quite a bit different from the production version. It was made with Reynolds 531 steel frame tubing, had a rear derailleur and six-speed rear cluster, shallow-drop touring handlebars, and weighed 28 lbs. The production model weighed 35 lbs., had a frame made from regular high-carbon steel, pressed steel forks, upright bars, and the same Sturmey-Archer S3B 3-speed with the miniaturized hub brake that had been developed for the RSW 16. Moreover, this was the only variant offered. According to Jim Bratby, whom TI transferred to Raleigh from another division in the late '60s, the failure of the Mk. III led to Alan Oakley's removal as head of production design, although he retained his position as director of domestic marketing, the beginning of a slow exit for the man who only five years earlier had been lauded as the father of the Chopper.[38]

In the last years of the F-frame, the majority of the units produced by Raleigh were the 14-inch variant, the Mini Moulton. Raleigh management made the decision to go down-scale in 1971 and withdrew the original Mini Moulton, substituting a model lacking a front suspension system, probably targeted at the youth market—a replacement for the RSW 14 and 11. Without the suspension, the joint between the head tube and main tube tended to crack–497 failures between 1971 and 1974. The factory issued an incredibly ugly retrofit gusset to bolster the joint on bikes already in customer hands, and suspended sales in mid–1974, scrapping 1,100 frames that had been X-rayed and cleared by quality control. The inability to make a price-competitive bike that looked like, was branded as, and worked something like, a Moulton was probably the precipitating factor in the decision to drop all Moultons in 1974.[39]

In 1971 Moulton presented Raleigh with a research report on design alternatives for a next generation bicycle to replace the Moulton-Raleigh Mk. III. Somewhat later he prepared a similar report outlining a concept for a replacement for the Twenty. For the 17-inch wheel bike, his strategy was to expand the basic Moulton concept into a broad-based platform that could be used to develop anything from an exotic sports cycle like the Speedsix to a utility bike. Some variants had dual suspension, some only a front suspension, some none. Designated the GP (General Purpose), the 17-inch concept used Raleigh's production costs for the Triumph 20 (the least expensive Twenty variant) as a baseline for comparison. The 20-inch wheel concept bike was called the Moulton Twenty and kept the basic rear suspension of the Moulton Mk. III, but had no front suspension.[40]

Moulton calculated that a 16/17-inch GP built with front and rear suspension and Moulton-level components would cost £10.97 ($26.50) to produce as compared to £8.57 ($20.50) for the Triumph 20 and about £13.00 ($31.25) for the Raleigh-Moulton Mk. III. A six-speed derailleur version with a retail price of £40.00 ($96.00) was feasible. As Tony Hadland notes, "The GP was probably the lowest-priced application ever of Moulton bicycle design principles. It showed that a Moulton bicycle need not cost much more than an inferior unsprung small-wheeler to produce."

There was little, if any, interest shown at Raleigh headquarters in either the 17-inch GP or in the Moulton Twenty, with its 451mm/20-inch wheels and squishball rear suspension. Although again, it is easy to read a conspiracy theory into Raleigh's behavior, a much simpler explanation is at hand: the American 10-speed boom. In 1963, Raleigh had exported about 700,000 bicycles to the United States, but as previously noted only about sixty to eighty thousand were Raleigh-branded. The rest were private label, sold at a steep discount, a legacy from the TI merger. After Raleigh acted on the Booz, Allen report and pulled out of everything except the specialty bike shop market, USA exports fell to 132,000 by 1966, but almost all were quality Raleigh or Carlton-brand bicycles.

But starting in 1969–70, ten-speeds really took off in the U.S., and in 1973 Raleigh exported 460,000 bicycles to the States. But now they were all lightweights, mostly 10-speeds. In 1965, Raleigh's planning department estimated that in 1970 it would be exporting £7.5 million ($17.5 million) worth of bikes to the world. In actuality, in 1971 it exported £18.5 million ($43.1 million). Most of the difference can be attributed to North America. In addition to these bikes, it was building 146,000 Choppers and 135,000 Twentys for the home market, which in 1972 totaled about 739,000, an increase of some 170,000 units over 1969. In 1970 the total market for adult small-wheelers (i.e.18 inches or less) in the UK was only 86,000 units.[41]

Between 1970 and 1974, the Raleigh works—especially Sturmey-Archer—was stretched to the limit. The Raleigh-owned Gazelle factory in Holland, and Raleigh plants in Ireland and Malaysia were re-fitted to supply the USA market. High-end work at the Carlton factory was abandoned and its machinery was used to make mid-level Raleighs for the North American market. The Handsworth factory of Lines Brothers, who made the Tri-ang line of wheeled toys, was acquired, originally with the intent to continue their existing product line for the domestic market, but it too was converted to bicycles, mostly children's versions of the Chopper called the Chipper and Chippy. The Tri-ang products, which included tricycles, pedal cars and pedal-powered go-carts, were almost certainly bigger money-makers than were the export bicycles that replaced them. Raleigh built the Enid, Oklahoma, factory in June 1974, only to shutter it thirty-three months later and give it to TI for another use. Overall, quality suffered, and many American models received scathing product reviews for mediocre component selection and uneven fit and finish.[42]

In 1974, figuring out ways to bring new and unique models to market was just not a front-burner issue. In 1960, Moulton had suffered because Raleigh was in disarray after the TI merger. It saw its problems then as essentially dealer-driven, not understanding that its point of sale weaknesses were inseparable from the products that it provided to them. In 1974, it viewed its problems as emanating from inadequate production capacity in the face of rapidly expanding worldwide market, not seeing that it was throwing everything into a flash fire that would flame out so quickly that there would be no new products to offer after it was over unless the company imposed self-discipline, rationed output, and explicitly reserved capacity for research and development.

It did none of these things. As Raleigh finance director Archibold McLarty put it: "History has repeated itself, once more management has mistaken the sales arising from a temporary boom as representing a continuing demand and built up productive capacity accordingly.... I do not believe that you can have both quantity and quality and the quest for increased output can only be at the expense of quality." He was ignored. Compare this to Schwinn, which deliberately capped output growth during the great 10-speed boom to 62 percent between 1969 and 1973, and chose to import Traveler, Voyageur, and Le Tour models from Taiwan's Giant and Japan's Panasonic, even though Schwinn's marketing department warned that Giant was offering aggressive pricing because they badly wanted an opening into the American market so they could later introduce their own bicycle brand, which turned out to be true.[43]

The Moulton Mk. III was dropped from the Raleigh domestic sales catalog in June 1973. Dealers were informed that the model was unavailable for purchase in mid–1974. In January 1975, the designated mid-point of the 15-year contract between Moulton and Raleigh, the parties exercised their option to terminate relations. Moulton had come up with a new design, the so-called "Y" frame, and asked Raleigh to license back his previous patents. Raleigh indicated that they would do so only if they received royalties; complete technical information on all aspects of Moulton's new designs; and guarantees that the new bicycle would not be marketed under the Moulton name. A few months later, to avoid litigation, Raleigh and Alex Moulton agreed that Moulton could use the patents (apparently without a fee) and that any bicycles would be sold under a new brand name, "Alex Moulton Bicycles." Moulton first tested the Y-frame in early 1977, but ultimately concluded that it was unacceptably heavy and development was abandoned in 1980 or 1981. Instead, Alex turned to the development of his spaceframe AM series. The earliest drawings of a spaceframe concept are dated 1977 and the first prototype was built in 1979. Moulton filed his patents in late 1982 and a new firm, Alex Moulton Bicycles, Ltd. was launched in May 1983.[44]

A prototype Y-frame Moulton. Made between 1977 and 1980, several were fabricated, virtually indistinguishable from each other except for their paint schemes. It is believed that none were sold (photograph by the author).

As the engineering of the bicycle jelled, so did Moulton's business plan. To avoid a repetition of the disastrous Raleigh experience, the AM would be a no-compromise, limited production model, with the price to be set only after an uncompromising design and fabrication process was established. Although a 1983 AM 2-speed initially could be had for £400, this did not last long; the spaceframe series was continually refined, and a Bradford-on-Avon built, stainless-steel, double-pylon version (with both the head and seat tubes replaced with spaceframe structures in addition to the main frame) is now in the range of $17,000 in the USA market. However, a less expensive all purpose version is built at the factory of the Pashley Cycle Company (since 2008 owner of the Moulton firm). Pashley also sells their own semi-handbuilt "retro" roadsters, path-racers, and delivery cycles. It was the supplier of the bicycles used by the Royal Mail until the service discontinued the use of cycles.[45]

In 1977, two years after the Moulton/Raleigh split, an engineer named Andrew Ritchie[46] wrote Raleigh about a folding bicycle he had been developing over the last five years. Ritchie studied engineering at Cambridge, graduating in 1968. Dissatisfied after working for a short time as computer programmer, he sold landscaping plants and material for five years. "It was neither a success nor a catastrophe," he recalls. "It just broke even." In the mid–1970s, his father, a stockbroker, had a chance meeting with an Australian named Bill Ingram. Ingram was associated with a group of financial backers that were raising money to begin production of Bickerton bicycles in Australia in the form of knocked-down kits that would then be shipped to Britain for assembly. "I met up with Bill who showed me the Bickerton design," Ritchie recalls. "After studying the designs, I felt it was a very handy bike to make ... [but] I thought the design left quite a lot wanting. That evening, after Bill had gone home, I started sketching out designs on scrappy bits of paper."[47]

Ritchie persuaded ten friends to invest £100 each. He deemed the first prototype unsuccessful, scrapped it, and built two more. They were assembled in a small flat overlooking a church known as the Brompton Oratory, across the street from the Victoria and Albert Museum, which is how the bicycle ended up with its name. "It was quite a knife-and-fork business," according to Ritchie. "The work went on in my bedroom, making a real mess."[48] Ritchie sent a letter to Raleigh describing his work. They asked to borrow a prototype to examine it. After looking it over for two weeks, they returned it with a polite rejection letter acknowledging its level of innovation, but asserting that it needed "a considerable degree of redesign" and probably wouldn't sell well.

One problem was that it was distinctly ugly, with a marked kink in the main monotube. Later, Ritchie figured out how to put a gentle arch in the backbone that made the bicycle quite attractive. He persuaded thirty friends to buy pre-production bicycles as a way of investing in his start-up and began

Andrew Ritchie at the first Brompton Bicycles factory, c. 1984 (courtesy Brompton Bicycles, Ltd.).

hand-building bicycles. This was in 1979–80. The initial batch required 18 months to complete. "I ended up making 50 rather than just the 30, thinking that maybe there were another 20 people out there who would be interested in buying one." They sold as soon as they were finished. Ritchie raised £8,000 from the shareholders. In two years, he had made 500 bicycles, all by hand, in batches of fifty. "There was me and one employee, who was doing all the brazing. It was hopelessly inefficient, but it broke even."[49]

The firm did get some breaks: in April 1982 the Brompton was featured in *Design* magazine. It received kudos from high-tech bicycle guru Richard Ballantine, publisher of the British magazine *Bicycle*. But by the time the Ballantine review came out, production had ended: the supplier for the main hinge quit making them and there was no practical alternative on the horizon.

In April 1981, the British Monopolies and Mergers Commission lodged a complaint against Raleigh Industries, Ltd., alleging that the firm's refusal to sell most Raleigh-branded models to discount stores on the same terms as specialty dealers constituted an anti-competitive practice under the Competition Act of 1980.[50] Raleigh, with the support of the Cyclist's Touring Club (CTC) and the British Cycling Federation, countered that it followed a stratified policy intended to preserve Britain's 1,500 specialty bicycle shops.

Raleigh argued that it did not withhold the sale of bicycles from discount stores or from mail-order suppliers. However, the bikes it did offer to these outlets generally carried Sun and Triumph labels, were spec'd slightly differently, and priced a pound or two higher or lower from Raleigh bicycles supplied to bicycle shops and other approved full-service dealers in order to prevent head-to-head comparison. The only identical Raleigh-branded bikes that discount shops or mail order houses could order were those at the low end—usually coaster brake kids' models.

The commission concluded that "Raleigh's policy of withholding supplies from some retailers therefore had, or was likely to have, the effect of restricting competition in the retailing of Raleigh bicycles.... We therefore conclude that the course of conduct pursued by Raleigh, as specified, is an anti-competitive practice."[51] When Raleigh pointed to the growing competitiveness of Asian firms, the commission sanguinely replied that "the increase in imports which has taken place

Andrew Ritchie's first semi-production Brompton, c. 1979-8 (photograph by Nora Quinlan).

at Raleigh's expense may be to some extent attributable to its existing distribution policy rather than in spite of it. If Raleigh had adopted a different attitude towards modern retailing methods ... we wonder whether it might be trading more successfully now."[52]

John Macnaughtan says that there had been significant investments in the 1960s in plastics and powder metallurgy in his division, but that no comparable efforts had been made at Phillips or Raleigh. The problem Sturmey-Archer ran into is that these technologies had such high front-end costs that they could never be recouped from supplying only the bicycle industry. So by 1970 Sturmey-Archer had expanded into components for the automobile and large appliance industries, which proved quite successful. However, within twenty years, England, like America, was awash in Asian cars, and unlike America, also awash in Asian washers, dryers and dishwashers, many assembled in Eastern Europe. At the same time, about 1989, Raleigh adopted a policy of purchasing parts from the cheapest source, regardless of identity. Raleighs regularly started to appear with Shimano components.[53]

Raleigh's post–TI merger strategy of selling millions of bicycles to relatively affluent Third World nations started to fall apart in 1979 when the Shah of Iran was overthrown and British holdings seized (and the American embassy staff held hostage for over a year). In 1964, exports had accounted for 78 percent of Raleigh's sales. In 1980 exports were 33 percent. The Nigerian market collapsed in 1982 when the debt-ridden government defaulted on its payments (on what may have been as many as half a million bicycle kits) and Raleigh walked away from its one-third interest in a factory in Kano, Nigeria that manufactured bicycles from knocked-down kits.[54]

In 1987 TI-Raleigh was sold to an investment group, Derby Investment Holdings, created for the purpose of acquiring it. Derby was headed by a London-based American lawyer, Edward Gottesman. A significant minority partner, Alan Finden-Crofts, was put in charge of the company. No price was announced, but it was widely known that TI was anxious to sell. Derby went on to make other bicycle industry acquisitions and by 1992 was the largest bicycle manufacturing group in the world. Revenues and profits were generally stable. As Melvyn Cresswell, who worked for Raleigh for thirty years in a number of supervisory and management roles recalled: "the move to profitability after the Derby takeover surprised many, and confounded the generally held view that Derby were only interested in asset stripping."[55]

In 1997, Grossman and Finden-Crofts sold 80 percent of Derby's assets to American investors, primarily Thayer Capital Partners and Perseus Capital, Inc. The firm was mismanaged, losses mounted, and from that time on asset-stripping did become the paramount strategy. By this point the firm's only three assets were the Raleigh trademark, the Sturmey-Archer division, and unrealized by almost everyone, the Brooks Saddle Company. As a maker of bicycles, Raleigh was no longer a particularly valuable asset, if it had any value at all, aside from real estate. The investors botched the sale of Sturmey-Archer in 2000 by selecting a buyer, Lenark, that was little more than a corporate Ponzi scheme, using rapid mergers and acquisitions to hide asset drains, but Sturmey was eventually sold (after the pension fund was drained by Lenark) to the Taiwanese firm SunRace for £750 million ($1.17 billion) who moved operations to Taiwan. The company was subsequently revitalized, with the quality and breath of products greatly expanded, including some of S-A's classic designs, which were updated and reintroduced.

Almost all of the remainder of the company, except for the Dutch Gazelle subsidiary, was purchased from the bankruptcy receivers by a team lead by Alan Finden-Crofts for $24 million in U.S. dollars. (Its actual price was $75 million, as about $50 million in debt—owed in currencies other than the dollar—was acquired along with the firm.) The Nottingham factory was razed in the mid–2000s and sold for a housing complex. Flying under the radar, former Raleigh executive John Macnaughtan engineered the purchase of the all-but-forgotten Brooks saddle division.

Macnaughtan and Adrian Williams, formerly a principal of Westland Helicopter, had already gone in together to buy specialty cyclemaker Pashley Cycles from the Pashley family. "Sturmey Archer was too big for me to do anything about," Macnaughtan later recalled, "but when Brooks became available, I thought 'well, there is some synergy here,' so I quickly got ahold of Adrian, and said, 'look, I think we could buy this thing.' An offer was being put up for it, small beer, half a million quid [£550,000], ridiculous." He quickly called Riccardo Baglian, the owner of Selle Royal saddles. "I said, 'Riccardo, I've got the chance to buy Brooks, will you come in on it with us?' He said no, he didn't want to be a part of something he couldn't control but then he said, 'John, buy it. Get the best price you can, and whatever you have to pay, I'll buy it from you.'" With that guarantee, Macnaughtan and Williams bid about a million pounds ($1.55 million) for it. They kept it for two years. With Macnaugton and Williams already running Pashley and negotiating to acquire Alex Moulton Bicycles, Brooks "needed more money and more time than Adrian and I could give it." Macnaughtan took Baglian up on his offer and sold it to Selle Royal, contingent on the promise that Selle Royal would continue its Midlands workshop. They have done so, and it continues to successfully produce leather bicycle saddles and other products.[56]

Raleigh designer Alan Oakley died of cancer in Nottingham on May 20, 2012. He was 85 years old. Riccardo Baglian died of a heart attack in early 2014. His daughter now heads up Selle Royal.[57]

7

Alternative Wheels

> I first met him [Banham] at a conference in London in 1969. He arrived there after peddling his way through dense London traffic on a collapsible bicycle with little wheels, which looked inadequate for his bulk. He was as enthusiastic about this little contraption as if he were a bike salesman.
>
> —Frank Dudas (1991)[1]

I

"I did it for a woman," Banham asserted of his sweaty-palmed attempt to pass the London driver's license test, "both the beard and driving." Although "I remain convinced that the examiner had taken leave of his senses; I wouldn't have passed me," he got his permit on the first try.[2] In spite of his attempt to palm the blame off on long-suffering Mary, the real reason for the license was the United States: he was starting to spend increasing time there, and as he was finding out, relying on public transit or friends wasn't the way to see America. Between 1968 and 1970 he stayed in Southern California for about 18 months doing research for what is probably his best-known book, 1971's *Los Angeles: The Architecture of the Four Ecologies*.

A follow-up to a series of BBC Radio 3 programs, followed by a BBC-TV documentary, *Reyner Banham Loves LA*, it is often considered the progenitor for Denise Scott Brown and Robert Venturi's book *Learning from Las Vegas*, but the two are really quite different. As writer Tom Wolfe once said, Versailles and Las Vegas were the only two completely architecturally uniform cities in the Western world. Brown and Venturi's main point was that Las Vegas was nothing but an assemblage of plain white boxes ("sheds") covered with megawatts of colored lights; the signage *was* the architecture.[3]

Los Angeles, on the other hand, was functionally different. Its hamburger stands looked like hamburger stands. Its freeways looked like freeways. The only exception was downtown. Downtown didn't look like downtown; it just looked like just another suburban activity center–Westwood; Irvine; Glendale. And besides, *Learning from Las Vegas* was almost in galley proofs when *Los Angeles* went to press.

On the other hand, Banham was following a trail blazed by his fellow Independent Group member Lawrence Alloway almost fifteen years earlier. After traveling across the United States in 1958, Alloway wrote up his impressions in a 1959 *Architectural*

Design article. "In Los Angeles close friends can live twenty or thirty miles apart, which gives a kind of privacy, but they are easily accessible by car on the freeways, which restores intimacy," he wrote. "It is a diffuse urban space with metropolitan occupants. It works and works well for the Los Angeles resident who uses his car like a cowboy used his horse, as a natural adaptive extension of his legs."[4]

Many were aghast when Banham did not adhere to the English intellectual-left party line, summed up by novelist Adam Raphael, that LA was "the noisiest, the smelliest, the most uncomfortable major city in the United States; in short, a stinking sewer." It is true that Banham played down the city's mobility and environmental woes, especially for those without a car, but his main point was never that Los Angeles was a great city. "The unique value of Los Angeles—what excites, intrigues and sometimes repels me—is that it offers radical alternatives to almost every urban concept in unquestioned currency," he wrote. In other words, "there are as many possible cities as there are possible forms of human society."[5]

Similarly, he argued in *Los Angeles* that just as architects had to abandon the concept of the "ideal building" in favor of "the right environment for the right activity," so too must planners. "The failure rate of town planning is so high throughout the world that one can only marvel that the profession has not long since given up trying," he wrote. It was a controversial conclusion. Francis Carney, in *The New York Review of Books*, called *Los Angeles* a "schlokology." Another LA-based reviewer hit pretty close to home when he noted that "the trouble with Reyner Banham is that the fashionable sonofabitch doesn't have to live here." And he never did.[6]

Banham had shifted from journalist to full-time academic when Richard Llewelyn-Davies, the new head of Bartlett School of Architecture at University College London hired him as professor of architectural history. "You have to remember that the first, and fortunately bloodless, student revolution at an architecture school was at the Bartlett," explained Mary Banham.

> [Hector] Corfiato was running the school along Beaux-Arts principles, fifty years after it was long dead, [so] the students, especially the graduate students, started an alternative series of courses. Nobody went to the set courses—they all went to the alternative courses where they brought in people like Peter and other people of his generation who were talking modern. And it worked! Corfiato was out, Llewelyn-Davies was in, and it was run as a modern architectural institution.[7]

After teaching alternative classes for a couple of years, first as favor to the student activists, then as a way of helping get Llewelyn-Davies's new administration up and on its feet, Banham accepted a full-time position in 1964. "One of the reasons he was prepared go into academia full time and give up being a journalist was because at the Architectural Press, he couldn't accept invitations to go off and lecture for any length of time," Mary explained. Phillip Johnson had invited him to New York in early 1961 for a weekend to participate at a debate at the American Institute of Architects, and by 1964 Mary recalled that he had been invited to a couple of different places as a guest-lecturer for a semester, "but he had to stay and see that the magazine got out every month."[8]

But as the 1970s approached, it seemed as fast as Banham ran to keep up with the times, it wasn't fast enough. His techno-pop optimism was falling out of favor. "Technology" now meant napalm, DDT, diethylamide, the Dalkon Shield, freeways, and miniaturized nuclear warheads atop autonomous, self-launching ballistic missiles. The 1970 Aspen design conference was almost the last.

The International Design Conference at Aspen (IDCA) had been founded in 1949 by Walter Paepcke, president of the Container Corporation of America, and his wife Elizabeth, who, as a pet project, were turning the isolated, all-but-abandoned former silver-mining town of Aspen into a world-class ski resort. Paepcke was led by his chief of design, Egbert Jacobson, to create a forum where business leaders and design innovators could exchange ideas. The conference never caught on with businessmen, but it became very popular with the international art, graphics and design community, a sort of designers' Milan Triennale held in a large tent pitched each year on the lawn of the Aspen Institute. Banham had been attending since 1964, and had chaired the 1968 conference. Mary recalled that the 1964 trip was her first opportunity to visit the United States with him.[9]

The program chair of the 1970 conference was William Houseman, publisher of the journal *Environment Monthly*, and the theme of the conference was "Environment by Design." Banham wrote Mary (who had remained in London that year) that "Bill Houseman really hadn't got the program together enough for it to gell, and [with] the kinds of people he invited it was a guaranteed communications failure." The vacuum left by Houseman was filled by Sim van der Ryn, an opportunistic UC-Berkeley professor who had been asked to represent the interests of students and environmental activists during the pre-conference planning. Van der Ryn had been invited because "I had a reputation of being someone within the establishment ... who had connections and sympathies with radical groups."[10]

Two years earlier, in 1968, radical students had taken over the Milan Triennale on its opening day, ejecting the speakers from the stage and smashing several installations, including that of the Smithsons. A Buckminster Fuller lecture in London in 1969 was similarly picketed. Peter Hodgkinson, writing in *Architectural Design* about the time of the Fuller protest, identified technology as the problem, not the solution: "System builders, throw-away utopians and plug-in idealizers will continue the trend to a worsening environment," he complained.[11]

As was the case in previous years, Houseman and Eliot Noyes, IDCA president and director of design at IBM, planned to invite a university team to Aspen and give them a stipend for travel, housing and fees so they could prepare an installation. Instead, van der Ryn recommended that IDCA distribute several small travel grants to various San Francisco area student and activist groups to come in buses and vans, set up tents and inflatable structures, and give informal and improvised sessions and performances. Noyes objected, as it seemed to him "to be in conflict with the Conference itself, almost as a counter-conference, or anti-conference." The Aspen Institute told Noyes that no shelters could be erected on their lawn that could be used for overnight habitation. Noyes passed this on to van der Ryn. Van der Ryn assured Noyes that "I knew them and had worked with them and they were all doing work I thought the Conference should know about." The stipends were issued as recommended by van der Ryn. As many as 400 students and activists showed up, erected their bubble enclosures, moved in, and gate-crashed the conference.[12]

In another tradition, IBM paid for a multi-disciplinary contingent from a foreign nation to attend. In 1970 France was selected. The delegation was headed by Roger and Nicole Tallon and included sociologist Jeane Baudrillard. There is a wide disparity of opinion about how the delegation approached the conference. Some claimed the French academics showed up, checked in, picked up their stipend checks, and immediately assured everyone within earshot that the whole set-up only reinforced their belief that America was a land barren in culture. Some members (probably Jean Aubert and François Barre) admitted that "they remember arriving in Aspen ready to fight with the Americans."[13] But member Odile Hanappe later wrote that "It was a breath of fresh air compared to the intellectual climate in France. Nothing is aggressive. To our great surprise, everyone listens. They were humanists." Gilles de Bure wrote "I really had fun and learned a lot ... we talked more about cinema and politics than we did about design."[14]

The trouble started early. Every retrospective of the 1970 conference agrees that the traditional program format, five days of paper-reading plenary sessions (with side projectors usable only at night due to the brightness of the tent) was worn out. The conference board knew this, but had procrastinated about changing anything, as Noyes's "anti-conference" reaction to van der Ryn's suggestions indicates. "It's curious to me that change is so long in coming to this design conference," said one design student." It's one speaker and a thousand people glued to their seats by regulation, or boredom, or both." Another added, "The format's outmoded. Nobody wants to sit passively and listen anymore."[15]

But in a 1961 article, Banham had written that while "Aspen has, to some extent, replaced the Triennale ... the findings of the congresses [at Aspen] don't constitute a body of literature comparable with CIAM."[16] For many of the old guard, possibly most of all Banham, generating formal, publishable papers was what Aspen was all about. In the nine years since that article had come out, Aspen had come to produce a body of work to equal that of the Congrès Internationaux d'Architecture. Moreover, CIAM had decided to disband after the tenth congress in 1959. By 1970, Banham was already at work editing an anthology of Aspen papers, what he hoped would be the first of a series. For the old guard, of which Banham had become a member, Aspen wasn't a place to learn exciting new things, it was a place to bounce ideas and draft papers off long-time friends and colleagues.[17]

On the opening day, Ecology Action founder Cliff Humphrey illustrated his "design" lecture with a pile of trash gleaned from the main tent's own dumpster. "If an item is made to be wasted, to be dumped in a dump, *then don't make it!*" he proclaimed.[18] So much for Banham's ideals of expendability and transience.

Banham chaired the last day's conference, in which a coalition of students, activists, and French structuralists used to read a number of position papers and resolutions, some critical of the conference itself. Tension and petty conflict had been rife all four days, and Banham later admitted that it had been his idea to turn the last day's wrap-up session into a soap box for the malcontents. The first part of the morning went reasonably well. The French statement, having been written by dedicated structuralists, was of course, incomprehensible to most listeners, although it was one of only two papers from the 1970 conference that Banham included in the Aspen Papers anthology. It was translated by Francoise Jollant-Braunstein. "At the end, they applauded very nicely. Some of us were humiliated; this was frankly not the reaction we had wanted," Jollant-Braunstein admitted sheepishly.[19] Following this, the "Black Statement" was delivered. It was, to quote Banham, "routine stuff, alas, just the usual threats—'we're together and we're here baby'—though effective enough when addressed to an uptight white audience."[20]

It got tough after the coffee break. Michael Doyle of the San Francisco–like Environmental Workshop read an eleven-part resolution concluding with a demand that all the participants pledge to refuse to produce any work intended to promote corporate profit or capitalism. It was clear from the audience response that this wasn't going to fly with the perennial attendees, most of whom were either university or corporate employees. Doyle hinted that unless it was ratified, there might be trouble. Banham turned to Jivian Tabibian, a loquacious, Lebanese-born, Princeton-educated political scientist. Banham and Tabibian turned the proceedings into a two man show, keeping the discussion going; getting the students (against Doyle's objection) to break down the resolution into individual clauses and to discuss and vote on each one separately "for the sake of inclusiveness"; and picking up each point raised from the floor and bouncing it around the hall.

This gave "all the frightened souls" a chance to slip out; the tent was less than half full by the time Tabibian and Banham ran out of steam and the votes were taken. "I shall not soon forget the hostility vibes that were coming up from the floor," he wrote Mary, "nor how uptight the students could get the moment they thought they weren't getting their own way." However, between Banham and Tabibian, a repeat of the 1968 Milan Triennale was avoided. That night, Banham wrote, "we didn't blow the conference, but I count it among the hollower victories of my public career."[21]

The day after the conference was, by custom, the annual board meeting of the IDCA. Eliot Noyes resigned as president of the organization, a position he had held for five years. Moreover, he recommended the conferences be ended. "I feel after this conference, battered, bruised, stale and weary. I am of the impression that we have come to the end of the line." Banham himself wrote that "as the chairman of that stormy last session in 1970, I could suddenly feel all these changes running together in a spasm of bad vibrations that shook the conference. We got ourselves together again, but an epoch had ended."[22]

Banham didn't return until 1973, and when he did, he said "he was somewhat of a stranger." He included no papers from the 1971–73 conferences in the *Aspen Papers* anthology, describing them as "years of participation and workshops and be-ins, and they emphatically did not produce papers in the classic sense." The conference had been fundamentally restructured to emphasize smaller simultaneous sessions; many panel discussions; workshops, and field trips. For example, in 1971 Bucky Fuller was the only keynote speaker who actually lectured on anything having to do with design, and was included only because Saul Bass and Elliot Noyes, fearful that nobody was going to register, insisted on him.[23]

Banham's preface and chapter introductions in the 1974 *Aspen Papers* anthology are a lament for a lost era. After 1974 Aspen was, for Banham, a place to get away for a week with old friends, if it could be fit onto the calendar. It was the past, not really a part of the present, certainly not the future.[24]

Alice Twemlow, who has written a history of the 1970 conference, concludes:

> Banham undertook a personal reevaluation during this conference. Hitherto he had been the hip spokesman for Pop and had prided himself on his ability to identify and characterize emerging trends.... Banham found himself, possibly for the first time, out of sync with the zeitgeist. Furthermore, his critical methods and approaches, which for the past decade had been considered unconventional, even revolutionary, were, in this new environment, no longer particularly relevant or effective.[25]

II

Similarly, Banham began to wonder if he would ever be, in the words of historian Robert Maxwell, "anything but a prophet without honor in his own country." His doubts had actually started well before 1970, at a 1959 visit to the Hochschule at Ulm, where he met "a community of intellectuals who treated me seriously. In Britain, people would pat me on the head and say 'Marvelous—now get lost.'" They grew after Ken Russell's film *Pop Goes the Easel*, a joint "day in the life" biography of four young London pop artists, Peter Blake, Derek Boshier, Peter Phillips, and Pauline Boty, aired on BBC's *Monitor* television series in March 1962. The four played pinball, drove American cars, worked, and met at an evening party. The Institute for Contemporary Arts, the Independent Group, nor any of its members were mentioned. Banham was so miffed that he began a long-term project to pull together a documentary on the history of the Independent Group. In 1979 he and Lawrence Alloway appeared in Julian Cooper's *Fathers of Pop*.[26]

He came to resent being dismissed as a "mere journalist" because he wrote so much, so often, using language that was something less ponderous than the usual formal academicese:

> Never having believed that journalism is a waste of talents and energy that should be reserved for more serious matters, I have treated whatever has come my way, not with levity (as some have claimed) but with the enjoyment of finding things out, and gratitude for having an audience to tell them to. Offense has been taken by those who insist that profound matters must be discussed only in "serious" language, but having seen the mess that a Marx, a Mumford, a Levi-Strauss, a Galbreath or a Freud (let alone a Hoggart) can make by trying to handle light matters with heavy equipment, I felt I had license to do the other thing—and a better chance of being understood.[27]

In fact, as several of his students pointed out after his death, his writings were always segregated: he never wrote a book on design. His editor at the *New Statesman*, Michael McNay, and Penny Sparke observed that "Banham thinks of his books as his serious statements and his journalism as the formulation of ideas that may be quite transient. This is not a denigration of the shorter pieces, but rather a description of their different function."[28]

Finally, to some degree, he had himself to blame for his declining influence as the era of pop modernism evolved into the era of semiotics-based postmodernism. He once wrote that "animus has long been the very breath of life to historians for as long as the tribe has existed ... bias is essential—an unbiased historian is a pointless historian—because history is essentially a critical activity, a constant rescrutiny and rearrangement of the profession."[29] But while crossing sabers with all comers back in the IG days, when he was 33, may have been great good fun, he was now approaching his mid–50s and starting to tire.

"Many have assumed that the United States was his natural home," notes officemate Adrian Forty, "Certainly his taste for bootlace ties and cowboy boots might suggest that America was where he really wanted to be, but dress aside, there was no sense in which he ever went native in either a personal or an intellectual sense." Penny Sparke agrees, suggesting that Banham's 1976 move to the United States was as much a case of "push" as "pull." "The rhetoric of the American environment—and growing dissatisfaction with the British—finally persuaded Banham to set up home there and after a period of about 15 years teaching architectural history at University College in London, he took the offer of a professorial post at the State University of New York at Buffalo."[30]

Architect Robert Maxwell concurs: "The International Expo '70 in Osaka marked the apogee of his influence.... When Banham decided to abandon his post at the Bartlett College and head out west, there was a palpable feeling that he had gone into voluntary exile." Photographer Tim Street-Porter recalls that "The fact that he made these moves, leaving London and that whole architectural scene behind, was really quite radical ... with Banham, there was always this conundrum—that here was a man so glued in to the center of things, yet he spends his later years on the edge, not even in LA."[31]

III

"We hadn't done our homework," Mary Banham rued many years later about their move to Buffalo. Banham had been hired to start a European-style tutorial-based program in architectural history at the State University of New York at Buffalo. But 1976 was the year that New York City functionally went bankrupt, and the state was in almost as bad a shape. "The chairman kept asking Albany about the new program and heard nothing. After awhile it became apparent it wasn't going to happen." Banham was put in charge of an existing program to while away the three years of his contract. "Things were very bad in England," Mary recalled. "Everyone was holding on to their jobs as hard as they could because they knew there were no others. So when this possibility turned up, it looked like a good idea. In retrospect, what we both came to learn, but had to go back to find out, was if the central places, New York, LA, Chicago, which did approach him at the end of the job in Buffalo, had known he was looking [in 1976] they would have outed him then."[32]

Banham and their son Ben went over in September 1976. Mary was working, putting together a retrospective exhibit at the Victoria and Albert Museum on the 1951 Festival of Britain. Banham and Ben returned home for Christmas, then came back with Mary after New Year's. The winter of 1976–77 was one of the harshest on record, which went back to 1835. "The wind roared off the lake like an angry animal," she recalled. With her prosthetic leg, it was three months before she could even venture out of the house. "The neighbors had to supply groceries using skiis ... in America, at least, they know how to heat their houses, so it was comfortable indoors."[33]

Ben, who had started attending SUNY-Buffalo, dropped out and flew home. "I thought he'd done the right thing, I'd have done it, too, if I could have known how to survive," said Mary. In retrospect, she believes it was (and is) a beautiful city with its necklace of Olmsted-designed parks, "but the people who love it are born there, and born into the climate, born into that rustbelt environment, and are inured to it and can see beyond it." The basic problem, "as Peter said, and he was quite right, was that we came ten years too late. We were already in our fifties. Earlier, we could have made friends, had our children to raise."[34]

Banham apparently did not take any of his Moultons with him to SUNY-Buffalo. But he did buy another high-tech folding bike, a Bickerton. "It looks like it was put together by someone who'd never actually seen a bike, only read a prose description of one by a visiting Martian, and then drawn the components from the wrong bin," he wrote."[35]

It wasn't quite that exotic, although it was, at that time, probably the smallest folding bicycle in the world, which is probably why he got one. Harry Bickerton was a British engineer, formerly employed by Rolls-Royce, who became a successful inventor after developing a mobile hospital bed. A serious traffic accident left him saddled with a three-year driver's license suspension. After some trial models he developed the Bickerton Folder. It was essentially a single box girder monotube frame with a short rear triangle. The frame hinged about half-way along the box girder. The hi-rise handlebars folded individually back and down, and the seatpost removed completely. The bike had two different-sized wheels, 14 inches in front and 16 in the rear.

Bickerton manufactured them himself as a sort of cottage industry starting in 1971. Over the next three years he sold somewhere between three and four hundred at $300 each. In 1974 he outsourced his production to the TCK Group of Birmingham, but they went bankrupt, orphaning the project. In 1976 the operation was acquired by Morris Vulcan (that's the name of a firm, not an individual) of Birmingham. Morris Vulcan slightly reengineered the bike and offered it as the Bickerton Portable at about $350 starting in January 1977.[36]

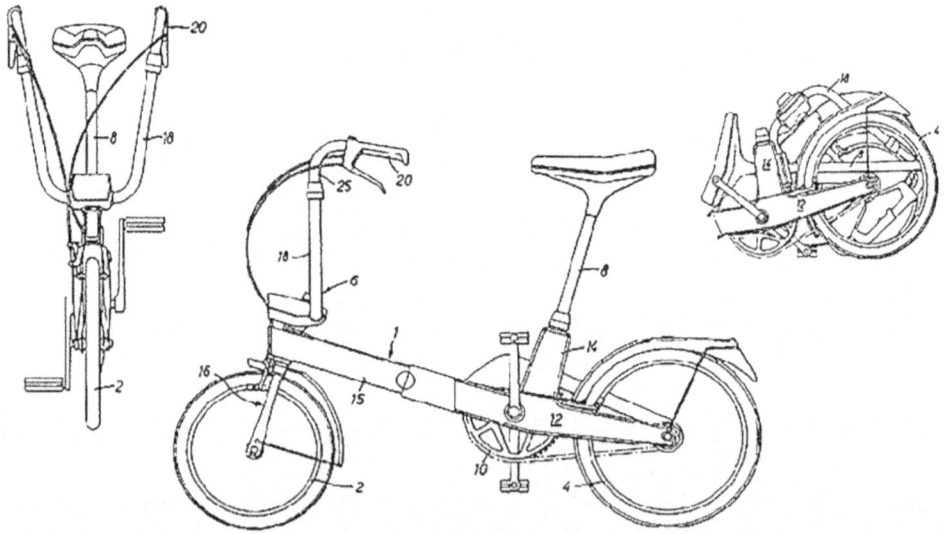

The original patent drawing for the Bickerton bicycle, c. 1970 (U.S. Patent Office).

Like the Brompton a decade or so later, it got a boost when high-tech bicycle expert Richard Ballantine raved about it in *Bike World* and *Richard's Bicycle Book*. "In all the word there is only one folding bike to consider—the Bickerton," he wrote. "The Bickerton, constructed of aluminum alloy throughout, weights only 18 pounds, (20 for the 3-speed) ... the portability of the Bickerton has to experienced to be believed. It is no trouble at all to take along in a taxi, train or bus ... in performance, a Bickerton will keep up with anything short of a flat-out racing or touring bicycle."[37]

It required no welding or brazing to manufacture. In theory, it could be made without power equipment. It appears the bicycles were, at least for awhile, manufactured in Australia and shipped to the UK as knocked down kits for final assembly. In 1982 the rights were re-acquired by the Bickerton family, along with financier Steve Rowlinson. They manufactured the bicycle in Welwyn Garden City until 1992. Production was permanently discontinued at that time, although some units were later manufactured under license in Taiwan.[38]

It was Banham's choice in Buffalo because it dropped effortlessly into the luggage compartment of his VW Golf (he couldn't bring himself to use its then-American name, Rabbit, without quotation marks), or he could take it as carry-on luggage on his now frequent trans-Atlantic flights. "From the days he lived in London and earlier too, I'm sure he was always riding a bike, just to get around," recalled photographer Tim Street-Porter. "It was no different [later, when he was teaching] at Santa Cruz. He lived near the campus and would cycle to work rather than drive. Cycling was something that he did naturally." Mary also recalls him "cycling around [Buffalo] looking for those old grain silos and factory buildings that appeared in the classic books by Groupius and Corbu from the 1930s."[39]

He was an early acceptor of the Bickerton, buying one of the semi-prototypes. Mary recalled that "our next door neighbor in Buffalo was an engineer of some kind and when he saw Peter riding this bike he said it looked a bit dangerous." In fact, Harry Bickerton had intentionally designed in a feature that can be described as, well, quirky: the handlebars were quite tall; shaped much like a Stingray's. Bickerton made them so they could be moved fore and aft when pushed or pulled firmly. As one reviewer put it, "at first this is a bit electrifying." The idea was that once you got used to it, the bars allowed for a number of riding positions: pulled back for climbing; pushed

Bickerton folding bicycle (courtesy Bickerton Portables/ Mark Bickerton).

forward for headwinds. There was a hard stop at the forward position so under hard braking your torso could only go forward a few inches. It was more popular with Harry than subsequent manufacturers, or for that matter, with most owners. After voicing his rather strident opinion of the bike's overall design, Banham's neighbor "without asking him [Banham] at all, one night he just got his tools together and tried to improve the bike, adding a brace to support the tilting handlebars, and actually ten years later this brace appeared as a standard part of the Bickerton," recalled Mary.[40]

Not everyone was as enamored of the Bickerton as he (or Ballantine) was. A few years later, in his office, Banham proudly unfolded it in front of his friend Buckminister Fuller. Fuller stared at it wordlessly for a few moments, then slowly picked it up and dumped it in Banham's wastebasket. Asked what attracted him to strange and unique bicycles, his son Ben replied "the technology." "Technology was an obsession in an almost romantic way, for him it was about optimism, the future, and immense possibilities, the astonishing capabilities of man."[41]

The Bickerton motivated Banham to ruminate in print in a December 1978 article in *New Society* about the fate of the Moulton and other (Banham's words) "advanced technology bicycles," the first time he had ventured onto the subject in several years. The early 1970s were the years of the "Great American 10-Speed Boom." Domestic sales

Left: Concrete Dreams I: Brunswick Center, Bloomsbury, London. *Right:* Concrete Dreams II: Barbican Complex, City of London (Chamberlin, Powell and Bon, architects, 1965-81) (photographs by the author).

shot up from 35 per 1000 in 1969 to 72 per 1000 in 1973. Almost all of this increase was in the form of adult, lightweight, multigeared bicycles. Between 1970 and 1974, over 40 million bikes were sold, a volume so impressive that Norman Clarke, head of the Columbia bicycle company in Massachusetts, called it "an unbelievable figure to those of us in the industry."

Yet these were not innovative designs; they were the three- five- and ten- speed lightweights that had been the bread-and-butter of the British and European markets for sixty years, and to a surprisingly large extent, the American market as well, at least since World War II. While Americans didn't ride many high-end ten-speeds before the 1970s, Clarke estimated that as early as 1965, a third of his firm's production was lightweight bicycles: "five-speeds, three-speeds, a few ten-speeds."[42]

There have been many theories advanced for the sudden boom of 1970 and its equally sudden bust after 1973–74. The most plausible comes from a 1978 strategic plan prepared for the Schwinn company and its vice-president, Jay Townley. It attributed the phenomenon to a combination of baby boom demographics and Stingray economics crashing into each other in 1969. Hi-rise bikes had all but pushed out middleweight and lightweight one-, two- and three- speeds as the bicycle of choice for subteens and teens by 1967. While middleweights and lightweights could be adapted for use by young adults—those between 16 and 20–Stingrays were strictly kids' stuff, so as their owners grew, they had to buy new bicycles if they wanted to keep riding. At the same time, these Stingray boys and girls, members of the peak birth years of the baby boom generation, were moving into young adulthood in record numbers.

Thus, a convergence of two factors: (1) young adults, who wanted to bicycle, were growing in record numbers; and (2) the bikes they already owned suddenly had turned, like so many Cinderella carriages, into useless pumpkins. Because they were young, more affluent than any previous generation of late teens, and open to new ideas, they were more likely to walk past the upright handlebar one- and three- speeds to the ten-speed section and ask, "what's this"? (It's also true, as Norman Clarke once said, "there were more bikes bought and never used in the '70s than you could shake a stick at.")[43]

But even a drop-bar 10-speed was still basically a seventy-year-old European design. The derailleur had been available since at least 1902, and affordable, practical derailleur gears (i.e. as cheap and reliable as the hub gears offered by the German Fichtel & Sachs or the British Sturmey-Archer firms) had been around since about 1924.[44] While 10-speeds may have been radical to Americans, they were old hat to Europeans.

Writing from the States, Banham described the "campus liberal bike-fancy" as "tall Peugeot 10-speeds," even a half-decade after the collapse of the 10-speed bike boom in 1974. Meanwhile, back in the UK, "Although a fair number of ageing Moultons continue to move substantial bodies of country folk about," he noted, "the bottoms on their saddles tend to be schoolteachers, social workers, and the like, not yer actual rural proletariat as normally understood. And when the Moulton was young, you'll recall, its typical rider was the young executive."[45]

Historian Tony Handland confirms this: "even in the early 1970s, on a Friday afternoon, a work colleague might remark to you, 'Looks like a nice weekend coming up, I think I'll get the car out.'" In other words, people generally did not use their cars during the week—they were cocooned in the garage for safekeeping. This was the residual suburban-exurban market that Banham was referring to. On the other hand, the RSW and (especially) the Twenty penetrated a somewhat different market segment—less upmarket, and overwhelmingly female. (One reason the Moulton Mk. III bombed so badly is that it was marketed as the "man bike," while the rapidly ebbing RSW 16 Mk. III was marketed—literally—as "the Dolly One.")[46]

But whether it was a just-out-of-university Salford social worker or a SUNY-Buffalo architecture undergraduate, Banham noticed a common link: when these new-generation liberals eschewed the automobile it was out of conviction, not a lack of money, and the bicycles they used were the same designs their grandparents were familiar with. At most, they rode Shoppers, which were just roadsters for the apartment-dwelling, bike shed-less era.

And it was true: there were no more bike sheds. Nor, for that matter, much of anything for cyclists any more. The first prewar new towns such as Letchworth and Welwyn Garden City had included only modest cycling improvements because nobody anticipated much car ownership, the towns had small neighborhoods, and they were built with shops and factories not far from the neighborhoods. The 1945–50 new towns, especially Harlow and Stevenage, included quite extensive, grade separated sidewalks and bikeways.

But by the time the thirteenth official new town, Milton Keynes, was started in 1966, no provision at all was made for cyclists, and one planner described the roadway system as designed on a "Los Angeles–type network." Increasingly, new towns like Cumbernauld outside Glasgow and Nordweststadt near Frankfurt incorporated gigantic city-center activity megastructures that combined shopping, recreation centers, libraries, hotels, government offices and transportation hubs. They resembled nothing so much as the fantastic "plug-in" cities or "megacities" that Peter Cook and Dennis Crompton were drawing for Archigram's eponymous magazine. Cumbernauld was nine levels high, not counting the hotel. The Nordwestzentrum had elevated sidewalks connecting it to the rest of town, but no bike racks. After the OPEC oil embargo of 1973 a "redway" system was grafted onto the then-existing roadway plan for Milton Keynes, but it was never well integrated, except for those parts of town newly built after the redways were introduced.[47]

In Britain, one of the most important factors was the adamant opposition to cycle tracks by the nation's largest bicycling club, the Cyclists' Touring Club (CTC, now Cycling UK). The CTC was a small organization, only 34,000 members in 1939, minuscule when compared to the UK domestic bicycle market of 1.4 million units that same year. However, its influence was outsized. It was unashamedly elite. Membership was expensive—in 1955, 15 shillings a year, plus another 10 shillings to join, at a time when a no-frills Hercules roadster for the workingman could be bought at a department store for £3 (60 shillings). Its ranks were made up almost exclusively of middle-class recreational cyclists on high-priced racing and touring machines.[48]

"CTC membership offered few clear benefits to the growing numbers of [prewar] working-class cyclists" observes British cycle historian Peter Cox. Although highly involved in cycle advocacy, when "viewed from the outside, if judged solely on what they appear to be concerned with," then the CTC campaigns "appear almost absurd." However Cox maintains that "only by understanding the actors involved and their social locations do they become intelligible."[49]

The CTC objected to laws requiring the use of red taillights after World War I and fought the construction of cycleways from 1934 to 1970. "After 1934 the CTC was dead against cycle tracks of all kinds," explains 2008 Cycling UK staff member Chris Peck, "we were very much of the mindset that we should try and recapture the road from the motorists." In 1965 the organization refused a request from the national Road Research Laboratory to help perfect a ventilated, lightweight helmet for cyclists. "Such helmets are neither appropriate nor necessary," wrote the club secretary, "they would rob cycling of its freedom, simplicity, and immediate contact with fresh air."[50] The CTC, deep into denial, would have gone under by 1970 had it not owned its headquarters on Craven Hill Road in London, which it sold in 1966 for fifty-five thousand pounds and moved into a house in rural Surrey. Thereafter, it rented a small storefront in London that, over time, eventually evolved into a travel agency for cyclists.[51]

Contrast this to the Netherlands. As historian Manuel Stoffers points out, "although the reputation of the Netherlands as a cycling country dates back to the interwar period, only after the Second World War did the Dutch deviation from the general European pattern of bicycle use become more marked." Ton Welleman, an Amsterdam road engineer who became coordinator of the Dutch Cycling Council points out (stressing "and this is important"):

> the bicycle was old fashioned, a vehicle for the poor, but cycling was recognized as a mode of transport that is part of life ... in the Netherlands, transport policy meant a pro-car policy, but in general not an anti-bicycle policy. That was wise at that time, as there were, after all, hardly any alternative modes of transport available for the Dutch. Mass motoring did not even start until about 1960, and public transport was minimal even then.[52]

Thomas Krag, a traffic consultant from Copenhagen, echoes this: "what distinguishes Denmark from many other European countries, with Holland being the main exception, is that cycling never disappeared as a normal transportation mode. The decrease was big, but the bicycle was still visible in urban traffic even when at its lowest."[53] Like the Netherlands, the bicycle was reduced to "the poor man's friend," but it was not stigmatized, nor was any deliberate effort made to eliminate it from the roads. It was tolerated.

This was not the case in Britain, with its emphasis on clearing out inner cities for ring roads, at least until the 1960s. The Smithsons prepared the "London Roads Study" in 1959, which proposed gutting most of Soho and Covent Garden in the center city, replacing them with motorways and ribbon-like high-rise buildings connected by skywalks and pedestrian overpasses. Asked by Ernesto Rogers at CIAM '59, "What, in fact do you conserve and what do you destroy?" Peter Smithson replied, "In the end, we would probably destroy everything, stage by stage."[54]

But in the end it was the Smithsons who were sent to the trash heap. As early as 1949, Arthur Ling, City Architect for Coventry (and a former assistant to Walter Gropius and Maxwell Fry) told a reporter for *Architects' Journal* that "to put several thousand people in one block of flats on stilts just because the designer thinks it is a good idea and gives scope for his designing ability is not a sure way of achieving successful results."[55]

The London Roads Study was not adopted. A decade after it opened, the Nordwestzentrum was considered out of date, and largely unused except for the shopping arcade and the government offices. The Cumbernauld Town Centre required massive infusions of maintenance money to keep it from falling apart and was more or less abandoned, much of it ending up a gigantic rubble pile. Both were rebuilt as conventional private shopping malls.[56]

There was growing evidence that the almost metaphysical belief of the modernists that the built environment could, in itself, modify "atavistic" behaviors and guide the development of "forward thinking" individuals without need to resort to counseling, social workers, or other forms of human-to-human intervention was just that, a kind of faith, a form of secular religion. Reporters for *Civic Affairs*, accompanying "Visitors" (social workers) to the tower blocks, reported that "It would be unfair to quote too many examples lest we be suspected of copying the comedians we have derided so many times, but coals *have* been found in the bath, the wash-basin *has* been used as a meat-safe, and recently a Visitor *did* discover that the cupboard to the kitchen dresser *had* been transformed into a chicken-coop."[57]

It's hard to say what finished off the concrete dreams of the modernists. Some say it was Ronan Point. It was a twenty-three story apartment building of 88 units, part of an ambitious plan to develop 650 flats in six towers in London's West Ham neighborhood. It was completed in 1967 using the rapid and economical Larsen-Nielsen system. The elevators and utilities were situated within a concrete core, and each floor's flats (four per floor in this case) were hung off this core, pinwheel fashion, using interlocking prefabricated panels. The Larsen-Nielson system had been developed in Denmark in 1948 for mid-rise buildings. Ronan Point was the tallest building to use it so far.[58]

On the morning of May 10, 1968, Mrs. Ivy Hodge, on the twentieth floor, turned on her stove. It exploded from a gas leak because the wrong type of flexible connector hose had been installed. One entire stack detached from the central core and pancaked down to the top of the lobby. Fortunately, most occupants had already left for work. Four were killed (Mrs. Hodge survived). All the West Ham towers had to be retrofitted with new brackets and beams. In 1964, 1175 high-rise flats were approved by the County Council for West Ham. In 1969, 50 were approved. The numbers were similar across London and most of Britain.[59]

Others say it was the Smithsons' own Robin Hood Gardens project, started in 1965 and finished in 1971. Located in Poplar, adjacent to the newly filled East India Docks, it was intended to provide council flats for the elderly and the families of dock workers. It presented many challenges. It was a difficult site. It was oddly shaped, a triangle pointed south towards the river, with the busy highway ramps to the Blackwall Tunnel on the east side. An already existing five-story apartment building sat along much of the north boundary, with no intervening road to provide a setback. The configuration changed significantly after the project started when the London County Council acquired and razed the Queen's Theatre, the largest contiguous parcel on the site. Previously, the authority wanted three buildings on three scattered parcels, the largest of which required a long, thin building. When the theatre acquisition opened up the triangle, the Smithsons took their building, stretched it, and duplicated it on the other side of the triangle, creating a single green area lined by 7- and 10-story curved slab blocks on the east and west sides, with the tall building on the Blackwall Tunnel side. The county, still with a wait list for council homes in 1965, also wanted the maximum density allowed, between 200 and 215 dwelling units. The final configuration had 213 units on five acres.⁶⁰

The layout made sense, in that it kept out traffic noise, but duplicating the ribbon-slab on both sides also effectively shut out the neighborhood. Also, the green space in the middle simply wasn't wide enough (less than three times the average height of the surrounding buildings at the midpoint of the triangle) to escape a looming, canyon-like effect, accentuated by the typical brutalist formula of heavy, plasticized cement surfaces. Robin Bowdler of English Heritage (the organization that confers architecturally protected status to buildings) notes that "on a popular level, it quickly acquired the nickname 'Alcatraz' because it was felt to be a rather grim, imprisoning environment." Peter Smithson recalls that "the week it opened, people would come in and shit in the lifts."⁶¹

Robin Hood Gardens, retouched to make it stand out in white. The view is from northeast to southwest. The entrance to the Blackwall Tunnel is on the near side of Robin Hood Gardens, crossing south under the Thames. The new Canary Wharf super-development is at top. The diorama is at the London Design Center (photograph by the author).

As early as the mid–1990s some architects suggested that Robin Hood Garden should receive protected status. The immense, 1000-unit Park Hill complex in Sheffied, designed by Smithson students Ivor Smith and Jack Lynn, built a decade before Robin Hood Gardens, was listed by English Heritage in 1997 and underwent a decade-long renovation by a private investment group starting in 2003. The debate over listing Park Hill was contentious, but as many residents supported preservation as opposed it.[62]

BBC architecture critic Maxwell Hutchinson reports that this was not the case with Robin Hood Gardens. "The campaign to save Robin Hood Gardens drew very little support from those who actually lived in the building, with more than 75 percent supporting demolition.... When we asked the conceriege what life was like here, his answer was expressed rather pointedly: 'Hell! ... I'm stuck here, nowhere to go. Some people stuck here twenty years.'" Roger Bowdler acknowledges that resident input was a factor in the decision to decline listing Robin Hood Gardens.[63]

There are many theories for why the high-rises and slab blocks failed so often: insufficient maintenance; inadequate tenant screening; bad design; bad planning; bad construction. Sheffield architect Sir Andrew Derbyshire says it really doesn't matter. "I think the system produces the outcome. The contractual system. The political system. I think they deserve most of the criticsm, but it's easier to give it to the architects because the built system is architecture. There's a building; architects do buildings."[64]

Edwin Heathcote, architectural critic for the *Financial Times*, believes that the New Brutalists, in a sort of Greek tragedy, brought their doom down upon their own heads:

> I've compared them to the Kitchen Sink authors, the "angry young men," John Osborne, Alan Sillitoe, men who were trying to upturn an existing social order, and in architecture this led to what is known as brutalism, which is kind of the architectural equivalent of a literary movement. It's a type of deliberate roughness, a celebration of process. But architecture is the one media you can't turn off, it's not like going to a play, reading a book, or seeing a film. It's with you all the time, it makes up your everyday environment, and perhaps it was a little harsh for that, because there's no turning it off.[65]

But in the end, much of it was simple economics. The high-rises were built because they were needed—badly. And then they weren't. Demographics had changed. Household economics had changed. Municipal finance had changed. By 1973 the government building boom was over. "The simple truth was that Britain couldn't afford to maintain the pace of redevelopment," concludes historian John Gringrod.

> Geologists can pinpoint global extinctions and climatic events by pointing to lines in the strata of rock where the geological record abruptly changes. An analogous hiatus occurred in our town centres, between the frentic activity of the fifties and sixties, and the mid-eighties, when widespread building began again. By that time styles had changed so much that the edifices of the postwar period seemed like relics from a distant age—a foreign country where they did things differently, where plans to demolish Covent Garden, put Newcastle on stilts or plough urban motorways through London's most desirable districts seem like alien curiosities, at which we can only scratch our head and wonder.[66]

Looking back, Bill Berrett, co-designer of Milton Keynes, says that in the postwar era "architecture became much more intellectual. We began to get the industrialized building. We began to look towards the early modernists, particularly Le Corbusier, which gave rise to things like the New Brutalism and heavy concrete, and I think some of the humanity began to disappear out of the younger and more capable architects, and I think that's a great shame."[67]

After completing Robin Hood Gardens, the Smithsons completed no more major commissions. "They take architecture, and themselves, completely seriously and have never been noticeably afflicted with false modesty," Banham once observed, "but it is one thing to be loved by students, and respected by architectural intellectuals, but it is something else to lead or even drive the profession."[68] In the words of Archigram founder Peter Cook, the Smithsons "deliberately cut themselves off" from the architectural mainstream and withdrew into solitude for the better part of a decade until Cook talked the University of Bath into hiring Peter Smithson for a part-time teaching post. They became, in the words of one critic, "the avant-garde in an era without an avant-garde." Alison Smithson died in August 1993, at age 65.[69] In 2000, Peter Smithson mused with a former student and old friend about life after the break-up of the Independent Group and the death his wife:

> Being jealous of nobody, I have become a terrible old bore. Without somebody like Nigel and Eduardo, we had no peers. Therefore you are entirely self-contained. That is the basic difficulty with Alison having gone. The relationship was more like a conspiracy. She and I against everybody else. Not literally of course, but you are more exposed when groups disintegrate. You have nobody to say ... [sentence drifts off] Also it's a question of aging. Gradually the people you can talk to disappear. You end up alone.[70]

He died less than three years later at age 81. In 2016 a private developer began dismantling Robin Hood Gardens to make way for a luxury housing project.

Postscript: At 12:44 am on the morning of June 14, 2017, a refrigerator caught fire in a fourth-floor corner apartment in Grenfell Tower, a 24-story residential building in the London suburb of North Kensington. The first emergency call was made one minute later. By the time the first unit of the London Fire Brigade arrived at 12:51, the fire had already spread over almost the width of one wall of the building and up more than ten stories. "How the fuck are we going to get into that?" one incredulous firefighter was recorded asking on his truck's cabcam. Two hundred firefighters did go in and 40 trucks fought the blaze, which continued to burn for the next twelve hours.

Robin Hood Gardens in decline, 2017 (photograph by the author).

Grenfell Tower was built in 1974, a 124-unit classic brutalist tower block with six flats per story. It had two elevators and one stairwell in a center core, connected to the flats via an H-shaped hallway arrangement. It had thickened cement firewalls designed to contain any fire within a single unit, but no sprinkler system. Each of the four corner units had L-shaped balconies; the two one-bedroom units in the center of the north and south sides were flush to the building exterior; again, a standard brutalist treatment, intended to give the outside of the building an alternating flush and sawtooth texture to differentiate it from the glass-and-chrome sleekness of the era's classic office buildings.

In 2016 a non-structural exterior cladding was added to give it just such a glass-and-chrome appearance. This included the balconies, which became sunrooms with slide-to-open windows. Sheet-foam insulation about two inches thick was added to the surface of the building. A one-inch air gap was left, then a cladding layer with integral windows was attached to brackets that had been bolted to the building before the insulation was hung. The cladding in this case was fabricated from Reynobond PE, made by Alcoa. (The European subsidiary changed its name to Arconic in 2016.)

Reynobond PE is a sandwich comprised of two thin sheets of aluminum with polyethylene in the middle to give the panel its strength. Polyethylene is flammable, and emits thick, black smoke as it burns. Alcoa-Arconic makes a fire resistant version of Reynobond, called Reynobond FR, and it is required in the United States and most of Europe for buildings taller than two stories. It is about thirty percent more expensive. The insulation affixed to the building was Celotex RS5000. Celotex RS5000 is made from PIR. PIR is an acronym that stands for polyisocyanurate. PIR is also inflammable, and gives off cyanide gas as it burns. Area hospitals treated as many as forty residents and firefighters with an antidote for cyanide poisoning the night of the fire.

A subcontractor, Omnis Exteriors, was brought in part-way through the job when the original subcontractor defaulted. Omnis Exteriors stated that if they had known that the insulation layer was comprised of a flammable material, they would not have used Reynobond PE. Celotex RS5000 comes in sheets, like plywood, and is covered with a foil-like vapor barrier. The name is clearly printed on this foil. However, the Celotex product identification sheet identifies RS5000 only as made of PIR. The word "polyisocyanurate" never appears on the

Construction fence being erected for the demolition of Robin Hood Gardens, 2017 (photograph by the author).

British product identification sheet, nor does it warn installers that anyone caught in the smoke from burnt RS5000 should be treated for cyanide poisoning; nor that RS5000 should not be used with flammable cladding. It notes only that RS5000 is approved for high-rise use if combined with a fire-resistant cladding. Similarly, the Alcoa product information sheet does not warn that that Reynobond PR cannot be used with a non-fire resistant insulation layer, only that it is acceptable for uses over 18 meters with FR insulation.

In the United States, Reynobond PE is not permitted on buildings over two stories. Polyisocyanurate insulation is not permitted for residential uses at all. If both an inflammable base insulation layer and an inflammable-core cladding are used, the air gap in between acts as a chimney. It burns, as one Grenfell witness put it "like a fire you put petrol on." If the base insulation is polyisocyanurate, it also issues the same gas that once was used in American gas chambers to carry out the death penalty.

It appears that the Celotex RS5000 insulating layer was so inflammable that it ignited from the heat of the refrigerator fire on the other side of the exterior cement wall. The original design of Grenfell Tower intentionally placed the kitchen appliances against the exterior walls to prevent the spread of kitchen fires from unit to unit. Virtually the entire surface area of the building was on fire or burnt out within two hours after the start of the fire. The building burned from the outside in. There are 79 known fatalities, although over 40 more are still missing as of July.

An unknown number of 1960s and early '70s era residential buildings across Britain, most of them of plasticized cement brutalist-type design, have been similarly modernized. Estimates range between 40 and 120. Similarly, it is not known how many have been treated with Grenfell Tower's combination of inflammable surface insulation and inflammable cladding. As of July 2017, the cladding and insulation are being torn off approximately thirty-six buildings in England, Scotland and Wales, apparently without waiting to determine the composition of the materials.[71]

8

Vanished into the Clouds

> Consider the Japanese cultural concept of *mano no aware*, which translates into something like "a pleasing sadness at the the transience of beautiful things." The literary scholar Motoori Norinaga coined the idea in the mid–18th century to describe the *Tale of Genji*, the great Heian-period novel.... When the protagonist dies late in the book, his death is never mentioned directly; it occurs in a chapter called "Vanished into the Clouds." Other than the title, the chapter is blank.
> —Brian Phillips (2014)[1]

I

Banham came to Buffalo to start a graduate program in architectural history at SUNY-Buffalo, but those plans never materialized, so after his three-year contract expired in 1980, he moved to the University of California at Santa Cruz, at the south end of what would later be called Silicon Valley.[2] Unlike in 1976, when he didn't realize that his talents were in demand, he received several feelers, but faced a new dilemma, according to Mary. "He had been, for a long time, an *enfant terrible*, and was expected, it was perfectly clear once we were in America, to behave like an *eminence grise*. He could not do it. He just could not do it.... He was already a man with a reputation, so he was expected to behave like an establishment figure, and he didn't know how."[3]

He was also both unfamiliar and uncomfortable with the American collegiate administration system, in which faculty members run their own departments and, in many cases, colleges. In fact, the involvement of non-academic provosts and administrators into their affairs is often seen as something of an affront. Banham was more comfortable with the British and European system where professionals run the schools at the university level, and, because of the tutorial system, there is less of an administrative structure below. "In America, academia is such big business, you know," observed Mary, "these universities with 27,000 students and all. What he enjoyed was the actual teaching, the actual lecturing ... most people who have worked in them for any length of time feel it's their duty to run a department or program, and he didn't want to do that." Moreover, starting at his time at Buffalo, "people who knew he was a famous person wanted to knob on to him, and get him to push forward whatever projects they had onboard. Other people did do that, but he didn't do that, and it disappointed a lot of people."[4]

So he ended up once again, in the words of Tim Street-Porter, "away from the center," at Santa Cruz. However, Mary believed that this time they had made a good choice:

> Santa Cruz was right for him in many ways, because it was a maverick campus. It had a lot of nutters, who had arrived from other campuses, thank you. Very beautiful, probably the most beautiful campus I have ever seen, not just in America, right on the edge of Monterrey Bay ... and there were some famous people there, who I got to know eventually, who were like him, in that they loved to be doing their jobs and working with students, and doing their own research, and writing their books or music or whatever. And they wanted to just get on with it.[5]

Shortly before leaving England he had noted that architectural history was changing: "We have had a good run for our money and there has been a lot for us to do, and I suspect that the bulk of it is for the moment done. The attention of historians will wander away from architecture, even in the architecture schools, into previously ignored fields like patronage, finance, land ownership, and the internal sociology of the profession."[6] He was right. At the Bartlett School, Richard Llewelyn-Davies had hired him to be Britain's first professor of architectural history. But nobody else had followed, at least in England. As Mary later pointed out, your career opportunities are fairly narrow when you are in a profession of one. But these new areas of focus were not of particular interest to him, and he took advantage of his time in Buffalo and Santa Cruz to produce two of his most personal books, *Scenes in America Deserta* (1982) and *A Concrete Atlantis* (1986).

Although the later published book, *Concrete Atlantis* was, from the point of view of research and preparation, the first. It was a study of modern industrial archaeology, examining the grain elevators, storage silos, and factories of the American industrial belt, which supposedly inspired Gropius, Le Corbusier, and the other European Modernists. He discovered, in his cycling perambulations around the Buffalo waterfront, that most likely, the Modernists had never visited the structures they emulated with such enthusiasm, but worked off photographs, often retouched photos that airbrushed away the practical engineering details of the American originals that cluttered their clean, platonic, ideal lines—things like roof scuppers, gutters, copes, jetty-brackets, or buttresses. That's why one encountered abandoned American factory buildings with flat roofs that still held tight forty years after they were built, and European apartment buildings with flat roofs that invariably started leaking after two.[7]

Banham also moved to California because he had fallen in love with the deserts of the American West. He was enthralled with the Sonoran Desert around Tucson, the Salton Sea in the bottom of California's Imperial Valley (the result of a 1905 breach in a temporary levee on the Colorado River that took two years to plug; originally 79 feet deep at its deepest point, it stabilized at 36 feet in 1920); and the Mojave. "I return to these landscapes in order in order to feast my eyes on visions that I am prepared to term addictive," he wrote, "I find myself driven closer than ever in my life to the idea that some scenes may be perceived as, simply, 'beautiful' ... there does seem to be something out there that communicates more directly to the pleasure centers of my brain than anything else I have ever encountered." Peter Hall believed that "what he particularly found and particularly valued in the American West was ordinary people, designing the way they wanted, untrammeled by the self-appointed thought-arbiters of good taste."[8]

Scenes in American Deserta provided the photographic bookend to the famous 1963 shot of him riding his Moulton in London. He and photographer Tim Street-Porter were at the Silurian dry lake in southern California in 1981. "Banham and I were paired up to do a few journalistic stories for a Californian magazine," Street-Porter explains. "Not long afterwards Banham contacted me about a book he was researching on the desert." They met at the Banhams' apartment in Santa Cruz, and headed south in Banham's 4-wheel drive.

> The whole thing lasted for three or four days during one of the winter months, I remember the softly modulated desert light being especially beautiful at that time of the year, with the lower sun, instead of the bleached, cloudless light ... the trip was great. We stayed in these little motels, and of course it was extremely low budget, but when you got out there, there aren't many other alternatives.... Banham had been to all the places already, scouting out the whole area and so he knew where he wanted to go.[9]

Banham was dressed in his usual California outfit of jeans, saddle boots, black cowboy hat and bolo tie with silver-and-turquoise slide. "Whenever or wherever we met he would always be wearing that same cowboy outfit—it was his signature uniform, and made him look startling and quite impressive ... the whole effect contrasting bizarrely with his East Anglian accent."

Banham had been making trips around the west since attending his first Aspen design conference in 1964. He had followed the original route of the transcontinental railway across the vast, empty continental plateau in south-central Wyoming. He had visited the ghost-white San Xavier del Bac church and the equally stark-white geometric angularity of the Kit Peak solar observatory, both near Tucson. A little farther north near Phoenix he toured Paolo Soleri's bizarre Arcosanti, looking like something out of a *Planet of the Apes* movie, one of the first attempts to actually build a true super-scale megastructure out of native material. Today, forty years later, it is half occupied, half-ruined, slowly returning to its desert elements of rock, rust and dust. He asked to order a meal at the mission-style train station at Kelso, California, only to be politely informed that no passenger trains had stopped there for close to thirty years—all of Kelso, including the village, the immaculate station and its café, was a company town for Union Pacific workers.

Finally, he climbed through Zzyzx (pronounced "Zye-zicks"), Curtis Howe Springer's abandoned health resort near the California-Arizona border. Springer was a radio preacher who got rich in the 1950s selling fire-and-brimstone sermons by tape. Somehow he discovered a forgotten, hot-spring-fed pool carved out of rock decades earlier by the lone-gone Tonopah and Tidewater Railroad on a ridge overlooking the Great Soda Lake. He built the resort practically with his own hands. But the government claimed he had no title to the land and evicted him. Everything—pools, buildings, guest cabins, electro-therapy tables—reposed in a state of ghostly suspension, still watched over by a caretaker paid by the Bureau of Land Management while the heirs of Springer, who had died years ago, tried to convince a judge, any judge, that he had in fact come into legitimate ownership of the land through a chain of title originating from a dubious Spanish land grant of ancient origin. And he wrote of the cyclists:

> I have seen a number of desert cyclists in the Southwest, and talked to some—the "California or Bust" type with bulging backpacks, or a memorable couple of tough-talking girls from Chicago fet-

tling over their ten speeds in a gas station forecourt outside Cortez, Colorado. And—something I would have dismissed as a hallucination did I not have a witness to bolster my grasp on reality—an elderly lady in a print dress and wool cardigan riding a tall dignified English bicycle that looked about the same age as herself, the sort that has an elaborate tracery of cordage to keep Edwardian skirts out of the back wheel and a wicker basket on the front of the handlebars. She was headed south about ten miles out of Shoshone, California, in a direction in which there was neither help nor habitation for another seventy miles. Too astonished to do anything else, we said, what was *that*, for God's sake?[10]

West of Laramie, Wyoming, atop the continental plateau, 1984 (photograph by the author).

He suggested that Street-Porter take some shots of him riding on the vastness of the Silurian lakebed. "We didn't set out with the idea of shooting him riding a bike across the desert—or at least I didn't—it was just there in the back, and at some point he fished it out for fun, really," recalled Street-Smith. However, he had to admit that Banham likely had thought out the whole thing in advance, as he had the rest of the trip. "He clearly loved his bike," he remembers, "almost like Keith Richards and his guitar, who in his pre-stardom days, slept curled up with his Fender." Colleague Thomas Weaver agrees: "He always seemed so amiable and chatty, talking to anyone, working in his office with his door open and engaging with conversation with anyone who walked by, yet he was also someone who carefully choreographed the defining image of himself as a man alone in the middle of nowhere, riding a bike across a desert."[11]

The Silurian lake was the only spot in the desert that you could actually cycle on; "everywhere else had huge crusts of dried earth which would have made it impossible," recalled Street-Porter.[12] He apparently used more than one camera; both color and black-and-white versions exist. Banham himself later wrote about what it was like taking long loops across the hard-packed salt as Street-Porter switched between lenses and cameras:

> Swinging in wider and wider circles or going head down for an ever-retreating horizon, the salt whispers under one's wheels and nothing else is heard at all but those minute mechanical noises of the bike that are normally drowned out by other traffic. Swooping and sprinting like a skater over the surface of Silurian Lake, I came as near as ever to a whole-body experience equivalent to the visual intoxication of sheer space that one enjoys in American deserta.[13]

"It was ultimately used on the cover, but at that stage he saw it just as something fun to do," says Street-Porter. "I never would have imagined that would have become perhaps the most lasting, the most iconic image of him."[14]

Banham and Mary had grown content in their little place near the campus, but in 1987 Banham was appointed professor of architectural theory and history at New York University, beginning with the spring semester of 1988. He never took up his chair; his ceremonial inaugural lecture, "A Black Box," was published posthumously. In late 1987 he was diagnosed with an undisclosed, but rapid, terminal illness. He and Mary rushed back to London so he could say goodbye to family and old friends. One recalls a last glimpse: fading fast in the Royal Free Hospital, he was still fascinated by all the medical gadgetry surrounding his bed. He died on March 17, 1988, at age 65.[15]

Fifty-four years later: the Carteret Street corner where Banham was photographed on his Moulton bicycle in 1963, taken in early 2017 (photograph by the author).

II

"It is ironic," points out Tony Hadland, "that in popular speech, the word 'Moulton' became for a time a synonym for 'small-wheeled folder'.... His original goal was not to design a portable cycle at all, but rather to create a bicycle for universal use which would be 'more pleasing to have and ride' than the traditional diamond frame machine." He concludes that the "aftermath" of the Moulton and the Stowaway was to create a new engineering horizon where "the aspiration of many designers was to produce a portable cycle that matches the performance of a conventional diamond-frame bicycle."[16] (He also points out that Moulton has never actually produced a folding bike. Both the F-frame Stowaway and the later AM are break-aparts, not folders.) Indeed, since the demise of the Mk. III in 1974, there has been a flowering of commercially successful designs, and, in one or two cases, firms that focus on originating new designs for licensing by others.

Andrew Ritchie's Brompton, as we have seen, went back as far as 1977, but his firm had gone dormant in 1983. In 1987 Ritchie was able to resurrect the company with the financial assistance of the late Julian Vereker, a British entrepreneur who had made a fortune in high-end audio equipment. Vereker guaranteed a £40,000 line of credit, and with his contacts and reputation in financial circles, helped Ritchie sell another £50,000 in equity. Volume production started under a railway arch in Brentford in 1988. "Private backing allowed me to be a perfectionist," Ritchie admitted. "In a way, I'm glad we didn't get a licensing deal with Raleigh. I'm not sure the business would have survived. The bike probably would have been dumbed down and cheapened."[17]

Thein lies two important differences between the Brompton and Moulton stories. The *second* difference is obvious: Raleigh *did* get their hands on Moulton Bicycles Ltd., and did try to reshape it to its own needs, which were simply not compatible with the goals and capacities of Moulton's firm. Given the choice of supporting a second prestige specialty house to go alongside Carlton or shutting down Moulton, Raleigh very quickly decided on the latter.

One of the great "What Ifs" in cycle history has to be: Would have happened if Raleigh, after acquiring the Moulton operation in 1967, had not moved it to Nottingham, but to the Carlton factory at Worksop, then given its director, Gerald O'Donovan, the autonomy of an independent firm? Founded in 1896 by Fred Hanstock in the village of Carlton, Carlton Cycles was bought by its general manager, Daniel O'Donovan, in 1939. At the time Raleigh purchased it in 1960 O'Donovan and his two sons were handmaking about 54 frames and bicycles a week. Up to 1974 Raleigh used it to produce all bikes with frames made out of Reynolds 531 and as Raleigh's custom-order department. The combination of handbuilt Carltons, high-end Raleigh Professionals, and Moultons could have been a formidable combination.

Alas, Raleigh management could never stomach such American-style divisional autonomy. Also, Raleigh refused to impose any production discipline during the great American bike boom of 1969–74 to maintain quality control in its premier products. By 1974, Raleigh management was pushing Worksop to crank out 1,450 Reynolds frames a week by applying some of the same mass-production techniques used in Nottingham, and dealers were returning too many of them for fatal alignment and brazing flaws. In response, headquarters moved the O'Donvans and the custom order shop to a tiny 7,000 square-foot shop in Ilkeston, renamed the Special Bicycle Development Unit (SBDU) to build team bikes, prototypes and other specials required by the marketing and engineering offices.

The Worksop facility was closed in 1981, when it became clear that the crash of the American bike boom would be permanent. After Raleigh's Dutch-based professional road racing team was discontinued, the SBDU was moved to Nottingham in 1986. It was closed in 1986. Former Raleigh marketing executive Melvyn Cresswell later wrote that he believed Raleigh killed off Carlton and the SBDU out of jealousy because enthusiasts wanted their bikes to the point where each had lengthy wait lists, while rightly disparaging Raleigh's front-of-the-catalog Nottingham models as obsolete and badly spec'd. Moulton later said that he had approached Gerald O'Donovan's son Kevin in the late 1970s with the idea of having Carlton produce Moultons, but by that time there was no more Carlton factory except in name, the Worksop plant having been converted to the mass production of Reynolds 531, and by 1978, Tange and Isawata, tubed bicycles.[18]

But more importantly, contrast the ways Alex Moulton and Andrew Ritchie chose to deal with the perils of success. Subtle, yes, but as different as day and night. Limited to 200 bicycles a week at Bradford-on-Avon when demand proved to be double, then triple that volume, Alex Moulton outsourced the majority of his production to Fisher and Ludlow, even though he had not taken into account the technical needs of mass-production in a non-bicycle factory with general labor of moderate skill given the Moulton's design.

The relatively crude fabrication techniques that Raleigh used on some parts of the later Moulton Mk. III (for example, using a cheap pressed front fork instead of a brazed fork with forged dropouts, and using heavy grade, regular carbon steel for the frame instead of Reynolds 531) shows the kinds of problems even an experienced bicycle-maker faced in producing a Moulton at a commercially viable price. On the other hand, Andrew Ritchie dealt with an excess of demand over supply by simply creating a wait list. "The main difference," says Tony Hadland, "is that Andrew Ritchie never deviated from his basic design. His entire thrust has been small incremental improvements.... Alex was the the opposite, forever tweaking and embarking on very clever but sometimes quite misdirected and arguably pointless projects. Yet both men have left companies that are making bikes in the UK at a profit and are exporting most of their production."[19]

By 2002, Andrew Ritchie realized that his physical plant was not up to the standards of his bicycle. He hired a new products manager, Will Butler-Adams, and in 2008 made him managing director, stepping down to the position of technical director. In 2015 Ritchie retired from the firm altogether to work on new projects. At the time he took over the helm of the firm, Butler-Adams was 38. "I'd worked for Nissan," Butler-Adams recalls of his first day at Brompton in 2002, "I'd studied engineering and I knew what world-class manufacturing was. I came in here and thought 'oh, my God,' I couldn't believe that this type of thing existed. It looked like a warehouse—there was no sign of any manufacturing going on." In 2002, Brompton made 6,500 bikes a year. In 2017, they will make almost ten times that number; all still in Britain, but in a modern, 46,000 square-foot factory in Greenford, twelve miles west of downtown London, where they moved in 2016. Brompton spokesman Nick Charlier states that the estimated the capacity of the Greenford plant is 114,000 units per year. That includes almost 10,000 square feet now reserved for the firm's latest development, an electric folding bicycle, which will be introduced in 2017.[20]

Ritchie did try outsourcing through a licensing arrangement with a Taiwanese firm for the Asian market in the late 1990s. (At the time Japan was their second largest export market, after the Netherlands.) "They were just crap," remembers Adams. "They [the licensee] outsourced various frame parts. Then the person they outsourced it to outsourced it to someone else and there was no coherent understanding of what they were trying to achieve. The thing was a disaster area—it didn't fold properly, it didn't sit properly, it was just shambolic and it was carrying our brand."[21]

Another highly influential design was the Bike Friday, developed by Hans Scholtz in Oregon in 1991. He and his brother started Green Gear Cycling in Eugene, Oregon a year later. Originally there were two designs, one with a single large monotube with the bottom bracket suspended beneath it, reminiscent of the Raleigh Twenty, and the

Assembling rear triangle and fender units at the Brompton factory, 2017 (photograph by the author).

other with dual parallel downtubes. Eventually the dual downtubes were phased out for everything except tandems, and all the models feature the oversize monotube layout. There are several different versions reflecting different purposes: road, touring, ATB. Many versions do not have suspensions. The assembly/fold-up time for the original Bike Friday takes longer than for most other makes. Its emphasis was on riding performance first, foldability second.

But Scholz was not satisfied with such specialization. "A bike is something that should be with you all the time," he told a reporter for an engineering journal, "it ought to be something that you can carry around, and you shouldn't get dirty using it." The problem, as always, was that the small, easy-to-fold bikes were flexible, and the rigid bikes were either heavy or not very foldable. Stoltz's solution was the Tikit, a single-hinge model that used the weight of the rider, acting in compression, to provide the needed rigidity across the joint. "A folding bike should flip out like a beach towel," he explained, and the Tikit does; if you turn it in the right direction, gravity will flip it open to the riding position. Within six months of its 2007 introduction, the factory was producing six a day, large-scale for a maker that had previously produced only on a custom-order basis. Prior to the Tikit, the high-end folding bike market was pretty much split down the middle: Europeans bought Bromptons; Americans, Bike Fridays; Japanese both equally.[22] (Jason Hon, president of Tern folding cycles, cautions that when it comes to statistics for the folding bike market "the numbers are skewed a bit because 15 to 20 percent of the market in China is folding bikes, all with a retail price of $100. There are 15 or 20 million of these made and that skews everyone else's numbers."[23])

Final assembly at Brompton's; one of 204 scheduled for this day (photograph by the author).

Harry Montague was a residential architect living near Washington, D.C. At 6 foot 3 inches, he wasn't interested in a small wheeler, but did want a folding bike he could take to the C&O Canal trail, the Potomac River trail, and other venues. He figured that with front-wheel quick release hubs already standard, all that was needed was a foolproof folding mechanism that would tuck the derailleur in behind the front fork when folded. He invented a full sized (26- or 27-inch wheel) bike that used the seat tube as one large hinge pin.

After it was patented in 1984, he had two batches produced by custom cyclemakers. They sold well, but not spectacularly. Then, in 1991, the U.S. government's Defense Advance Research Projects Agency gave him a two year grant to develop an electrically assisted, folding mountain bike for the Marines. It is not known if any of the bikes were ever used in combat, but many were used behind the lines for basic transport. The firm introduced a non-electric version, the Paratrooper, in 1997 and it was their most popular model for many years. Harry Montague died in 2011 at age 77, but by the time of his death the firm had been incorporated with control jointly shared between the Montague family and outside investors, and both the production of existing products and development of new models continues.[24]

A final contender was Dahon, out of California. David Hon and his brother Henry introduced their Hon Convertible in 1982. Due to trademark conflicts with the Hon office furniture manufacturing company, the bicycle company agreed to change its name to Dahon (David A. Hon) in 1983. The company originally contracted out production, but opened its own Taiwan factory about 1984. There were originally two models, a regular 3-speed and a marine model with a stainless steel frame. The bicycle was somewhat reminiscent of the Bickerton: very compact, but somewhat flexible and limited in its riding capabilities. About 1986 a five-speed derailleur model was introduced. At the time it was the most successful folding bike on the market: by 2002 it had sold two million units.[25]

However, the original design grew dated and had some drawbacks. The most significant of these was its use of a long, narrow A-shaped lateral brace running from the folding hinge at the base of the seat tube up to the head tube, just behind the handlebars. By using this lateral brace, Hon was able to forego a positive locking mechanism for the long head tube when it was in the up (riding) position. The lateral kept it in place and absorbed the rider's fore-and-aft loads on the handlebar, so the very long head tube could be relatively light. But if the rider was not careful to make sure that the spring-loaded pin that dropped into the hole behind the handlebars actually snapped into place, the lateral could fall away during riding and the handlebars would flop around. Also, in an unrelated problem, the snap clamp that locked the frame into the riding position was just in front of the bottom bracket and hard to see, and if ridden with the snap clamp not fully locked, it could bend the frame hinge so it wouldn't mate properly from then on. Nevertheless, the design worked very well and could be produced at an affordable price, and they sold well from the start, mostly under private labels.[26]

Alan Issacson, who today lives in Florida and works as a transit analyst, sold the original Dahon in Colorado in the 1980s. "We sold them out of a booth at RV shows and boat shows along the front range, from Laramie to New Mexico," he recalls. "They worked well if you showed the customer how to fold, and especially, unfold them." The biggest problem was that "they rattled unmercifully when you rode them because the handlebar snap pin was a loose fit, even though they were really quite strong." They "were priced right" and sold well, but the market became saturated after a few years,

and you started to see them in RV catalogs, boating catalogs and so forth, and people started buying them mail-order. The whole point in selling at RV shows was that you had the bikes right there—customers bought them on impulse; these were people who had a boat or RV that cost as much as my house. But that meant you had to lay in an inventory, and the Hon brothers only sold to dealers for cash on delivery. If you had a bad show you could go home with a trailer full of bikes worth several thousand dollars, and four months until spring and the next show.[27]

David Hon says that "Our natural market turned out to be boaters, private pilots, and RV owners. That has been our core. We advertise in magazines like *Yachting*, *Sail*, *Southern Boating*, *R.V. News* and *Flying*. We're just trying to

An early Hon Convertible, one of the scarce marine models in stainless steel (photograph by the author).

keep the public aware of the product and hopefully drive business to retailers." However, the retailers he referred to weren't bike shops. "I tell IBDs [bike shops] not to worry about big box retailers," Hon told a reporter, "don't butt heads with the big boxes, you can't beat them." In return, Issacson says "I don't ever recall seeing a Dahon at a bike shop before the new generation bikes in the early 2000s. They just didn't carry them."[28]

The original Hon/Dahon is still one of the smallest-folding compact bicycles with regular (that is, 12-inch or larger) tires yet devised (photograph by the author).

The original Dahon was about as compact as the Bickerton, but like that supersmall bike, it too had compromises, with at least one, the diagonal brace, that eventually had to go. Dahon abandoned its reliance on the original platform, diversifying its in-house offerings and occasionally adding bikes designed by other firms. It began significantly re-working its original concept in 1995, and in 1999, brought out its first all-new design, a dual suspension 20-inch model that sold for about double the price of the original. By 2003 Dahon was marketing a range of over a dozen different designs. Most of these were manufactured at Dahon's two factories, in Taiwan and on the mainland in Shenzhen, but others were made under contract in factories in the Czech Republic, Macau and Taiwan. The firm introduced new designs and variants with great rapidity and if they did not catch on, they withdrew them with equal swiftness. In 2005 alone Dahon sold 285,000 bikes.[29]

In 2010 David Hon's wife, Florence Shen, and their son, Joshua Hon started a new firm, Tern Bicycles, that competes with Dahon using the same basic business model. The parties gave vague explanations for the split, but it appears that the family's disagreement started after the firm opened a new replacement factory in Shenzhen in 2008. The firm had been courted by the economic council of Tianjin, like Shenzhen a bicycle-making center, but David Hon chose to remain in Shenzhen. But in 2010, David Hon decided that the original factory in Taiwan would be closed, a second mainland factory built in Tianjin, and all design and headquarters operations moved to Shenzhen, where he was now living.[30] But Florence Shen and Joshua Hon had been splitting their time between Shenzhen, Taipei City and California, and Dahon's engineering and senior management team all lived in Taipei City. They did not want to transfer to mainland China and instead convinced Shen and Joshua Hon to break away. Complicating matters was the fact that Shen and Joshua Hon held significant equity shares in several of Dahon's various corporate components.

8. Vanished into the Clouds

Tern's founders articulated a vision for the two firms resulting from the dissolution much different from David Hon. Shen and Joshua Hon wanted to see Tern and Dahon function as sister entities, with a common logistics office, an exchange of design ideas and interconnected product lines, and coordinated retail distribution channels. David Hon and his brother Henry (Dahon's co-founder, who returned from retirement) on the other hand, asserted that there should be a full corporate divorce resulting in two fully independent firms, with a formal division of the assets of the original Dahon corporation arranged through either a court order or a court-supervised consent decree.[31] What appears to have emerged is closer to the latter.

III

The first two Alex Moulton AM models, introduced in 1983, were a two-speed with a Fichtel & Sachs Duomatic rear hub (£400 retail, $840), and a seven-speed with a rear derailleur (£490, $1030). (Again, the dollar amounts are just simple currency conversions. Add about 30 percent for American retail prices.) Production took place in half of the 1962 factory building. After a few months demand stabilized at about fifteen units a week.[32]

The AM was built with standard Reynolds 531 tubing, but in an entirely new way. The head and seat tubes were normal, but the rest of the frame was built up from small tubes, each about the diameter of a seatstay or less. The frame was not only much stronger and lighter, but it could be taken apart in the middle. For the rear suspension, the squishball principle of the Mk. III was retained, but in a far more refined way (it's actually a cupped hemisphere incorporating two different densities of material), and the front suspension was again buried in the head tube, but this version didn't require a rubber bellows to keep out grime and it was adjustable for weight and preference of stiffness.[33]

Some idiosyncrasies remained. Moulton insisted on keeping the 369mm bead-diameter wheels (also known as 17 × 1¼). Wolber of France volunteered

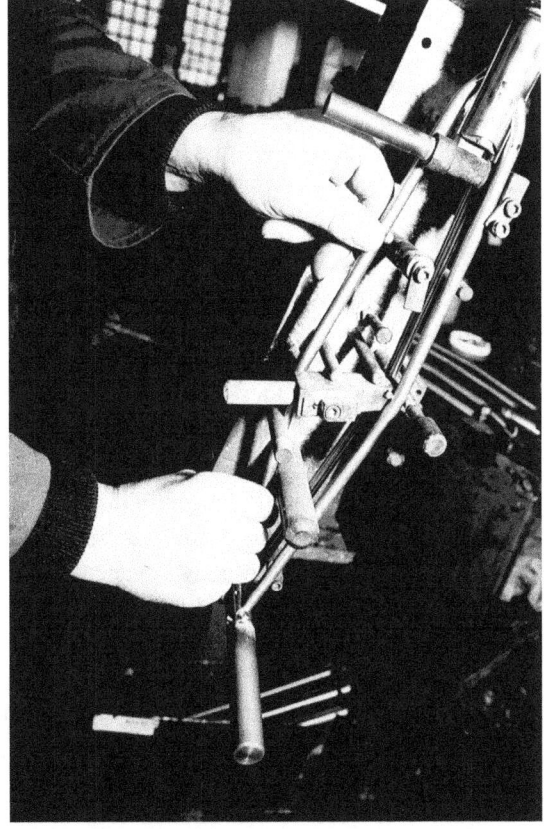

Setting up stringers for a double-pylon Moulton AM. This is the seat tube assembly. When finished, the seat post will slip into the tube at top right. Over forty tubes and cross-supports will eventually be brazed into this one assembly. The cost of perfection: US$17,000 (photograph by the author).

to produce tires and tubes that were narrower and higher pressure than the 65-psi Dunlop Sprite–like tires that John Woodburn had complained about. At first, Moulton was forced to produce his own rims for the AM in-house, making them from raw stock under the supervision of famous British cyclemaker and wheelbuilder Jack Lauterwasser. At one point, Moulton had a prototype version of the AM built with 406 mm/20-inch wheels[34] and Lauterwasser recommended that Moulton go ahead and use it for the production version, as he believed it worked better, sacrificed little (in fact, it fit in the same shipping box), and solved many supply problems. However, Lauterwasser recalled that Moulton was "sold" on the smaller wheel and nothing could dissuade him from using the smallest practical wheel size.[35]

The smaller wheel also required that Moulton fabricate its own 9-tooth rear sprocket that was added to the outside of a standard freewheel in order to achieve an adequately wide range of gears. This, of course, became less of a problem after the 1990s when component manufacturers introduced cassette freehub units with 7 sprockets, which eventually grew to 8- and 9-cog units by 2008, and later even more.

"He's very much an individualist," explains historian Tony Hadland of Alex Moulton.

> He's said himself that he's unemployable. He's not someone you could fit into the modern corporate world. He's really from sort of the culture of the Victorian entrepreneurial engineer, more in the cut of a Bruniel, if you like. Someone who has a personality that they stamp on a design, which is very difficult in the modern world, because industry and finance and the means of production just aren't set up that way.[36]

Dan Farrell, Moulton's Director of Design, says that Alex's aims changed as a result of his disputes with Dunlop over bicycle tires and the Hydrolastic/Hydragas auto suspension, then with Raleigh in the 1960s and '70s over the F-frame, the Mk. III and the stillborn Moulton-Raleigh design exercises. (Some kind of long-term disagreement with Dunlop appears to be the reason he turned to Wolber for the tires on the early AMs.)[37] The goal of the F-frame had been to make a bicycle "that was more convenient and pleasing to use." But with the AM the goal shifted to (in Moulton's words) "concentrating on producing the highest quality machine for those who want the best available." "Alex's intent to make a Moulton bicycle 'suitable for all the family to use' was not fully met," according to Farrell, mostly because its wheelbase was too long. On the other hand, the "AM series was never intended to be 'suitable for all the family' but did retain the one size fits all ideology."[38]

The company experienced a crisis in 1985. Moulton describes it as an accounting failure, but fundamentally it appears that the firm was selling each bicycle below their true cost. The volume of production was so low, and indirect costs so high when compared to labor and materials, that normal accounting procedures had difficulty capturing what it really cost to make a Moulton.[39] To cure immediate cash flow problems, Moulton sold the 1962 factory to a former employee, Anthony Best, who was already renting half of it to run a business dedicated to motor vehicle suspensions. He moved the bicycle factory into the stables of The Hall, developed new models, became more of a custom-builder, and increased prices.[40]

The development of the AM went in two different directions. By 2008, Moulton had pushed the spaceframe concept to its ultimate end with the double pylon. In the original AM, only the main triangle was trussed; the head and down tubes were normal Reynolds 531 tubing. In the double pylon, even these were replaced with triangulated structures. Reportedly, a double pylon frame had over a hundred separate tubes and stringers, took 48 hours to braze, and could theoretically support a rider weighing 3,000 pounds. The stainless steel version costs about $17,000, depending upon equipment.[41]

But Alex Moulton also went in a second direction. In 1988 the firm introduced an All Terrain Bicycle (ATB) version—with 20-inch wheels. It varied in some details from the AM, mostly to adapt the suspension to the needs of off-road riding. But the biggest difference was that these changes facilitated the development of an "everyone's AM," the Moulton APB, the first Moulton since the Mk. III intended to be built in another factory. Production started in 1992 at Stratford-on-Avon at the factory of W. R. Pashley Ltd., an old-line maker that survived the onslaught of Asian imports by moving into a niche market of English retro-nostalgia bikes: deluxe roadsters, delivery cycles and pathracer designs from the 1920s.

The Moulton-Pashley APB had some minor engineering and material changes (primarily the substitution of less expensive generic cro-moly steel tubing for the ATB's Reynolds 531), but the biggest change was that it was manufactured at Pashley's semi-mass production factory. When introduced, it sold for £500 ($900), compared to the

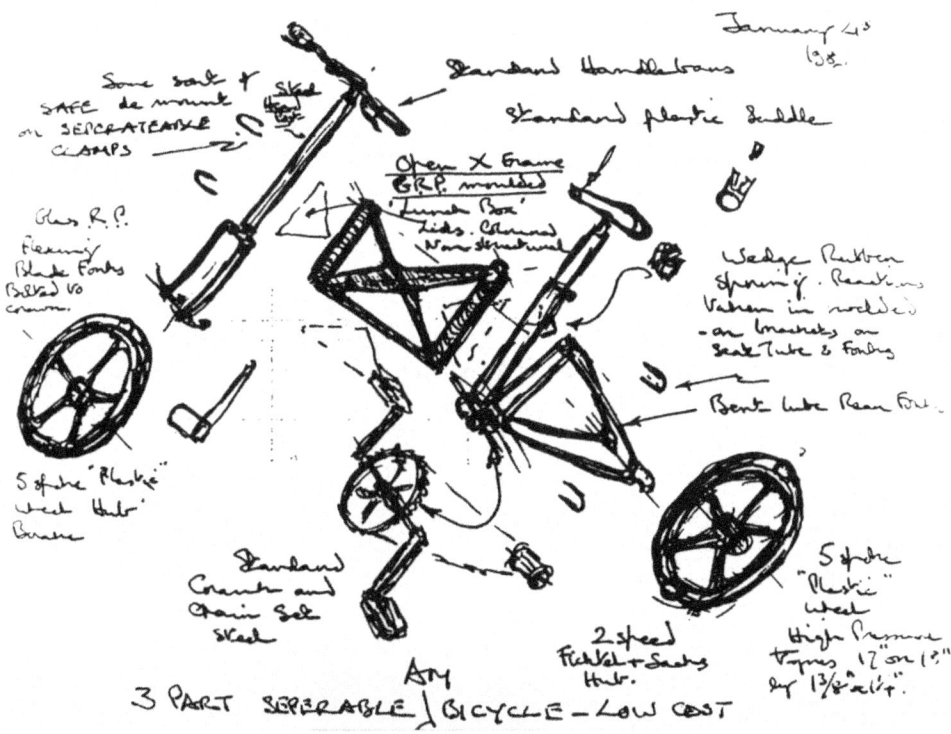

An unrealized design concept by Alex Moulton for an X-frame G.R.P. molded bicycle. Like the Y-frame, passed over in favor of the spaceframe. Dated January 21, 1980. Original drawing in the Science Museum, London (photograph by the author).

ATB's £1,295 ($2,330). Tony Hadland and John Pinkerton noted that by replacing £200 in components, the weight of the Pashley-Moulton could be brought down to within a pound of the Moulton AM-ATB. Most of the Pashley-Moulton APBs were, for many years, sold under a marketing agreement with Land Rover in which the bicycles were labeled with their name. (Owned at the time by BMW of Germany, Land Rover was later acquired an Indian conglomerate.) In 1999 Moulton licensed the Bridgestone bicycle company to make a modified aluminum version of its original F-frame design, available only in Asia.[42]

In 1984, Alex Moulton revived the Cowl project that had been abandoned shortly before John Woodburn's Cardiff-to-London record ride in 1962. He had patented the Cowl in 1963, but had done no additional development work. In the late 1970s, Moulton Developments was involved with General Motors-USA in a research project involving automobile suspensions. As an intermediary, Alex Moulton turned to an American research firm, Milliken Research Associates (MRA), and its owner, William F. "Bill" Milliken, based in Ithaca, New York. In 1980, Moulton sent a handbuilt F-frame-based special to MRA for use and evaluation by Bill Milliken's son Doug. The Millikens had access to an advanced wind tunnel at Cornell University, and Doug Milliken designed and tested twenty Cowl variants, sending a 50-page report to Alex. One significant finding was that under the right conditions, a crosswind would interact with the fairing like a sail, helping propel the bike and rider forward.[43]

In 1984 MRA was sent a new spaceframe AM7. A partial fairing with Plexiglas nose and spandex tail was built for it by Doug Milliken and raced at the International Human Powered Vehicle Association meet at the Indianapolis velodrome in September 1984. A full fairing was constructed over the winter of 1984–85 and a year later, at the Indianapolis Motor Speedway, the AM7 achieved 50.21 mph for the flying 200 meters, a record for bicycles with a traditional riding position. In 1986, in Vancouver, with a more aerodynamic and lighter fairing and 82 × 10 gearing (140 inches), the AM7, piloted by Jim Glover, achieved an improved 51.29 mph, a record for traditional-position bicycles that still stands.

In 2008, Alex Moulton Bicycles was acquired by Pashley Cycles. Previously, the Pashley firm itself had been sold by the founding family to two partners, Adrian Williams, who had earned a fortune at Westland Helicopter, and John Macnaughtan, the former Raleigh and Sturmey-Archer executive who saved the Brooks Saddle Company. Williams owns 75 percent of Pashley and Macnaughtan 25 percent.[44]

Although the offices of the Moulton Bicycle Company are still at The Hall, only the most expensive and custom-order AMs are made at Bradford-on-Avon; the standard-specification models are manufactured at the Pashley shops at Stratford-on-Avon. At the time of the buy-out Alex Moulton Bicycles had a 15-month wait list and even Alex Moulton had to admit that the situation had become untenable. "In a year's time, I'm hoping we can significantly reduce that," he told a reporter when the new organization was announced. "But," he added, "we are not downing quality to make ourselves more commercial."[45]

In late 2012 Alex became increasing infirm. Normally, his bedroom was on the second floor of The Hall, adjacent to the famous Oak Room, his oft-photographed office and studio. Both to save himself from having to traverse The Hall's Victorian-scale staircase, and, more importantly, to save the now-elderly Toby the Cat from having to do the same, Alex moved onto a chase bed in his study on the ground floor. In his final weeks, Toby became his constant and reassuring companion. On December 9, 2012, the 50th anniversary of John Woodburn's record-setting Cardiff-London ride, Alex Moulton passed away at age 92. A few weeks later Toby followed him. The cat door in the kitchen was grouted in.[46]

IV

Macnaughtan came out of retirement to run operations at The Hall. "I can blame Adrian for this," Macnaughtan light-

Brazing spaceframe Moultons in the Paddock at The Hall, 2017 (photograph by Nora Quinlan).

heartedly told an interviewer in early 2016. "I did not intend to go back to work full-time. He asked me if I would like to get involved, and I didn't realize what 'getting involved' meant. You can't just dip your toe into the bicycle business—it's total immersion or nothing."[47]

On a more somber note, Macnaughton recalled that he had barely begun work when Moulton passed away. "It's very sad, because he died just at the time I started," he recalls. But he was forced to admit that, when seen from the point of view of the business, it was not a calamity:

> In terms of working it made it easier, because, you know, there are a hundred and one stories coming out of the factory here. One is that when Pashley bought Moulton in 2008, one of the biggest problems we had is that he would come wandering down to the work force and say "Timmy, I'd like you to clean the Bentley, please," or "Timmy, I've just knocked up these drawings to the tool room, could you get a unit out," and of course we were paying Timmy and he was the one driving the Bentley or devising some new invention.[48]

Actually, Macnaughtan's still somewhat light-hearted comments gloss over what threatened to become a major dispute between Alex Moulton and the Pashley firm only a year after the Pashley purchase. In April 2010, Moulton Developments (the firm Alex Moulton had owned since the mid–1950s) unveiled the M Dev. At first glance it looked like a Moulton AM, the one with regular seat and head tubes, but with the Double Pylon's front fork. On closer inspection, it was narrower and had none of spaceframe's usual zig-zag cross bracing.

From the head tube forward, and from the seat tube back, it was a Moulton AM. The main truss, however, was radically different. The goals were create a main frame that required no brazing, and that could be shipped to dealers in the form of a knocked-down kit of parts and assembled at the point of sale. Each frame member was an aluminum tube. Inside this tube was a "spoke" which could be rigid or made of cable, including carbon fiber. The head of the spoke was set into recess in special lugs attached to the head tube and the top and bottom of the seat tube. The frame tube was then slid over the spoke into its own recess in the lug. The process was repeated at the lug on the other end of the tube. However, the other end of the spoke was threaded. A nut was screwed on these threads and tightened to a specified tautness. The tube and the two lugs on either end were now held rigid by the spoke. When all the tubes were placed in all the lugs and all the spokes tightened, the frame would be locked up and would have all the properties of a regular brazed frame.[49]

The M Dev raised immediate concerns with the Pashley firm. For one, it was, essentially, a patent for an improved way to make what was now their bicycle. Second, it was developed using their personnel and machinery. In late 2010 Alex Moulton announced, in a press release to the Moulton bicycle owner's club, that he intended to find a firm to make the M Dev, the implication being that this would not be Pashley. It does not appear that Alex Moulton had the opportunity to do much more developmental work after the end of 2010 before ill health interfered with his plans. The M Dev was not in itself successful, according to Dan Farrell, for reasons he prefers not to go into. He acknowledges that the firm is exploring some of the concepts contained in the patent, which he sees as more of a manufacturing process than an actual device.[50]

Macnaughtan, sitting in Moulton's old office at one end of the drawing studio, is, surprisingly, far from sanguine given that Alex Mouton Bicycles exports 90 percent of its output and still has a backorder list that exceeds nine months. "Mouton is not in the bike business, as presently conceived," he explains. "We are selling jewels, artwork ... selling jewels is jolly hard work, and the amount of time we have to spend with each customer is totally out of proportion to the return."[51]

John Macnaughtan, the former Raleigh and Sturmey-Archer executive who saved the J. B. Brooks Saddle Co. from the collapse of Raleigh Industries, and who now runs Alex Moulton Bicycles (courtesy Alex Moulton Bicycle Co., Ltd.).

Asked to look into the future, Macnaughtan has two goals. The first is to resolve the uncertain relationship between the bicycle company and the new owners of The Hall. It is currently being maintained by a trust established by Alex Moulton. He intended for it to go to a permanent preservation trust, but it is unclear whether, and to what extent, Alex Moulton

Bicycles, Ltd., will be able to remain in its premises in the stables and the drawing studio. The 1962 factory on Holt Road was sold in 1985 and "we don't know right now if we will have permanent tenure," Nacnaughtan candidly admits.[52]

With the much bigger Pashley factory only 75 miles away at Stratford-on-Avon, the danger of being evicted from the stables would not seem that much of a crisis, but Macnaughtan argues that The Hall is a large part of Moulton's identity. Tony Hadland agrees. He compares the Moulton, not to a jewel, but to a premium single-malt Scotch whiskey. Both are expensive luxury items; both rely on export, mainly to Asia; and both are inextricably tied to both a production process and a specific location.

Macnaughtan's second goal will likely be even more controversial: "We're being asked, 'Well, Alex has died, what's next?' We really haven't come out with something yet that really puts someone else's stamp on it. That's overdue." This will come as a surprise to many, who assumed that Alex Moulton Bicycles would become a repertoire-technology firm, tweaking older Alex Moulton designs or producing designs that were brought to fruition, but never produced, such as a Reynolds 753 version of the six-speed Mk. III prototype that Peter Knottley tested in 1969; or an aluminum version of the Y-frame that Alex considered, but rejected, in the late 1970s; or even a domestic version of the aluminum F-frame, until 2016 manufactured under license by Bridgestone in Japan.

Dan Farrell, director of design at Alex Moulton Bicycles. "People said, 'You couldn't fly to New York faster than a bullet,' but we did. They said, 'You can't improve on the diamond frame,' but we did. He did" (photograph by the author).

Macnaughtan does not reject this, admitting that they are exploring alternatives based around the Y-frame and the Bridgestone. But the spaceframe is still the crown jewel. Macnaughton emphasizes that "we need to take control of the Moulton's design and take control of the Moulton's quality in a way that both continues the Alex Moulton tradition *and* moves on to new innovations." However, this probably does not mean a clean-slate design. "It's a risk to go clean-slate," he admits. Dan Farrell says that "I think that if Alex were here, he would say 'Keep making small changes, keep improving it.'"[53]

9

Yesterday's Tomorrow Is Not Today

> The recent past and the austere present was something we did not like to dwell on. Tomorrow had to be "new and improved".... In those days, no one thought about posterity and who should get credit for what.... We did not begin to suspect the importance of the movement until a later generation began to ask about it.
> —Mary Banham (1990)[1]

I

Alan Oakley's RSW "family of bicycles" concept proved to be an early example of what would soon grow into a widespread manufacturing phenomenon: the "platform": a hardware device that serves as base upon which accoutrements are installed to create a variety of different useful and marketable products. With 16-inch wheels, enclosed wiring and a plethora of accessories, the RSW was a versatile, high-fashion, sports accessory, well suited for upscale department stores; with 20-inch wheels it became a solid, reliable bicycle, the replacement for the 70-year-old roadster; with chrome, bright paint, and a fancy saddle it could be recycled as a faddish kid's toy at an outlandishly high price.[2]

Banham first mentioned the idea of "clip-on" technology in 1958 in, of all places, an article on science-fiction robot detectives. He again discussed the idea briefly in 1960 in *Architectural Review* in a review of a plastic-composite living unit designed by the French firm Coulon and Schein. The units, small enough that four or five could be loaded on a flatbed truck, could be used individually to provide minimal shelter in emergencies or to assist the socially disadvantaged; installed as prefabricated bathroom units for otherwise traditionally constructed homes; or combined on-site into multi-room dwellings. These could be arranged as free-standing homes, or structured in racks into larger apartment-style buildings. Banham thought such structures could be assembled entirely out of modules, but more likely they would serve a more limited purpose as plug-in service units to house kitchens or bathrooms. "In either case, Banham observed, "the result should be that service-rooms, which need to be connected to the public mains, might be treated as expendable clip-on components," thus obviating the need to rip up the apartment building in order to upgrade its shared appliances. Just yank the whole kitchen or bathroom out and replace it.[3]

But it was the mid–1960s work of the architectural firm Archigram that really focused Banham's attention on the idea of plug-in, clip-on modular design. Established by architects Peter Cook, Warren Chalk, Denis Cromption, David Green, Ron Herron and Mike Webb (all but Chalk and Herron under thirty), it is not true, popular mythology aside, that they never actually built anything, but even Banham noted that they frequently had problems wrestling with such abstract concepts as gravity. Archigram was better known (read: infamous) for their mega-watt light-and-sound presentations. In the early days, many audiences were left slack-jawed, others fought back. At a 1966 architecture conference in Folkestone, Peter Cook, along with his sound-and-light team, was slow-clapped off the stage. In Paris, in 1971, an entire audience simply walked out, down to the last seat. For many years they were banned from the MIT architecture building. (Cook: "they really hated us there in those days, you know.")[4]

Archigram's other major activity was its irregular magazine, also entitled *Archigram*. In *Archigram* 4, the team had outlined a giant "plug-in city" comprised of publicly owned racks dozens of stories high containing privately owned "capsule homes." Banham was so enthused about the plug-in city (and Archigram's whole "zoom wave" approach to architecture) that he wrote about them in two 1965 articles in *Design Quarterly* and *Architectural Design*, plus a briefer story the following March in *New Society*. According to Banham, plug-in city's appeal was that "it offers an image-starved world a new vision of the city of the future."[5]

"It was all about Pop art, really, Pop architecture" recalls Mary Banham. "It was so much about imagery. They were all wonderful draftpeople—incredible draftspeople. They did these simply unbelieveable images."[6] But, as Banham admitted, "even Archigram can't tell you for certain whether Plug-in City can be made to work." Instead, what turned his head was the application of their "plug-in" idea to what he called the "hot-rodder attitude to the elements of building." A few years earlier, a British group of architects had developed the CLASP (Consortium of Local Authorities Special Program) system of prefabricated small neighborhood school buildings, which they displayed at the Milan Triennale in 1960. In the USA, John Entenza, editor of *Arts & Architecture*, created the Case Study House program as a way to enable architects to design (and find clients for) innovative, low cost modern houses using donated materials from building suppliers and other industries. Starting in February 1945, 31 homes and two apartment buildings were featured, of which 26 homes and one apartment building were built, all except the apartment building in the Los Angeles metro area.[7]

The noted husband-and-wife designers Charles and Ray Eames designed Study Houses #8 and #9, the latter with the assistance of Eero Saarinen. Number 8 was built for the Eames and #9 for editor Entenza. Both were alike in that they were built out of a steel frame and roof kit designed for farm and industrial buildings, but without the standard stamped-steel walls. This frame was then fleshed out by selecting standard building components from various suppliers' catalogs, adding in hardware and material from suppliers outside the building trades, and "kit-bashing" everything together to make unique, but reproducible designs that could be mass produced.

The difference between the two was that in the Eames House, this "industrial-ness" was left open and exposed, something like the Smithson's Hunstanton school, but in the Entenza House, every affordable effort was made to make the house seem "normal": the steel truss roof girders were covered with a tongue-and-groove wood dropped ceiling, fewer jalousie windows were used in favor of tall sliding doors, and so on. The two houses were next to each other, so the contrast between them could be easily compared.

The result was more sophisticated than Buckminster Fuller's gain-silo Dymaxion Deployment Unit, but less complex than his Wichita House, made by Beech Aircraft using re-worked aviation materials. None of the Wichita House components were aircraft parts, but on the other hand, only an airplane manufacturer had the necessary tooling and experience with sheet and extruded aluminum and stainless steel to fabricate them. The Wichita House looked rather like a shiny hamburger (with wrap-around windows being the hamburger patty) levitated a foot or two off the ground, suspended from a center pylon by cables.

The Eames houses and the CLASP School on the other hand looked like a typical plain '70s-modernist civic building—a small town city hall or a recreation center in a park. Banham liked the idea of "ingeniously mating off-the-peg components, specials, and off-cuts from other technologies ... the great virtue of the hot-rod method—and this cannot be said too often or too loud—is that it demonstrates how to make architecture out of what is available."[8]

What made Banham's enthusiasm for the hot-rod "plug it in and see how it works" culture was his visceral, negative reaction to the adoption of the idea of the "bricoleur" into the growing architectural movement of the early 1970s. A leading figure in the postmodernist movement was Banham's own doctoral student, Charles Jencks. It's hard to define exactly what postmodernist architecture is, other than possibly a denial that architecture can ever be anything more than architects exchanging puns, jokes, inside messages (often commentary on the clients who commissioned them), and insults to each other and their critics through their buildings. Modernist architecture claimed to be substance without style; the post-modernists asserted that architecture could never be anything but style, re-labeled "symbolism." Postmodernism did, however, highlight a longstanding weakness in modernist architecture. It was dedicated to the idea of "form follows function." What happens if the function of a building was not very interesting? You ended up, Banham had to admit, with a dumb building. But he could not bring himself to approve of the idea of postmodernist artifice: "these are 'silly' buildings like the 'silly' buckets on the Ancient Mariner's deck—empty."[9]

As opposed to the constellation of professional clients, architects, engineers and contractors (all bad guys) stood the bricoleur, derived from the work of the anthropologist Claude Levi-Strauss. Now, Levi-Strauss himself used it in a rather narrow, technical sense. The point he was attempting to make, seemingly self-evident today but radical in 1962, is that the members of very pre-industrialized cultures were fully capable of conceptual, abstract thinking, and that they applied such thinking when presented with the objects of modern material culture. They may not understand what a pair of pliers is, how they work, or what they are used for, in the same way as someone from an industrialized society, but they do abstractly think through what a pair of pliers is, and how they can best be used.

Over the years, the idea became something of a cliché for the make-and-do patcher and mender, a nostalgic fixture of a vanishing, pastoral, agricultural way of life, as forensic scientist, cycling author and George Washington University professor James Starrs once recounted:

> In Washington, Kansas, in the summer of 1971, my daughter broke two spokes in her rear wheel, of course, on the freewheel side. At the time three of my children and I were traveling by bicycle from California to our home in Washington. That is, Washington, D.C. ... Washington, Kansas is small in size and remote. The local machine shop was our only hope for assistance in removing my daughter's Atom freewheel.... They puzzled over what they might have that would fit inside the freewheel snugly enough to break it loose from the hub. After casting about in boxes of rusted metal, they set about locating bolts with heads large enough to fit the center cavity. But one bolt after another broke after its shaft was twisted in an effort to loosen the freewheel. When the last of the bolts was about to be used, the owner told the younger one to heat it under a torch and then immerse it—in short, to temper it. He did; the bolt held; the freewheel dislodged and the spokes replaced.
> Even today this home-crafted freewheel puller is cherished by me as a formidable example of prairie-land bricolage, defined by Claude Levi-Strauss as "the art of the do-it-yourself handyman who must solve problems given only limited tools and his own ingenuity."[10]

But about the time Starrs's daughter was getting her bike fixed, critical (read: French) sociologists such as Jaques Derrida took hold of the idea and turned it into what architect Patricio del Rio calls "a model for creativity and a critique of dominant culture in general.... The construction of marginality is perhaps one of the single most important ideas that traverses the important discussion of bricolage." This was picked up by a group of Americans, including Charles Jencks and Nathan Silver, and broadcast in a 1972 book, *Adhocisim*. They jumped on the Derrida bandwagon with a vengeance. The highest, purest form of architecture was the third-world slum shanty made from modern building materials begged, borrowed or stolen from high-rise construction sites and put to use in ways never intended by their inventors or fabricators.[11]

"The bricoleur is set up as the archetypical make-and-do-and-mend patcher and improviser and thus the patron saint for their new gospel of improvisatory design," complained Banham. "However, the bricoleur has also come in patron-handy to eco-maniacs and alternative technologists, to architectural dissidents reacting against the sophisticated rigor mortis of 'system building,' to Popperian proponents of piecemeal planning as against comprehensive redevelopment, and *haute bricolauge* has even been detected in the work of Marcel Ducaump and Robert Rauschenberg." What had turned it into such a popular academic fashion statement was its new formulation: "it works *only* with the offcuts and discards of engineering," Banham explained. No raw material or from-scratch fabrication allowed. Hence, "it proves to be just another colonial dependency of the dreaded technostructure."[12] By definition, as used by Jencks and Silver, bricolage must use as its raw material the gleanings from the garbage heap of a dominant (in both senses of the word—higher-tech and overbearing) culture.

Architecture was a particularly difficult place to take a stand for a Derridean interpretation of Levi-Strauss' original ideas. As del Rio succinctly put it: "slums cannot, however, be reduced to romantic, antimodern or idealized characterizations, to vernacular spaces of rural sociability and precapitalist exchanges or microcommunities that escape global capital.... For a world that sees bricolage simply as the commerce of contemplative experiences or as a tool for critique, the stench of the slums has yet to be excised." In other words, an improvised, patchwork residential building isn't a piece of folk art, it's a shack, and if you put five or ten thousand of them together, well, you've got a fetid slum, a *favela*.[13]

If, on the other hand, if you weren't talking about big, obdurate, fairly simple things like buildings, but relatively small, portable, sophisticated gadgets, Banham asserted that the concept just became a joke: low-high tech. "The bricoleur is now established as kind of third-word alternative to science and engineering, pottering about in his *bidon* in the *barriada*, improvising, as like as not, a wheelchair out of Coke cans and dismembered rollerskates ... this, however, is not the role for which Levi-Strauss intended him ... if you go back to the text, you will find that the bricoleur is not part of any anthropological statement or argument." Banham wrote off Jencks and Silver's spin on Levi-Strauss's concept as "only a symptom of the general ignorance about engineering among those who profess to tell us what's wrong and what's right with western culture—and, above all—what's wrong and what's right with western engineering."[14]

The concept of bricolage worked its way into the cycling vocabulary in the mid–1990s with urban bicycle messengers and their stripped-down, brakeless, single-speed "fixie" bikes. Bicycle messengers are nothing new, of course. Western Union and Postal Telegraph Service bicycle messengers were common in the 1920s. Their numbers peaked during World War II, when gas and auto tires became hard to come by. But by 1974, Western Union employed only 779 messengers, of which 116 worked in New York. The modern bicycle messenger started in New York City with "can carriers" in 1979 to transport film for the advertising industry. Their numbers topped out in the mid–1980s. The only reliable estimates are those for New York City: 500 in the late 1970s, about 4,000 in 1985. It appears that no other city had as many as a thousand except for San Francisco.[15]

Moreover, these figures are inflated because firms deliberately overhired to create a two-class system. A small core of long-time employees worked full-time and was dispatched to a steady flow of desirable runs, usually repeat customers downtown, while a much larger pool of short-term hires was used to handle peak loads, outlying addresses, and bad weather. They were considered expendable, treated poorly, and had high turnover and injury rates. Kickbacks to dispatchers were often needed to gain preferential treatment. The second class "on-call" riders were often "undesirable" minorities, the identity of which could vary from location to location.[16]

Although messengers and firms continually complained about police harassment in the late 1980s, the industry was doomed the minute regulatory agencies stopped ignoring their own health, safety and workers' compensation laws and forced firms to treat bicycle messengers the same as teamsters, package delivery drivers and other transport workers. In 2006, the State of New York issued new rules that categorically determined that a bicycle messenger was an employee, not an independent contractor, if the firm "set the order and priority of delivery," including promising a specific deadline for delivery.[17]

By 2002 even the nontraditional documents that had been the bread-and-butter of messenger services could be transmitted electronically: blueprints, photographs for retouching, and legal documents. "It's a dying way of life," complained one messenger as early as 1993. In the late 1980s a top rider could pull down $1,000 a week and an

A Postal Telegraph messenger boy, Birmingham, Alabama, c. 1918 (photograph by Lewis Wickes Hines/Library of Congress).

experienced, hard worker could make $700. By the mid–1990s, $350 was about the most anyone could expect to take home. "In the old days business people would pay $80 just to get something across town in an hour. Cost was no object. Now they just say to hell with it, tomorrow's soon enough," said one messenger firm dispatcher.[18]

The final blow came with the September 11, 2001, attacks on the World Trade Center and Pentagon. Security was bolstered in every American city, and the idea of permitting a bike messenger to pass beyond the lobby became unthinkable. Even leaving an unattended bicycle became next to impossible. In 2009, one researcher claimed that a thousand were still working in New York City, but many were doubling as walking couriers or second-seat riders on delivery trucks to assist with heavy boxes or prevent theft while the driver was in a building.[19]

But in the '80s and '90s, the bicycle messenger gained a reputation as the archetypical "noble savage," and it seemed at one point that there were as many doctoral students crowding the grungy offices of bicycle messenger services as the break rooms of university sociology and anthropology departments. "Why does everybody want to interview bicycle messengers?" responded one exasperated courier to an academic's request for research subjects."[20]

It appears, oddly enough, that the trend was literally a case of life imitating art. In December 1980, Karl Kim Connell, a creative writing instructor at Northern Illinois University, published a short story, "The Bicycle Messenger," in *The North American Review*.[21] Richard, the story's narrator, jams down the mad streets of New York for the Mobile Messenger Service, taking his tips in coke and closet sex while he contemplates the offer of his (until very recently) girlfriend to abandon a city clearly collapsing into open anarchy for the promise of hope and happiness in Paris. But on the other hand, what's hope and happiness compared to the thrill of smashing in some Jag's windshield with a U-Lok?

The story was passed around, samizdat-style, among messengers and urban cycling advocates, and more often than not was mistaken for (or deliberately palmed off as) an actual memoir, creating the myth that a community of testosterone-saturated, drug addled, thrill-seeking cycle primitives lived in the shadowlands of America's large cities. As a result, some cyclists started to seek out courier work as a form of urban extreme sport, thus creating the very culture (or pseudo-culture) they went in search of.

The first of these academic groupies, Dr. Jack Kugelmass, published an article in the journal *Natural History* a little less than a year after Connell's short story appeared.[22] It set the template for most that would follow: (1) the author was a cycling aficionado, although usually not up to competition level; (2) the author tuned out the vast majority of itinerant messengers that do not like the job, focusing solely on a small group of "lifestyle" messengers—essentially careerists; (3) the author played up the flamboyantly anti-social behaviors of messengers, even though this group constituted a further subclass within the lifestyle messengers; (4) the author emphasized the importance that messengers place on independence and autonomy, valorizing the exploitive and often blatantly illegal practices of the messenger firms that hired them; thereby turning corporate criminality into an open-minded accommodation to the extremist-lifestyle messengers' desire to avoid (or inability to accommodate themselves to) a traditional employment relationship they saw as "restricting" and "depersonalizing."

Jeffery Kidder, a Ph.D. candidate at the University of California–San Diego claimed that "messengers are who they are, wild-riding madmen living the life many readers may have already dreamed of." Elizbieta Drazkiewicz prefaced her Lund University master's thesis, with the description of the bike messenger as "the toughest, looniest cat around because he gets paid the fat bills to ride a fat bike through hells nine circles ... the glamorous transgressor, the free man who laughs at slow humanity huddled under umbrellas at bus-stops." Ben Fincham, at the University of Cardiff, was only a little more clinical when he wrote that "there are many couriers that behave in ways that would be considered 'deviant' in sociological literature. There is a fair amount of recreational drug use and a level of self-organization." He goes on to quote one interviewee, Dee, as saying that "I think that [in] most places people who are couriers enjoy cycling, at least I would like to think that. I honestly think that generally speaking people are in it for the cycling and the beer."[23]

All three authors were motivated to study couriers because it gave them a chance to get their master's or doctoral degrees by, mostly, spending a lot of time riding their bikes. However, none made it clear in almost everything they published—the typical exception being the thesis or dissertation itself—that they were not interested in the vast majority of delivery cyclists, and explicitly excluded them from consideration. Fincham found that less than a quarter of the couriers he interviewed had any interest in cycling outside their jobs, yet paradoxically concluded that for "most" of the messengers he spoke with "being a courier … is much more than earning a living—it's a way of life."[24]

Kidder reported in his dissertation that for 85 percent of the messengers he studied, "being a courier starts and stops with the workday," but then disregarded this group as "largely outside my analysis" because they were insufficiently "deviant." In subsequent papers he downplayed this detail, giving the impression that his fifteen percent represented the general bike courier community. Similarly, Drazkiewicz also drew a dichotomy between "just a job" couriers and "ideological image" couriers, which, like Kidder, were the only subtype considered for study, but Drazkiewicz did not attempt to estimate the proportion of riders that were "just a job" or "ideological image" riders in Toronto, Copengagen or Warsaw, the three cities she studied. Nor did she consider food delivery riders to fall within the ambit of "bicycle messengers."[25]

Likewise, both Kidder and Fincham excluded food delivery riders, even though at one point Kidder estimated they outnumbered document couriers, and observed that food delivery work served as an alternate—and sometimes preferred—form of employment for document couriers, depending upon prevailing wages and conditions at a given time and place. Moreover, the number of bicycle document couriers was only a very small proportion of the total workers in the business of package, prepared food and purchased-goods delivery in all cases studied. Fincham at one point noted that bicycle document couriers "are, in fact, a small number of workers in a much larger industry—an industry where work is not determined by age or lifestyle," then proceeded to ignore this baseline observation.[26]

Another graduate student, Sam Shupe of Boston University, who has done research into the bicycle technology of messengering, notes that

> all the bicycle messenger research has been extremely narrow and focused only on one class of messenger. My own work made me realize that there are indeed two types of messengers. Ones who do it just because it's a job; using a bike as a means to live. Then there are the messengers (like the ones talked about in Kidder's articles) who are messengers because messengering allows them to do what they love to do all day (biking) without having it feel like a real job. I think sociologists like Kidder see the latter as much more interesting and thus focus on them for study.[27]

A bicycle courier in London, winter 2017 (photograph by the author).

The tendency for firms to deliberately overhire, the ever-mounting competition from electronic communications, and the low barriers to entry for firms (and the ability of existing firms to dissolve in the face of fines or legal liability with hardly a paper trail) means that pay and working conditions are chronically poor. After an investigation, *New York Times* reporter Robert Lypsyte called bike messengering the "sweatshop of the streets." While the well-compensated, dashing, go-to-hell rock stars were mostly young white males with a smattering of African-Americans and African-Caribbeans, the rapid-turnover "occasionals" were almost always Hispanic or of mixed-race origins. "We are expendable," Lypsyte quoted 21 year-old Tony Rubirosa as saying, "the city takes us for granted; the companies grow rich off our blood because they can. There are hundreds more of us, hungry and gullible, to take our place for no benefits, no future, no respect."

Rubirosa is the son of a Dominican father and a Puerto Rican mother. His brother, Danny, 18, is also a messenger. Danny Rubirosa described his job as "death on the installment plan." Tony added that the certainty that he would eventually go down one day frightened him the most: "I get scared when I see the old riders, the ones with all the body armor. They have given themselves to the job; they have become the robots the company wants." Danny, on salary, earned $237 a week. Tony, on commission, averaged about $300.[28]

The same trajectory of academic myth-making was true with the messengers' alleged self-expression of technological bricolage, the brakeless, single-speed "fixie," ideally a purpose-built track-racing machine. In reality, actual track bikes were (and are) rarely used by messengers, for two reasons. First, the typical track bike comes from the factory equipped with wheel rims made for glued-on, ultra-light, very expensive tubular tires; hardly the thing to handle the grungy mess of the typical urban street gutter. Second, the wheelbase of a track bike is so short that the toe clips on its pedals (or the front of the shoes of the newer, ski-type "step-in" clipless pedals) overlap the sweep of the front wheel. Turn your wheel at the wrong time to dodge an obstacle and forget to point your toe down and you'll jam your foot into the back of the front tire, leading to a very sudden stop. Many "track bikes" are actually road machines converted into "single speeds," (one gear, with a freewheeling rear hub) or "fixes" (one speed, no freewheel, which allows backpedaling in place of a brake—sort of).[29]

There are conflicting accounts on the origins of fixes in messenger use. In 2008, James Norton, a Boston bicycle shop owner and former courier, told the *Boston Globe* that when he started as a courier in 1991, there were only "two or three other kids" riding bicycles stripped down into fixes, but by 1995, they predominated within the business, at least in the Boston area. The reason for the evolution was not esoteric: the couriers wanted to avoid downtime from mechanical breakages. However, there is much evidence to contradict Norton's assertion. As far back as 1981, Jack Kugelmass said that he saw "dozens of messengers on track bikes," although his subsequent description makes it clear that he was actually seeing a mix of true track bikes and road bikes stripped down into fixes. A 1987 article in the *New York Times* reported that "the word on the street is that messengers are abandoned their traditional 10-speeds in favor of faster track bikes." The author, Sarah Lyall, appeared to know bicycles and it was clear that she meant genuine, out-of-the-box track bikes, not modified road equipment.[30]

But much of the mystery could be revealed from a visit to Central Park in Manhattan on any given weekend in the 1960s. Many premier-grade European bicycles, both road and track, were single-speed machines. "Historically, British track bikes (unlike European ones) tended to be road/track bikes," notes Tony Hadland, "designed to be ridden quite long distances to and from the race, as well as in it. And proper track bikes were as rare as hen's teeth, and usually made to order." Similarly, many of the better racing machines available in the United States in the 1950s were road/track bikes, because most American racers competed in track events—at least for their state championships, which were often held on auto race ovals or grass tracks. Hence, single-speed road machines were far from an oddity in the 1960s, especially along the East Coast. In New York, Tony Avila ran a bike shop in East Harlem, near the northeast entrance to Central Park, that specialized in track bikes and single-speed road machines.[31]

Avila's family started the shop before World War II, when there were twice-yearly professional six-day races in Madison Square Garden, which lasted until 1939. Up through the late 1920s, there was also a regular summer circuit of outdoor velodrome racing run by promoter John M. Chapman on tracks he owned or controlled in New York (Bronx), Coney Island, Newark, Providence and Buffalo, with a week-long winter racing meet in Miami.[32]

The writer Henry Miller, asked late in life about his best friend, answered:

> Believe it or not, it was my bike. This one I had bought at Madison Square Garden, at the end of a six-day race. It had been made in Chemnitz, Bohemia and the rider who owned it was German, I believe. What distinguished it from other racing bikes was that the upper bar slanted down towards the handle bar. I had two other bikes of American manufacture. These I would lend my friends when in need. But the one from the Garden no one but myself rode. It was like a pet....
>
> There was a period when I spent most of my days job-hunting, presumably. Suddenly I stopped doing this, and did nothing. Nothing but ride the bike. Often I was in the saddle, so to speak, from morning to evening. I rode everywhere and at a good clip. Some days I encountered some of the six-day riders at the fountain in Prospect Park. They would permit me to set the pace for them along the smooth path that led from the park to Coney Island.... I must have given it to one of my cronies [when I moved to Paris], but to whom, I no longer remember. Years later, in Paris, I got myself another, but this was an everyday sort, with brakes.[33]

Miller himself was a bicycle messenger, around 1920. A few years later, while working as a manager at a utility company, he wrote his first book, "a study of twelve eccentric messengers." Rejected, it too was lost when he moved to Paris in 1930.[34]

Track bikes were popular among East Coast bike messengers because they were popular among urban enthusiast cyclists, and had been for a long time. (The same was true in Chicago, where the legendary Oscar Wastyn, Sr., ran his Schwinn bike shop. Wastyn had been the race mechanic to the Paramount six-day race teams of the 1920s.) They were less popular elsewhere, for obvious reasons. Jeffery Kidder quotes one Seattle messenger as telling him "this is a stupid city for riding fixed. It's a city with seven hills."[35]

Sam Shupe, who found 15 percent of the messengers he observed adhering to a "strong messenger culture," and that "the other 85 percent simply do their job and go home," found that this same breakdown carried over to the type bike being used.

> The 15ers ride fixed gear bikes without brakes. Some of them ride with brakes, it all depends on who has had too many close calls ... some of them also ride single speed free-wheels, but I would

say the majority of the people I met rode fixed gear bikes. The 85 percent largely ride whatever bikes they can find—mountain bikes, old road bikes, anything. I think this illustrates the difference between the two. One cares deeply about their bikes, the other just has whatever bike they can find.[36]

This dichotomy punches a hole in the whole "bricolage" theory behind the bike messenger's fixie. For the 15 percent, it was a conscious choice, but not an act of technological rebellion. By the mid–1980s, the urban velo-cognoscenti had been riding single-speeds of various types for thirty years. On the other hand, for the 85 percent, it was simply a case of removing all that fragile, complicated stuff before it broke, or more typically, strip it off *as it broke* and get the bike up and running again with the tools, bailing wire and duct tape you could scrounge up at the closest hardware store. Hopefully, before the whole machine broke down for good into a heap of scrap metal, you'd have a new job. That may have come close to what Levi-Strauss originally meant by bricolage, but it probably lacked the necessary ambiance required by the postmodernists.

Cycling historian Jonathin Bodkin describes the popularity of fixes, both among messengers and, starting about 2005, with general cycle consumers, as a fad, simply the latest in a long line of cycling crazes: the Stingray boom; the 10-speed fad; the mountain bike craze of the late '80s and '90s. "Aesthetics play a large role in the allure of fixed gear bikes for many riders," he notes. "All one needs to do is find a cheap bike, strip off unneeded parts, and put on a track hub…. When cycling magazines discuss $1,500 bicycles as 'entry-level,' a custom fixed-gear fashioned from a garage-sale bike frame offers freedom at minimal cost."[37]

In "marked contrast" to the prior Stingray, 10-speed and mountain bike crazes,

the fixed gear fad does not stem from the lower classes, or the less wealthy, emulating the upper classes. Instead, the ideal of this fashion craze is the bicycle messenger: a lowly, overworked, underpaid servant to the large corporations who require his services. In this case the attraction is not the freedom of the upper class that comes with wealth and leisure. It is instead the minimalist messenger … he eschews all the trappings of wealth and achieves enlightenment by reducing his lifestyle to the basic necessities, which in this case, often include a Brooks saddle and a carbon fiber fork.[38]

Charles Jencks answered Banham's critique with a sharp reply in *New Society*. "Does Banham really believe that the managers are really creating new combinations of off-the-rack parts every day?" Banham's answer was yes. Look at the Mustang. Before becoming John Kennedy's Secretary of Defense, Ford chief Robert McNamara had ordered the design of a small, simple car for basic transport, the Ford Falcon. Introduced in 1958, it bombed. After McNamara left, Ford recycled the chassis, engine, and rear suspension, gave it three new bodies, re-worked the front suspension so it could handle an optional V-8 and developed the Mustang. Banham saw bricolage. Jencks scoffed. The project cost millions and took three years. That wasn't bricolage; that was technostructure.[39]

So far, the debate was close to a tie. Then Jencks threw in a ringer: "We should all peddle with Banham, but even his mini-cycle is Purist, integrated, and made by specialized parts. Look downwards, sirrah, you are sitting on the truth every day!"[40] The Mustang was debatable, but it was hard to argue that the Moulton was a mere re-arrangement of existing parts, especially after you had described it as the biggest thing "since the invention of safety bicycle." Beaten, not so much by facts as by style, by the ability to take on and absorb, marshmallow-style, the assault of any rhetorical argument, Banham folded his tent. He later rued that "the life of an *enfant terrible* is tough—I know, I used to be one. Quite apart from the constant pressure to be more and more terrible (or infantile) there is the certainty of being over-run from behind by younger and more terrible *enfants* before you've had the time to enjoy the role."[41]

The new *enfants* had finally caught up.

II

While the demographics, economics and politics of bicycle transportation may have ebbed and flowed back-and-forth across the years, one thing has remained fundamentally the same: the technology. At least the technology of the cycling masses. The bicycles being used in England and America when Alex Moulton introduced his original F-frame were little different from the Flying Pigeons in China or the 1930s Raleigh roadsters made in Nigeria. On Planet Bike, time may as well have stopped the day before the Earl's Court Cycle Show of November 1962. And in the decades since, they have evolved only in a few technical details, and mostly for the benefit of the cognoscente.

"It may seem odd at first, this failure of the progressive bicycle to link up with the betterment of the labouring poor," Banham wrote in 1978, but as he said more than a decade earlier about his hometown, "the working class don't ride bicycles anymore." Instead, they had their Poplars, Cortinas, and Minis. But something else now struck Banham:

> Even odder, at first blush, must seem the failure of the seemingly radical bicycle to link up with political radicalism. After all, it was the alternative community who pioneered the revival of the bike, but it has always avoided technologically radical models—from the Provos communal White Bikes in Amsterdam to the Cong's jungle-cargo carriers in Vietnam, it has always been Old Diamond-Frame that has borne the burden, the message and the imagery of the committed left.... However, reflection will show that the radical position is not really paradoxical. A world-wide "People's" movement needs a bicycle that is as common as the great unwashed themselves.[42]

Banham may have just been smarting from his rough joust with the Situationists and the radical enviros at Aspen '70, but he did have a point. As early as 1965, Harrison Salisbury of the *New York Times* told Senator William Fulbright that without their tens of thousands of bicycle porters, the North Vietnamese Army, led by General Vo Nguygen Giap, "would have to get out of the war." Modified with basket panniers, shoulder slings, carrying handles, and a push-tube lashed across the handlebars, the bicycles functioned more as single-track carts than bicycles. The average load for a bike porter was 120 kilos (260 pounds); this could go as high as 225 kilos (500 pounds) in an emergency.[43]

The roadsters also had one additional advantage: almost all were equipped with dependable, efficient, front wheel dynohubs. Fill a frame with explosives, run the wiring to the dynohub with an arming switch in between, leave it partially blocking a sidewalk or building entry, close the switch and get away—fast. The dynohub was so good that turning it a quarter of a turn was enough to set off the detonator.[44]

But even sixty years after the 1962 cycle show, you still needed a trained eye to see what had changed in the bike world. True, they looked different, but like the giraffe's neck or the peacock's tail, what had changed was that the bikes on the periphery, the extreme bicycles, had just gotten even more extreme. However, they were now being treated as if they were the norm; the everyday bike for the everyday rider. They were taller, or thinner, or leaner, or lighter, or merely more outrageous.

Grant Petersen, former president of Bridgestone Bicycles (USA), notes that while bicycle racing may have been responsible for some technical improvements, the crossover from racing to utility bikes ended years ago: "racing may have been responsible for some improvements up through about the 1950s, maybe even the 1960s, but soon after the practical improvements stopped, [and] the impractical refinements kicked in." As a result, "the modern race bike has become too specialized, a one-trick pony, a disposable, fragile flyweight that isn't suitable for anybody who doesn't race. Yet it has become the standard road bike of the day, even for non-racers."[45]

What Petersen is describing isn't innovation, it's hypertrophy—the overdevelopment of one or two details to a Rococo level of refinement good for only one thing—or for no good purpose at all. The traditional road bike is probably the worst possible technology upon which to base a general-purpose bicycle. "If car design had followed the same [design] path," says Leigh F. Wade, "we would be driving Model A cars with titanium frames and the hand crank would be carbon fibre."[46] A European would scoff at the idea of riding a "tall Peugeot 10-speed" to school every day. It wasn't made for the weather, for the potholes, to be left chained outside all day. But, Tony Hadland notes,

> whilst what Petersen says is true of some European countries, 10-speeds were exactly what many British males who still used a bike for commuting or local trips took to using in the late 60s until the coming of the mountain bike. It was ludicrous—the riding position was daft, the braking poor, the gear shifting awful and the range often no more than what a 3-speed hub offered. But a pseudo racer was seen as "manly." An otherwise liberal and well educated female friend once described the Raleigh 20 to me as being "a poof's bike".... Small wheelers generally were much less likely to be ridden by the working class. In their eyes, real men didn't ride "pushbikes," and if they did, the bike had at least to be a (pseudo) racer.[47]

Cycling author John S. Allen says "carbon fiber is not a forgiving material. It's easy, for example, to damage a carbon-fiber seatpost simply by tightening the binder bolt ... the torque wrench is mandatory with carbon fiber. Calibrated hands used to be good enough." The modern team-issue racing bike, according to Peterson, will have an average lifespan of eight months and 3,000 miles, will never go more than 300 miles without being completely disassembled for service, and if it has a frame made of one-piece ("laid out") carbon fiber, will likely be retired if it sustains a single major crash. And finally: "the average stage racer weighs 138 pounds. And you?"[48]

Much the same is true for the mountain bike. By 2000, it was impossible to find one without front and rear suspensions. They make for great off-road riding, but aren't all that helpful for city riding. However, if you want a mountain bike, you have to buy them. That means you either have to spend a lot of money for good suspension systems you don't need, or buy an inexpensive mountain bike with cheap ersatz suspensions that don't work well and break a lot. It's impossible to find a mountain bike with high-end components and no suspension system.

The point that Allen, Hadland and Petersen are making isn't that these bicycles are bad; it's that they had become either/or extremes of a one-size-fits-all world in the Anglo-American market. A large, specialized bike shop may have more than one commuter or city bike to choose from in a range of prices, but all that most consumers are probably going to find in a sporting goods mega-retailer or department store outside of a racing bike or full-out ATB is a low-price one-speed cruiser and (maybe) a mediocre 3- or 5-speed roadster.

On the other hand, highly innovative designs that initially appear odd, but that make sense, have been proposed often, but just as often rejected, mostly by component makers who do not want to tool up for new products that serve only one category of bicycle, the compact commuter bike.[49] Mike Burrows, the bicycle developer who invented Chris Boardman's Lotus carbon-fiber "superbike" to take on the one-hour record in 1992, was asked three years later by Jim McGurn, editor of the British magazine *New Cyclist* to illustrate an ideal city bike. Instead, Burrows built the "Amsterdam." A monotube, 20-inch shopper, with monoblades both front and rear. Monoblades (one-sided forks) with cantilevered hubs were a trademark of Burrows's advanced recumbents, but in this case, were included for a very practical reason—so you could fix a flat without having to unbolt the wheel.

However, on his advanced recumbents, Burrows had radically reengineered the standard rear freehub to place the gear cluster on one side of the monoblade and the wheelhub itself on the other. This helped keep the centerline of the wheel reasonably

International delivery service DHL showed off this prototype package delivery cycle at the Voyages à Vélo exhibit in Paris, summer 2011. Intended for use in congested northern European cities, the layout is known generically as a "Long John," and has been around for almost a century. However, new materials and production methods allow its weight to be cut almost in half. It's not known to what extent DHL put them into service, but several cyclemakers now specialize in such cargo machines (photograph by Nora Quinlan).

close to the centerline of the bike. On the Amsterdam, he did the same, but because the rear stay was an extension of the monotube beam, and because the seven-speed internal rear hub was itself rather fat, the centerline of the rear wheel was offset 6 cm (2.25 in) from the centerline of the bike.

Reportedly, it felt odd at first, but after a few minutes rode about the same as a normal bike. But it did look strange. Over the next five years, Burrows worked as a consultant for Giant Bicycles and produced several more prototypes, less radical, but still with front and rear monoblades. Almost always, the idea never got farther than the component makers.

From these efforts, Burrows became convinced that the ability to fold front-to-back was less important to storability than the ability to become "thin." He then developed the "2D," a monotube, 20-inch bike with a carbon fiber chaincase that doubled as the rear monoblade. It had folding cranks and handlebars. It was a brilliant concept, but could not be produced by hand for a profit, and only four were made, the last one going to Tony Hadland. After 2003 Burrows turned almost exclusively to recumbent superbikes.[50]

In 1971 Banham complained that "The old Safety norm—a classic not to say a platonic ideal—of industrial design, emerged as long ago as the early [nineteen] nineties and like a spectacularly successful natural species, it overran all competitors and bred out all deviants. For three-quarters of a century, bike design was so stabilized it was practically fossilized.... North American bike manufacturers were even more somnolent than British ones."[51] Forty years later, that was still true. There was again this same stabilization, the same narrow mainstream of choice, only now it was, from a technical standpoint, survival of the least fit. Spindly road racing bikes and hyper-complex mountain bikes had become the dominant duopoly only because they were easiest to sell.

Yvonne Rix, Raleigh's marketing director in the 1990s, put it simply: "we're a marketing company." Ultimately, as both British and American bicycle manufacturers discovered, marketing bicycles that had (or at least appeared to have) the same technological features as team-issued equipment meant outsourcing to China, both because of lower labor costs and because of lax Chinese intellectual property laws.[52]

The market has become inverted: the bikes best suited to general daily use actually require more knowledge and sophistication to find and evaluate, while the most accessible bikes, available in the widest variety and in the greatest range of price and quality, are all targeted at two narrow functions, neither of which has any relationship to what the majority of purchasers actually need.

It started to occur to people: there were perfectly good older bicycles, such as the Raleigh Twenty or Dawes 500 around, in the tens of thousands. They were still delivering good service because their basic design was so well thought out, and because they were so well built. They cost a pittance. You could buy them at yard sales and in thrift shops. For decades, sports riders, even those who were not super-good mechanics, had been taking older ten-speeds and swapping out cheaper steel parts like handlebars, rims and seat posts and substituting lighter alloy components. Why not do the same with a shopper and make a compact commuting bike?

Heck, if you got really ambitious, you could probably modify almost everything down to the frame. Look at what hot-rod car guys had been doing forever with Chevy V-8s and Volkswagen Beetles. How hard could it be?

10

Clip-On; Plug-In; Burn-Out

> The Splendor (and misery) of writing for dailies, weeklies or even monthlies, is that one can address current problems currently, and leave posterity to wait for the hardbacks and Ph.D. dissertations.... The misery (and splendor) of such writing, when it is exactly on target, is to be incomprehensible by the time the next issue comes out—the splendor comes, if at all, years and years later, when some flip, throw-away, smarty pants, look-at-me paragraph proves to distill the essence of an epoch far better than subsequent scholarly studies ever will.
> —Reyner Banham (1981)[1]

I

By the early 1990s, the new generation of compact, folding and separable bicycles was an established fact. In Britain, the Folding Society was formed in 1993. Within two years it had a thousand members, and a year later held its first Folder Forum in Weymouth. The organization went international in 2001, dispensing with the need for dues by putting its newsletter on-line. An American version took place in May 2003 when Michael McGettigan, a Philadelphia bike shop owner, organized a Small Wheel Festival to celebrate the 40th anniversary of the original Moulton. Most of the best-known names exhibited. "One bike can't do it all," McGettigan later explained. "If you are near a bike path, you are going to sell lightweight Dahons and Bike Fridays—bikes geared to a higher performance level. If you are next to a major transit network, like we are, you are going to sell a ton of Bromptons because they are so easy to fold. Small wheelers sell in urban environments. Full size folders appeal to suburbanites who are less interested in commuting and more interested in taking their bike with them overseas."[2]

In 1987 Andrew Ritchie's Brompton won the Best Product at the Cyclex show in London. Twenty-five years later, now in a state-of-the-art 46,000 square foot factory, the firm he founded was producing over 50,000 a year. Dahon was the 27th most popular brand in the U.S. among all bikes, not just compact or folding bikes.[3]

The market was still highly bifurcated. On one hand there was the very expensive Moulton AM, running between $2000 and $10,000, and the less expensive, but still pricey Brompton and Bike Friday, which cost about the same as a high-end sports bike. At the other extreme there was the Bickerton, which cost about $450 and the original

Dahon, by the 1990s in its twilight, at $300 to $400. There were also many inexpensive U-frame folding bicycles, predominantly from Asia, but they were not extensively imported into the United States due to import tariffs and problems with product safety regulations. Also, the U-frames were more properly considered compact bicycles, as they were not much smaller folded than open, just shorter.[4]

Then a third alternative cropped up on-line courtesy of a Newton, Massachusetts, repair technician and cycling writer named Sheldon Brown: take a stock Raleigh Twenty shopper, change it a little, and make it a decent medium-range commuting bike, or change it a whole lot, and make a Moulton-style pocket-rocket out it for less than half the price of the original. Some of his creations looked like they belonged in hot-rod magazines. In fact, a couple were in magazines.

Sheldon Brown (1944–2008) was an intensely interesting character. He is probably the only modern bicycle mechanic (certainly the only American) to have a 2000-word obituary in the *Times of London*. His father, George M. Brown, was an electronics engineer who worked for General Electric and the New York Central Railroad. His mother Madalyn was the daughter of a widow, Helen Joyce, who was a successful Boston real estate investor up to the Great Depression. After losing most of her properties in the crash of 1929, Helen Joyce moved to Hollywood to become an actress. That apparently did not work out, as the family returned east and Helen started an antiques store. Madalyn Joyce was working at General Electric in Schenectady, New York during World War II when she met George Brown, whom she married in 1943.[5]

George Brown was an enthusiastic light-plane pilot. He owned a 2-seat Ercoupe, one of the first all-metal light aircraft built, and later a Piper Cub. A member of the Massachusetts Civil Air Patrol, he was asked to fly a search and rescue mission after a fishing boat had been struck by an ocean liner. The airfield was substandard and wind and weather conditions were not good. He crashed on take-off and was killed. He was the only occupant of the plane. Sheldon was then nine years old.

After George's death, Madalyn and her three children moved briefly to Schenectady, then to Marblehead, where Helen Joyce lived and was running her antique shop. In high school, Sheldon started scavenging bikes and parts from the town dump, building up bicycles and selling them. He entered college, but didn't stay

Sheldon Brown on his Moulton Stowaway, c. 1970 (courtesy Dr. Harriet Fell/estate of Sheldon Brown).

long. He worked selling shoes and stereo equipment, then drove a cab. In the 1970s he co-founded the Boston Bicycle Repair Collective with four others, including Stan Kaplan, inventor of the Kryptonite brand bicycle U-lok. "The next generation of people after the founders were Maoists," recalls author John S. Allen. "In the end [about 1981], they purged the founders, including Sheldon. He was very bitter about it. The founders' concept was to establish a fix-it-yourself bicycle shop, where people could learn how to work on their bicycles, not lead the charge in a Velorution." (Fortunately, the shop, in somewhat altered form, eventually survived the revolution.)[6] Brown became a camera repairman, and always remained a proficient amateur (and occasionally freelance) photographer.

In the 1970s, Brown began writing for *Bike World* magazine, then continued for *Bicycling* after it bought *Bike World* in 1979. He also wrote for the trade magazine *American Bicyclist*, and penned the "Mechanical Advantage" column for *Adventure Cyclist*, the members' magazine of the Adventure Cycling Association (formerly BikeCentennial) from 1997 to 2007. In 1990 he was hired at Harris Cyclery in Newton, Massachusetts just outside Boston. He became a regular contributor to the then-new world of internet newsgroups. In 1995 Aaron Harris let him set up a website for the shop. It was initially intended to sell specialty parts, such as small parts for older Sturmey-Archer internal hubs. But over time it morphed into one of the earliest, and even now, years after his death, still one of the most popular, sites on the technical aspects of cycling, especially regarding older or less common bikes. At the time of his death in 2008 it was recording half a million visitors a month, and was answering an average of 250 emails a day.

At one point, Brown owned two of the original F-frame Moultons (an early Deluxe and one of the break-apart Stowaways), and later a Raleigh-Moulton Mk. III. The Deluxe had the curved rear forks, and Brown recalled that it was "quite easy" to twist them out of alignment with hard pedaling because they were cantilevered out so far from the bumper plate. The bike had a "billiard cue" paint job, indicating that it was a very early unit, not one of the bicycles shipped to Huffy's Azusa, California facility for assembly in 1964 and 1965.[7]

As was true for most of the F-frame Moultons, his other bike, the Stowaway, came equipped mostly with steel components meant for roadsters, and Brown replaced them with aluminum components, taking off at least four or five pounds, especially as he dispensed with the Sturmey-Archer rear hub (one of the dreaded S3B combination three-speed and compact hub brake units designed for the RSW 16), turning it into a single-speed. He still had it in his collection at the time of his death.

He recounted that the Moulton Mk. III was the "first new bicycle I ever bought," but that "I didn't keep it stock very long." Like Banham, he found it stiffer than the F-frame Stowaway, but heavy and thought that the rear suspension was too soft. He cured the suspension by putting an adjustable band clamp around the squishball to stiffen it. Like the Stowaway, Raleigh had equipped the Mk. III with the S3B, which Brown found "pitiful," so he brazed on his own brake bosses and added a dual-pivot centerpull rear brake.[8]

His comments on his three Moultons are fairly perfunctory, but when it comes to his Raleigh Twentys, they fairly bubble over with enthusiasm. "I've had a love affair with these bikes," he admits. "I bought my first one in 1973 ... and did many modifications to it over the years."⁹ The early changes were fairly modest: platform pedals and toe clips meant for touring bikes, a leather Brooks saddle, and the substitution of a Sturmey-Archer 5-speed rear hub for the original Sturmey-Archer "AW" (3-speed wide ratio) hub. "For a while I had BMX tires on it, and I did quite a lot of off-road riding on this bike before real mountain bikes became available." He also noted that "one nifty use for such a bike is hitch-hiking."

> Back when I was a starving hippie, I used this bike to visit friends on Cape Cod, during the winter off season. I hopped on the bike at my commune in Allston (an outlying section of Boston) and rode in, perhaps 6–8 miles to the main north-south highway that runs through Boston, and up an on-ramp. I then folded the bike and stuck out my thumb. A hitch-hiker with a crumpled bike next to him looks less threatening than a normal hitch-hiker and I got a ride, almost immediately, all the way down route 6, about 10 miles from my destination. There I was, on a dark November night, on a deserted two-lane in Cape Cod, with nobody going by. If I had been purely hitch-hiking, I'd've been S.O.L., but since I had my trusty Twenty, I just unfolded it, turned on my Elite headlight and had a pleasant ride to my friend's house.¹⁰

Later, Brown began to modify the bicycle more extensively: "it acquired aluminum rims, Cinelli handlebars and stem, a Campagnolo Nuovo Record crankset, Phil Wood bottom bracket, and other goodies." These were all top-end components, most intended for racing bicycles. Unlike Brown's earlier modifications, adapting the Twenty to accept some of these parts required a fairly extensive reworking of the frame, because the Twenty used a propriety Raleigh 26-tpi (threads per inch) threading system, not the standard 24-tpi "British" (and Japanese) threading, nor the 25.4-tpi (1mm) "Italian" threading. Moreover, some parts of the frame have non-standard dimensions. The bottom bracket, which holds the crank axle and bearings, is 76 millimeters wide, whereas on standard European bicycles it is 68 mm wide, and on many mountain bikes and some tandems, 73 mm.

Thus, it is necessary to either mill down the frame's bottom bracket shell or buy an extra-long sealed axle and bearing unit that is quite expensive. Aaron Gross, at Aaron's Bicycle Repair in Seattle, recalls that "back in the 1980s when I worked with Andy Newlands and later in the 1990s with Dan Winn, we used to fill the bottom bracket threads with brass and re-tap to English [threading]," but John S. Allen adds that "this was before cartridge, Shimano-type bottom brackets and long axles were available." One maker, Phil Wood, also manufacturers a bottom bracket cartridge and set of lock rings that will fit an unmodified Twenty bottom bracket. It is fairly expensive, but competitive when compared with the cost of paying a framebuilder to modify the bottom bracket.¹¹

The headset, the two-part (upper and lower) set of matched bearings that permits the fork and handlebars to turn within the frame's head tube, is also very non-standard. Originally, it was almost impossible to replace the stock setup with a touring bike headset without replacing the front fork. In fact, to accommodate the quick release feature on the stock bike, it doesn't even have an upper set of bearings; instead there is a nylon bushing that fits into the head tube of the frame.

But recently, there have emerged an amazing plethora of innovative solutions to work around the various problems. Many involve the use of so-called "threadless" headsets first developed for mountain bikes, then transferred to road bikes, because of their increasing use of carbon-fiber front forks, which cannot be threaded. On the other hand, some modifiers go in the other direction, keeping things super-simple by replacing the factory caged ball bearings in the lower cup with loose balls, and then just greasing the bushing well. The lower set of bearings carries almost all the weight of the bike and rider, so tuning it up by making sure that the race is properly aligned and isn't pitted, and by using loose balls, which doubles the rolling surface area, makes a big difference.

Brown became convinced that taking the extra trouble was worth it because "a stock Raleigh Twenty offers performance comparable to that of a 26-inch wheel 3-speed 'sports' bike, but due to its robust, well designed frame, it lends itself to being 'hopped up' in performance by upgrading components. With suitable equipment, it can approach the performance of a Bike Friday at a much lower price."[12] Brown's friend, cycling author John S. Allen, concurs:

> The very nicest thing about the Twenty is, as I think Sheldon was the first to state publicly, the frame is nice and stiff, requiring no compromises in riding style. That is what makes it worth all the trouble to customize. Standing on the pedals does not result in any more flex than with the diamond frame. This bicycle can be ridden at high speed with confidence.[13]

Allen adds that the Twenty "is a workhorse. Its small wheels entail some slight increase in rolling resistance, but it has a normal riding position and good handling qualities. With the small rear wheel placed well behind the seat tube, the wheelbase is conventional, and a large amount of baggage can fit on top of the rear rack."[14]

Fortunately, the easiest components to swap out are those that make the biggest difference in performance. These include the pedals, rims, saddle, handlebar, stem and seatpost. Replacing them with aluminum alloy components will cut the stock bike's 35 pounds down by four or five pounds. Just replacing all the hand-tightening levers on the folding version with regular bolts and clamps saves two pounds. Everyone agrees that the change that makes the single biggest difference in performance is converting the stock bike's steel rims to aluminum alloy, both because it saves weight and because it improves its rather marginal brake performance, given the loooong brake arms.

Sheldon Brown's "Twenty No. 1" underwent many permutations. After his daughter Tova was born, it sprouted a baby seat for awhile. It was modified as a commuting bike for his wife, Harriet Fell, a professor at Northeastern University:

> It was a bit of a Q-Ship. Harriet had a lot of fun blowing off posers on thousand-dollar bikes. When you're a poser with a thousand-dollar bike and you pass a middle-aged woman on a small-wheel folding bike with a baby seat on it, there's precious little glory (especially as the only chance these worthies had to pass her was while she was waiting for a red light to change.) On the other hand, once the light changes, and the middle-aged woman on a small-wheel-folder-with-a-baby-seat catches and passes you, you know you've been *passed*![15]

Banham had run into the same thing back in 1963 when he started riding his F-frame around London:

> Members of the outgoing big-wheel culture clearly recognize the Moulton as a threat of some sort: the first real hairy-knee'd cyclist to see me on mine promptly reacted with a tremendous virility bit, standing up in his toe clips, thrashing the pedals up and down with his calf muscles coming out like reef-knots and the frame of the bike whipping this way and that as he smoked off along North Carriage Drive. Happy days, mate![16]

Brown admits that "in a moment of weakness" he sold off Twenty No. 1 for $300. However, by this time he had gone on to at least two other projects. Twenty No. 2 was a non-folder (something of a rarity in the U.S.; it was only listed in the Raleigh in America

The manufacturing floor at Brompton's new factory, March 2017 (photograph by Nora Quinlan).

catalog for two years, 1969–70). He turned it into a dedicated speed machine with dual disc wheels. He found riding it in a crosswind with the front disc "an interesting experience." Twenty No. 3 was, in his words, "my piece-de-résistance," an ultra-light, ultra-fast, single-speed, fixed-gear. However, as he began to suffer health problems in his later years, he replaced the single-speed rear hub with a Sturmey-Archer 8-speed hub and added a BMX suspension front fork so he could continue to use the bike's then-existing high-pressure rim/tire combination, which was fast, but very harsh.

II

Brown and Allen's "hop-up" work was an example of what Banham called "clip-on" or "plug-in" technology. Leaving aside Archigram and their gigantic "plug-in" city, more pragmatic applications included Smithson's "House of the Future" (1956) and "Appliance Houses" (1957–58); Fuller's grain silo-based "Dymaxion Deployment Units" (early 1940s); the Eames Case Study Houses #8 and #9 (1946–48); and the British CLASP school project (1960). "I first remember using the phrase 'clip-on' in connection with architecture," Banham later recalled. "It must have been in circulation for at least a year before I used it in print in the *Architectural Review* in February 1960, and its meaning was, to some extent, fixed by conversational usage before then…. The epitome of the clip-on concept, at that time, was the outboard motor ... given an Evinrude or a Johnson Seahorse, you could convert practically any floating object into a navigable vessel."[17]

The Smithson homes put all of the utilities in a single integrated unit, which was replaced completely as its stoves, dishwashers and washing machines started to wear out or became obsolete. The Eames Case Study Houses and CLASP structures were assembled from standardized components in the anticipation that some parts of a building (doors, windows, roof) wore out faster than others. Why repair? Unbolt, scrap and replace.[18] In fact, Banham mused, "when your house contains such a complex of piping, flues, ducts, wires, lights, inlets, outlets, ovens, sinks, refuse disposers, hi-fi- reverberators, antennae, conduits, freezers, heaters—when it contains so many services that the hardware could stand up by itself without any assistance from the house, why have a house to hold it up?" And his article contained a drawing of an "unhouse"—a home comprised of nothing but mechanical, electrical, and HVAC systems, exposed, out in the open, needing nothing but a curtain wall, such as one of Bucky Fuller's geodesic domes or even an air-supported bubble, to keep the rain out and the warm air in.[19]

"Too much should not be made of this distinction between extreme forms of the two concepts" notes Anthony Vidler. "Admittedly, Archigram had reversed the idea of clip-on by adopting that of plug-in, but Banham was ready to fold this concept into his theory ... the aesthetic tradition overruns niceties of mechanical discrimination."[20] Translation: clip-on and plug-in are just labels for the same idea: (1) a basic platform; (2) holding interchangeable modules; (3) that are easy to replace as they become worn or a different function is required. Today, the idea is almost banal. In the early 1950s it was really revolutionary, especially when it came to buildings. "Modular structures" were still usually thought of as "trailer houses."

In addition, Banham was not, by this time, thinking only about clip-on or plug-in as an architectural concept, but also as a way of increasing the usefulness of consumer products, as his reference to outboard motors suggests. He had often taken advantage of his "Not Quite Architecture" column in *Architects' Journal* to discuss his first love, car culture, and in July 1961, he extolled the virtues of the American Warshawsky's mail-order hot-rod car-parts catalog. It was full of "odds and sundries to mop up the differences between what Detroit supplies and what the heart desires." And in another column, he referred to the Eames Case Study houses and CLASP School as examples of the "hot rod method" of architecture.[21]

Most of the components in catalogs such as Warshawsky's and J. C. Whitney's weren't for exotic or expensive cars; the typical customer was a shade-tree mechanic just trying to keep the family station wagon going. But in addition to factory-spec replacement parts, both offered speed kits, again not for exotic machinery, but for the rugged, lower-end cars acquired and modified by the young and not-so-rich: Chevys, Dodges, Ford V-8s, and later VW Bugs. They could be bought cheap, and built-up to suit the owner's style: smooth-gliding boulevardier, screaming quarter-miler, or in the case of the VW, cheap sports car.

This was precisely the kind of thing that Banham had been talking about when he called the Warshawsky's and J. C. Whitney's automotive catalogs "a guide to the first folk-art of the do-it-yourself epoch," evidence of a "live technological culture." With hot rods, "the vehicles are largely assembled from selections permuted from a wide range of ready-made components, standardized (but highly specialized) accessories and ingenious bolt-on or drop-in adaptors." As opposed to flat-out racing cars embodying "absolute originality of design as well as absolute craftsmanship":

> We aren't all endowed with absolute originality, we have different talents differently arranged: a situation like this enables you to concentrate on your areas of talent and get the rest done by experts. You can make up as much of an original life style as you want and conform where you feel the need.[22]

Unlike Levi-Strauss's and Jencks's bricolage, this was not some form of noble savagery; nobody was, to use today's language, hacking the system; nobody was pretending they were beating the man at his own game. That is, nobody was asserting a rejection of the mores and boundaries of modern society. Instead, these were the new happy warriors of the Second Machine Age (the first coinciding with the era of the high-wheeled bicycle); taking the best mass-produced goods, then using them in ways their makers never had intended to kit-bash products that approached the semi-customized, high-technology products of the craftbuilders.

Moreover, Brown and his cohorts were not stripping *down*, they were building *up*. Unlike the bricoleur bicycle messengers on their fixies, they weren't seeking a form of urban primitivism through velo-minimalism. To build up a fast commuting bike or pocket-rocket out of a shopper was to create something more than was initially there at the start. It was an act of creation, not negation. It also took more persistence, patience and a tad more skill than simply stripping down a bike.

Moreover—and probably most importantly—the hot-rodders had no disdain for the craftbuilders. If you were to ask Brown, Allen or their colleagues, they would probably express nothing but admiration for Moulton, Brompton's Andrew Ritchie, or Bike Friday's Hans Scholz. (In fact, Allen has a Bike Friday New World Tourist, and admits "The Bike Friday in its hard suitcase/trailer is the real deal for air travel.") They might have suggested some changes, but they would not have rejected their basic premise, that there is always a better bicycle out there, that the technological state of the art is never played out. Brown started as a bike-building kit-basher because that's all he could afford, and in the case of the Twenty, kept at it long after he could buy anything new because he liked to do modification projects and the Twenty was a favorite project platform.

David Lucsko, who has studied hot-rod culture, agrees with Banham that its essence is the purposeful alteration of stock machinery, but argues that his "speed catalog" technique is only one of three ways to approach hot-rodding. The others are to alter a machine's existing parts (or an entire system) for improved performance or appearance; or to replace parts or an entire system (such as a drivetrain) with used parts from another machine, typically by modifying the transplanted parts, the recipient machine, or both. These three strategies (upgrade, modify, transplant) are not mutually exclusive, and a single hot-rod will often incorporate all three.[23]

"Speed equipment was not always cheap or plentiful, nor was it ever available for every conceivable need," Lucsko observes. "By the middle of the 1930s, a number of established Midwestern speed equipment companies had gone out of business…. Although this gave a handful of entrepreneurs in Southern California an opportunity to break into the business … for many hot rodders it simply meant a dearth of relevant and affordable speed equipment."[24]

But there was another option: "parts cars," wrecked or worn out autos, sought not for their intrinsic value as automobiles, but for their mechanical guts. The Chevrolet Vega subcompact of the 1970s, for example, is usually considered a terrible car. But it had a front steering system that was uniquely self-contained and compatible with almost anything you could build. Thus, the steering column, rack-and-pinion assembly, and right and left steering linkages, as a unit, was eagerly sought after for hot rods and dune buggys. In another example, a well-kept secret was that the engine in the VW squareback and hatchback (and after 1969, the van), while outwardly identical to that in the Beetle, was actually a different model, and about 15 percent more powerful. Some enthusiasts risked a visit from the code enforcement bureau by storing parts cars in their back yard, but most hot rodders just went to a salvage yard.[25]

In 1990, Unocal (the Union Oil Company of California) announced plans to buy up 7,000 older, large-engined automobiles from the pre-emission-control era in the Los Angeles metro area and scrap them. The average price offered to owners was $700, so most of the cars were expected to be near the end of their useful life. Unocal's "cash for clunkers" program was part of Southern California's air quality management program, intended to offset Unocal's non-conforming smokestack emissions in the short-run until the company could refit or replace polluting refinery equipment.[26]

The firm was promptly bombarded with complaints from motoring clubs and enthusiast magazines. Its executives were flabbergasted. "We're not crushing classics," said a company spokesman. For seven hundred dollars each "there aren't going to be any Ferraris in this program." According to Lucsko, what Unocal "failed to realize was that even the least-desirable sedans of the 1960s and 1970s nonetheless were of value to 1990s enthusiasts, especially hot rodders and restorers, as sources of major drivetrain parts and other components."

Local zoning laws were slowly squeezing out junkyards. New American steel micromills had increased the demand for shredded cars, and shredding technology was better and cheaper. In addition, EPA rules now made the commercial harvesting of pre-emission-control power trains from scrapped cars a paperwork-intensive hassle. This had turned clunkers into a sort of rolling warehouse of parts for big iron aficionados, but without the awareness of their owners, who simply saw them as cheap, rugged cars that could be fixed in the driveway without fancy computer diagnostic equipment.

Unocal's argument that the purchase price offered was so low that no rational person would cash in a classic (or even a potential classic), was irrelevant. Those driving these cars didn't know they had something of value. In fact, it probably wasn't valuable. Yet. Only when that rusty, wheezy heap became sought after as a source of cannibalizeable parts would its role change. The important part, as far as the hot-rodders were concerned, was to keep them going as somebody else's cheap, basic transport until destiny called. The hot-rodders claimed it was a symbiotic pairing of car aficionados and shade-tree mechanics; the California Air Resources Board (CARB) called it parasitic, with poor clunker-owning families subsidizing the hobby of rich hot-rodders. Ultimately, only three states ever adopted ongoing "cash for clunkers" programs, and two dropped it within a few years.[27]

(Interestingly, Alex Moulton recalled late in life that with the exception of the motorcycle he had in his first year at Cambridge, "I never was much interested in the modifying or tuning of existing cars and engines. It needs to be done for the best performance and it gives so much interest and pleasure to many, but I was more interested in creating something new."[28])

"I know of some collectors who have these incredible boneyards of bike parts in their garages," says Florida bicycle shop owner Eric Hodges. Alan Issacson, the former Colorado Hon salesman, recounts the story of a friend he knew in college in the early 1980s. "His dream was to own a bike shop, a Schwinn shop. He had been

A modified Raleigh twenty at Gainesville-Hawthorne Trail, Gainesville, Florida, 2007 (photograph by the author).

through the Schwinn factory training program as a teenager and thought the sun rose and set on Schwinn." He made repeated offers to buy the town's Schwinn shop. "It was in bad shape, an obsolete location, the owner was old, tired." The owner sold and Issacson's friend moved it, expanded and modernized it. "It was the best place in town during an era when you couldn't say that often about Schwinn stores."

But Schwinn went bankrupt in 1992 and was sold to the Scott Sports Group. Scott couldn't make it work and bankrupted it again 2001. It was bought by Pacific Cycles, a leading importer of department store bikes, who announced that they were pulling out of bike stores in favor of big boxes. Schwinn became just another department store brand. "He was on the point of bankruptcy himself. We went down into the basement of his shop. It was filled to the brim with valuable new-old stock he had moved over from the old place and sorted. Stingray handbars, Solo Polo saddles, grips and gear shifts. All three versions of the Bendix kick-back rear hub. Dragster slickx rear tires for Stingrays. New Departure Model 'D' coaster brakes for old cruisers and the overhaul kits for them, in Schwinn boxes. I told him, 'Man, you've got to get a price list out for this, or hire a kid from the university to set up a web site. You're sitting on ten or twenty thousand of the quickest dollars you'll ever earn. He wouldn't do it. He couldn't part with the stuff. He chose bankruptcy.'"[29]

III

When asked if he would still have embarked on his Raleigh Twenty project had there been the wide variety of compact and folding bicycles in the '70s and early '80s as today, John S. Allen replies, "Yes, I like to build up my own bicycles. And at the time,

I had more time than money. I am a parts scrounger as Sheldon was and so I can find most parts for my bikes at flea markets, as closeouts, etc."[30]

Allen Issacson has two, one stock ("Greenie") and one heavily modified ("Root Beer"), which he has spent "twenty five years more or less continuously tinkering with." Asked the same question, he gives a very different answer: "No, if I could have gotten my hands on a Brompton or Bike Friday in 1991, I probably wouldn't have gotten the second one, Root Beer. In fact, if I had bought a Brompton or Bike Friday, I probably wouldn't have bought anything since then." He says he was motivated to hop up his Twenty by what he calls a "very specific problem": the need for a bike that could be ridden for 30 or 40 miles "but could be thrown in one of our (relatively small) cars for a trip on a moment's notice and not get banged up." In 1991, "such a bike didn't exist at a price I could afford. The original Dahon had the size, but not the range. The Moulton had the capabilities, but it was very expensive, and very few dealers in the U.S. carried it back then."[31]

John S. Allen says that "it reigns supreme as my daily utility bike, giving me a freedom in my commuting habits which I never knew before. I can ride to work without having to ask the question, 'Will I want to ride home tonight?'" British urban transportation specialist Peter Cox points to this as the most important characteristic of the folding bicycle. Overall, the transport potential of the bicycle is somewhat limited as a solo mode, but this changes if the bicycle is easily and universally transportable as a personal possession—that is, if it can be folded or broken down and carried along on a bus, train or taxi without the formalities of being "checked baggage" or placed in a distant, inconvenient or unguarded location. "The relevance of the use of folding cycles," he says, "is primarily in their multi-modal capabilities."[32]

In many European cities, commuters use two cheap roadsters—one ridden in the morning from home to the suburban transit station and the other picked up from the bike rack in front of the downtown station and ridden to work or school. "A familiar sight in cities where these patterns are common are ranks of apparently unused and rusting bicycles, which, being of minimal value in the first place, are abandoned without ceremony when movements change."[33] Not only do unclaimed and possibly abandoned bicycles create a form of mechanical littering, but after the Charlie Hebdo and Bataclan attacks of 2015 and 2016, pose a clear security risk. The bicycle bombs of 1960s Saigon suddenly do not seem such a distant memory anymore.

Allen notes that "the [Twenty] doesn't rate too well on compactness or cleanliness when folded. For one thing, "it folds the wrong way, to the left, leaving the chain exposed."[34] Issacson believes his modified Twenty works better as a break-apart than a folder, something he didn't discover until after he replaced the 3-speed rear hub with a Fichtel & Sachs Duomatic kickback 2-speed/coaster brake, which eliminated both the cables running to the rear of the bike. He says that's typical: "I haven't really felt compelled to change things simply for the sake of changing them," but hasn't hesitated to do so when there was a good reason. "Removing all the quick-release handles, taking off the chainguard, installing alloy rims, seatpost, stem, dropped bars; all that happened right away. Everything else was incremental. Something wears out; upgrade it. I had been looking for a Duomatic for years. It was standard equipment on the AM-2 Moulton for a long time but is now out of production. I came across one, new-old stock, and had a wheel built up around it. A gradual process."[35]

In 1979 Brown located a Twenty for Allen, his Boston-area friend. "He found my Twenty at a flea market and sold it to me for $35. It sported an extended seatpost made from the seatpost of another bicycle, a 3-foot high handlebar crowned by an automotive steering wheel, and gold paint from a spray can. I can only begin to imagine how it was used before Sheldon acquired it."[36] Allen extensively re-modified it for use as a performance commuter, installing dropped bars, alloy racing-bike cranks and other improvements.

> My ambitious project began with adding several brazed-on fittings (brake and water bottle bosses, mudguard eyelets, etc.) to the frame and then painting it with ugly-but-durable marine finish, aluminum alloy 20-inch motocross rims, a Sturmey-Archer five-speed hub, and an ordinary aluminum hub for the front wheel. A Stronglight 49D crankset with Phil Wood bottom bracket, Ava drop handlebars, Atom pedals, and a Wrights leather seat were other choices. I tinkered with a Pletscher rear carrier so it would sit as low as possible and attached the brake cables so they extend down, not up from the brake levers. This protects them from damage when the bike is transported.[37]

Allen was already a contributing author to the largest-circulation American magazine of the era, *Bicycling*, and in 1981 its publisher, Rodale Press, published a guidebook to commuting and urban riding he wrote. It featured several photos of Brown's and Allen's Twenties, and Allen recalls that he was asked about them "pretty often" in the years after the book was published. "The Twenty, when customized, works for much more than short range urban transport," Allen notes. "I have ridden mine on a half-century (a 50 mile club ride) and found it entirely acceptable. Sure, it was a bit slower than my touring bike, but it didn't have any glaring deficiencies."[38]

He adds:

> The bike's riding characteristics are excellent, especially in city traffic. Although the bike is slower than a large-frame bike, it does not *feel* slower. The drop handlebars and low center of gravity give a feeling of speed, as does the ease of steering of the small front wheel. The quick steering and short wheelbase give excellent maneuverability; and it has it has proved its worth on rough and soft surfaces. I've commuted through slush and snow of a Boston winter and felt secure doing it.[39]

Issacson takes exception with a little of this.

> Its handling is generally very good, but unless you change the front fork, put on a much straighter one like John did, it has an envelope you need to stay within. It *is* quite sluggish in turns at some speeds. The rear brake *is* inadequate. The bottom bracket *is* low, and if you install narrow profile tires and road bike cranks, it *is* easy to dig in a pedal on even shallow turns. No matter how much you alter it, you have to keep in the back of your mind that this is not what this machine was

A heavily modified Raleigh Twenty, used for club touring (photograph by the author).

designed for, and you are asking it to do things it wasn't meant to do. Under the wrong set of circumstances, it can be unforgiving[40]

Issacson bought Greenie, a 1969 folder, "in remarkably good condition" in Colorado in 1984 and took it with him when he moved to Florida. He still has it, and it is still stock, down to the original pump and lamp. "Only the tires, handlebar grips and that godawful mattress saddle have been changed." He found Root Beer, a 1975 folder, in Florida in 1991. "I was definitely influenced by seeing the photos of John Allen's bike in his book on bicycle commuting, that Twenty of his with the French racing cranks and the dropped bars. I had the word out that I was looking for one and a mechanic in a shop around the corner said that they had taken a couple in as trade-ins. I had my choice of root beer or green. I took root beer. They were $50 each. I should have snapped up both. I have been slowly upgrading it ever since."

Eric Hodges points out that while fewer Dawes Kingpins were imported into the United States, it is just as good, if not better, than the Twenty for use as the basis of a hop-up job. "Dawes, unlike Raleigh, didn't make their own components," he explains. "They had to outsource them from Phillips, GB, Nicklin, and so on. Because of that, things like the bottom bracket, headset and front hub are less proprietary, more standardized. People turn their noses up at Kingpins simply because they aren't Twentys, but I think the frame is better, at least on the non-folder, because the brazing wasn't automated." The hinge is not as good on the folder, though. Tony Hadland believes the shopper made by Royal Enfield and sold about the same time as the Twenty may be the best of the breed, but it was not produced in numbers close to the Twenty or Kingpin, nor was it ever imported into the United States.[41]

There was another reason mechanically inclined cyclists began taking older city bikes like the Raleigh Twenty and the Dawes 500-series and upgrading them: dissatisfaction with the standard products of the bicycle manufacturers. A $2000 Brompton or Bike Friday may be a nearly perfect city bike, but it was hardly something you would leave chained to a parking meter while you dashed in for a loaf of bread. As Issacson related, he needed something that he could throw in the back of a small car on short notice for road trips to various places in and around Florida. "Breaking the bike down makes it almost theft-proof. Undo two bolts, split it apart, and it doesn't look like a bicycle. It doesn't look like anything recognizable. Nobody would want to steal it."

IV

There is a lot of debate about how many Raleigh Twentys and Dawes Kingpins are still rolling. The Dawes KP500 Kingpin was introduced in 1964, followed later by the NP20 Newpin, the same bike with narrower tires. Folding and take-apart versions of the Newpin (NP20AH and NP20TA, respectively) were introduced in 1965. In the mid–1970s the Newpin name was dropped and the name changed to Folding Kingpin. The take-apart did not sell well, even in England, and was dropped by 1967. The hinge of the folder was not as strong as that of the Twenty and was criticized by a consumer research organization twice after testing, in 1972 and 1975, and as a result the folder may have been discontinued for a time. In 1984 a new model with a revised hinge was introduced. The non-folder was discontinued in 1986 and the folder was dropped in 1988. Total production of the Kingpin series is believed to be in the range of 600,000 to 750,000 units.[42]

The Twenty was introduced in the UK in 1968 under the Triumph logo. The Twenty and Twenty Folder were first listed in the Raleigh in America catalog in 1969. (In 1969 the RSW 16 was dropped from the Raleigh in America catalog and the Twenty was apparently intended as its replacement.) Both the Twenty and Twenty Folder appear in the 1970 USA catalog. From 1971 to 1976 only the Twenty Folder is listed. In 1977 no small-wheeled bicycles are included in the USA catalog. *Bicycling* magazine reported in March 1981 that Raleigh would "soon" begin importing a 16-inch folder from Japan into the U.S. The description matches that of the Bridgestone Picnica, which, in the end, was only imported by Raleigh into Australia starting in 1982. In 1984 Raleigh began importing a Nottingham-built U-frame folder called the Compact. Like the Twenty, it was available in both folding and non-folding versions in the UK, but only as a folder in the USA.[43]

Tony Hadland reports that in 1975, 140,000 Twentys were sold in the UK alone. The numbers are certainly much smaller in the U.S. It was only offered for eight years, and for six of these only in the relatively more expensive folding version. It was listed at $129 in 1976, its final year in the United States. (As a comparison, the nation's most popular entry-level imported 10-speed, the Peugeot U0–8, carried a list price of $139 at the time.)[44] Twentys were also made in New Zealand and South Africa. The production of Twentys at Nottingham was discontinued in 1983 after approximately sixteen years. It is believed that the Nottingham production was in the range of 900,000–1,400,000 units.

This brings up an interesting difference between Britain and North America. In North America, Raleigh Twentys have become a hobby onto themselves, and as a result they have become (at least those in good condition) surprisingly expensive—$300 to $500 is the norm. But in England they are so ubiquitous that the idea of agonizing over the decision whether to preserve or modify a Twenty is faintly ridiculous—rather like agonizing over the preservation of a Model T in 1927 or a Volkswagen van in 1970. On the other hand, John S. Allen, in Massachusetts, notes that "I keep seeing Twentys come out of the woodwork. The minister at my church gave me one only a few years ago which she had put aside shortly after purchase some thirty years ago. It was all stock and I customized it ... eBay and Craigslist increase the likelihood that a Twenty which had been disused will be sold to a person who will fix it up."[45] An overview of the bike photos submitted to the Raleigh Twenty website is about equally split between preservers and modifiers, but the submissions come from both Britain and North America, and it is not possible to tell with any specificity which is which.

While England has its Folding Society, with a website and regular "Origami Rides" around the country, there is nothing comparable in the United States, other than some informal websites. "The only thing I know of that's well organized for collectors, with newsletters, decals, parts exchanges, shows, retail stores for restored bikes and parts is for Stingrays," says shop owner Eric Hodges. At one time that used to be true for classic cruisers like the 1950s Schwinn Black Phantom, but prices got so high that the market was taken over by specialists, who only deal in those bikes and their parts.[46]

Hodges points out that Stingrays are collectable, but not especially rare. "There were so many Stingrays as opposed to only ten thousand or so Phantoms, or 25,000 Elgin Blackhawks from the 1930s, and if you are willing to settle for a plain, coaster brake Stingray in a stock color, you can pick one up in good shape for two or three hundred, about the price of a Twenty," Hodges says. "I see original Stingrays every month. I saw more Twentys a decade ago, before there were so many compact performance bikes."

Hodges adds:

> I think the rising price of used Twentys and the shrinking pool of available bikes in the States have a lot to do with it. Also, a big factor is the price and availability of replacement parts. Stock parts, I mean. They are much more available in the United States for Stingrays than Twentys. That's important. Once a bike becomes modified, it has a finite life. If a Stingray or a Twenty is kept showroom stock, it can always find a buyer, even if the type goes out of fashion. It doesn't have to be pristine, just a solid, well cared-for unit. But if you decide to turn your Stingray into a lowboy, or your Twenty into a pocket-rocket because you can't get stock parts, or because they are too expensive, it now has a very narrow repurchase audience. It may go from one friend to another, or maybe through a flea market or even ebay. But in general, it has largely ceased to be an article of commerce. The demand is too thin. People want to build these bikes, not buy them. So when the owner loses interest, moves, dies, whatever, they have a good chance of being lost or destroyed.[47]

Issacson, listening in on the conversation, asks, "then do you think it's a bad idea to hop up old bikes?" Hodges replies, "No. These aren't Black Phantoms or billiard-cue Moultons or Cinellis. There were two million Stingrays made, and what, a couple hundred thousand Twentys? The world will not be a worse place if you take even a shiny, pristine Twenty and go after it with a tap and die set and a brazing torch. And if you pull a Twenty or a Stingray off a scrap pile, then you are keeping valuable resources from going to waste no matter what you do. All I'm saying is that once you've put all that time and money into that pocket-rocket you've always wanted, don't be shocked if nobody else wants it, at least enough to write you a check for it. You're not Ed Roth with some custom-car creation ready for the auction house."

The Raleigh Twenty proved to be much like David Lucsko's "hidden in plain sight" rolling parts-cars of the Unocal "cash for clunkers" fiasco: the well-built little bike was so rugged that, as Tony Hadland said, it could hold together with minimal maintenance and attention for twenty years, as successive generations of roadsters, ten-speeds and cheap mountain bikes chained up beside it on the bike rack lived and died. Then, seemingly out of nowhere, people started to get excited about them. As Eric Hodges said, the secret was parts. If enough units are out there to support a market for either new and/or recycled repair parts, then a given model has a better chance for a second life as a classic, independent of its original value or price. And as John S. Allen's story of his minister's Twenty points out, compact bicycles have an edge: they are more likely to get folded, stored and forgotten, something difficult to imagine happening with a 1968 Dodge Dart.

On the other hand, this logic is upside-down for high-end imported racing, touring and track bikes made since the 1960s. They likely aren't rare, but they aren't being spotted, either. "It is a great mystery what happens to higher-end bicycles, and especially, their components, here in the States," observes Hodges. "They don't resell today in anything approaching the numbers they have sold over the last, say, forty years. But they don't get used, and don't appear to have been worn out or thrown away. They may be getting stripped down. Parts sell better than bikes, especially since the ebay era. I wonder if a lot of them have been crashed and stripped. I can't believe they've all had their braze-ons ground down and been turned into fixies, not the good stuff."

Dan Farrell, Director of Design at Alex Moulton Bicycles, recalls being phoned by a woman whose husband had recently died. She wanted to know if his two Moultons were worth trying to sell. He asked her to describe them. They were high-end machines, even for Moultons. Farrell gave her a general price for them and recommended that she contact the Moulton Preservation Society to assist her in finding a reputable shop to broker a sale. After a long pause, the widow told Farrell, "I'm glad he's gone, because if I had known he spent that kind of money on a couple of bicycles, I'da killed him myself!"[48]

Issacson asks: "I'm starting to read on-line about guys who are motorizing Twentys in England." Hodges interjects: "A Twenty is stiff enough, and it would look cool, but the small size would work against you. And price. Here [in the States], you can go to Wal-Mart, get a retro-look balloon-tire cruiser from one of the big names, and there is probably already a pre-made gas whizzer kit you can buy for it on-line."

Issacson agreed that "this whole motorized bike deal" is probably "going to be the next big thing. Take a gas motor out of a leaf blower, drop it in a cruiser bike with a 7-speed rear derailleur, damn thing'll go 40 mph. Poor people, people with suspended driver's licenses, undocumented workers, just about anyone afraid to apply for a driver's license, that's what they are turning to, not push-bikes. A clip-on [motor] is hardly more expensive than a Wal-Mart or Costco bike."[49]

Hodges adds that he can't treat them in his shop as bicycles: "If it's a gas whizzer, you can't even bring it in unless you drain the tank first. Customer asks, 'how can you fix it if you can't even start the engine?' But zoning and fire codes don't allow me to do small engine work. I'm a bike shop, not a lawn mower place. But go to a small engine repairer, like a lawn mower or a motor scooter shop and they'll tell you: 'go to a bike store, you can't afford us.' And they can't. If they had any money they would be on a Vespa. Or in a Chevy."

Issacson asks Hodges: "Have you seen a Twenty with a motor?" Hodges replies, "No, nothing but cruisers. Wait. I saw a Stingray. But he was an enthusiast, not a basic transportation guy. The bike was real nice. It was a Krate, with the 16-inch front wheel and the extended front fork. He said it had a shimmy above 25 or 30."

Issacson: "25 or 30? If he lived to describe it, he was doing good."

But with the interview concluded, standing in the doorway of his ten-foot wide shop, door already open, Hodges adds one final thought: "I did hear something about a guy that was talking about going to England, buying up a bunch of Twentys, putting them in a container, shipping them here to the States, so who knows?"

In the end, one has to ask, which is really "Banham's Bicycle"–the high-tech Moulton, or the lowly, ubiquitous, infinitely tweakable Raleigh Twenty?

Chapter Notes

Introduction

1. The origins of the label "Historian of the Immediate Future" are murky. According to Anthony Vidler, Robert Maxwell first used it in an introduction to an article by Colin Rowe titled "Thanks to the RIBA—Pt. 1." Maxwell wrote: "Where Banham invented the immediate future, Rowe invented the immediate past." Vidler does not fully cite the Rowe work and I cannot find it. Based on another essay Maxwell wrote, "The Plenitude of Presence: Reyner Banham" in *Architectural Design* (included here as "Reyner Banham—Historian"), I suspect it was about 1981. "Reyner Banham—Historian" is in Anthony Vidler, *Histories of the Immediate Present: Inventing Architectural Modernism* (Cambridge, MA: MIT Press, 2008): 107–112.
2. Lawrence Alloway, "The Independent Group: Postwar Britain and the Aesthetics of Plenty," in *The Independent Group: Postwar Britain and the Aesthetics of Plenty*, ed. David Robbins (Cambridge, MA: MIT Press, 1990): 49–53.
3. Barbara Penner, "The Man Who Wrote Too Well," *Places Journal* (September 2015), https://placesjournal.org/article/future-archive-the-man-who-wrote-too-well/. "Bibliography," in Reyner Banham, *A Critic Writes*, ed. Mary Banham, et al. (Berkeley: University of California Press, 1996): 301–479. Bibliography updated in: "Bibliography," in Nigel Whiteley, *Reyner Banham: Historian of the Immediate Future* (Cambridge, MA: MIT Press, 2002): 477–479.
4. Peter Hall, "Introduction," in Reyner Banham, *A Critic Writes*, ed. Mary Banham, et al. (Berkeley: University of California Press, 1996): ix–xii.
5. Reyner Banham, "The Atavasm of the Short-Distance Mini-Cyclist," *Living Arts* 3 (1964): 92, also in Reyner Banham, *Design by Choice: Ideas in Architecture*, ed. Penny Sparke (New York: Rizzoli, 1981): 84–89. Emily King recollection: Mark Haworth Booth, "Camera Lucida," *Frieze* 99 (May 2006): 22–23.
6. Bicycle production and Raleigh's aborted interest in Moulton: Tony Hadland, *Raleigh: Past and Presence of an Iconic Bicycle Brand* (San Francisco: Cycle Publishing, 2012): 205–212. Banham quote: Reyner Banham, "A Grid on Two Farthings," *New Statesman* (November 1, 1963): 626, also in Reyner Banham, *Design by Choice: Ideas in Architecture*, ed. Penny Sparke (New York: Rizzoli, 1981): 119–120.
7. Chris Mansell, "The Rallying of Raleigh," *Management Today* (February 1973): 83–92; Thomas J. McNichols, "Raleigh Industries" in *Policy Making and Executive Action* (New York: McGraw-Hill, 3rd ed., 1967): 485–524.
8. Dissatisfaction with traditional bike shops: Mansell, "The Rallying of Raleigh": 89–90. Targeted at women: electronic mail from Tony Hadland to the author, January 28, 2010, and January 31, 2010.
9. Reyner Banham, "Had I the Wheels of an Angel," *New Society* 18 (August 12, 1971): 463–464.
10. Hadland, *Raleigh: Past and Presence of an Iconic Bicycle Brand*: 218–219.
11. Banham, "A Grid on Two Farthings": 626.
12. Ben Banham quote: Pedero Ignacio Alonso and Thomas Weaver, "Deserta," *AA Files* 62 (2011): 20–27, quote at 21; Hall quote: Peter Hall, "Introduction," in *A Critic Writes*, ed. Mary Banham, et al.: xi–xv, quote at xiv.
13. "The Bike Show from Resonance FM: The Moulton Story (Part 1)," http://bikeshow_20080929_64kb_M3U (2008).
14. Mary Banham quote: Corinne Julius interview with Mary Banham, Tape 10 (January 2002), British Library Oral Histories, http://Sounds.bl.uk/oral-history/Architects-Lives/021M-CO467X0067XX-0010V0. Banham quote: *Fathers of Pop* (Julian Cooper, dir., Arts Council of Great Britain, prod., 40 min., 1979).
15. David Kynaston, *Austerity Britain 1945–1951* (New York: Walker, 2008): 357; David Kynaston, *Modernity Britain: A Shake of the Dice, 1959–62* (London: Bloomsbury, 2014): 102–103, 161.
16. Nigel Whiteley, "Pop, Consumerism, and the Design Shift," *Design Issues* 2, 2 (Autumn 1985): 31–45, quotations at 33–35.
17. Eleanor Bron, *Life and Other Punctures* (London: Andre Deutsch, 1978).
18. Hamilton quote: Whiteley, "Pop, Consumerism, and the Design Shift": 37; Seales quote: McNichols, "Raleigh Industries": 506.
19. Banham quote: Banham, "A Grid on Two Farthings": 626; Raleigh executive quote: Mansell, "The Rallying of Raleigh": 83–92. "C2-D-E" refers to the bottom three economic classes of the UK's National Readership Survey. "C2" means skilled working class; "D" is working class, "E" is laborer or dependent.
20. Reyner Banham, "Alternative Wheels?" *New Society* 46, 846–847 (21/28 December 1978): 712–713.
21. Banham quote: Reyner Banham, "A Grid on

Two Farthings," in Reyner Banham, *Design by Choice: Ideas in Architecture*, ed. Penny Sparke (New York: Rizzoli, 1981): 120 n1. Moulton Mk. III: Hadland; *Raleigh: Past and Presence of an Iconic Bicycle Brand*: 201–212.
  22. Moulton quote: Alex Moulton, "Innovation: An Address," *Journal of the Royal Society for the Encouragement of Arts, Manufacturers and Commerce* 128, 5281 (December 1, 1979): 31–44. Banham quote: Banham, "A Grid on Two Farthings" (1981 version): 205n1. Maxwell quote: Robert Maxwell, "Reyner Banham-Historian," in *Sweet Disorder and the Carefully Careless: Theory and Criticism in Architecture* (Princeton: Princeton Architectural Press, 1993): 163–175.
  23. Mary Banham quote: Corrine Julius interview of Mary Banham, Tape 10 (January 2002). Banham's contribution had been: Reyner Banham, "Vehicles of Desire," *Art* 1 (September 1, 1955): 3ff.
  24. Reyner Banham, "Arts in Society: Zoom Wave Hits Architecture," *New Society* 7, 179 (March 3, 1966): 21.
  25. "Softer Hardware," *Ark* 44 (1969): 6–7; Whiteley, *Reyner Banham: Historian of the Immediate Future*: 347–350.
  26. Tony Hadland and John Pinkerton, *It's in the Bag! A History in Outline of Portable Bicycles in the UK* (Cheltenham: Quorum, 1996): 44–47.
  27. Whiteley, *Reyner Banham: Historian of the Immediate Future*: 4; Martin Pawley, "The Last of the Piston Engine Men," *Building Design* (October 1, 1971): 6; "Alex Moulton—Pioneer From a Stately Home," *Director* 19, 1 (July 1966): 42–43; Moulton, "Innovation: An Address," 36–37.
  28. Moulton, "Innovation: An Address," 38; Tony Hadland, *The Moulton Bicycle: the Story from 1957 to 1981* (Coventry: Author, 2000): 14–23.
  29. Banham, "A Grid on Two Farthings," *New Statesman* (November 1, 1963): 626 (reprinted with 1979 postscript, in *Design by Choice*, 1981); "Back in the Saddle," *Architects' Journal* 138, 19 (November 6, 1963): 109, 929; "Easy Rider" (letter), *Design* 181 (August 1964): 59; "The Atavism of the Short-Distance Mini-Cycle," *Living Arts* 4 (1964) (reprinted in *Design by Choice*, 1981); "Had I the Wheels of an Angel," *New Society* 18 (August 12, 1971): 463–464; "Alternative Wheels?" *New Society* 46 (December 21/28, 1978): 712–713; *Scenes in American Deserta* (London: Thames and Hudson, 1982). The Moulton also puts in a brief appearance in the documentary *Reyner Banham Loves Los Angeles* (1972), in London, not LA.
  30. Banham, "The Atavism of the Short-Distance Mini-Cyclist" (1981 version): 84.
  31. Gillian Naylor, "Theory and Design: The Banham Factor," in *The Banham Lectures: Essays on Designing the Future*, ed. Jeremy Aynsley and Harriet Atkinson (New York: Berg, 2009): 47–58.
  32. Banham, "The Atavasm of the Short-Distance Mini-Cyclist" (1981 version): 84.
  33. Penny Sparke, "From Production to Consumption in Twentieth Century Design," in *The Banham Lectures: Essays on Designing the Future*, ed. Jeremy Aynsley and Harriet Atkinson (New York: Berg, 2009): 127–141.
  34. Moulton, "Innovation: An Address": 37.
  35. "The Bike Show from Resonance FM: The Moulton Story, Part 1" (2008).
  36. Banham: "Atavism of the Short-Distance Mini Cyclist" (1981 version): 84.

Chapter 1

  1. Robert Maxwell, Reyner Banham—Historian" in *Sweet Disorder and the Carefully Careless: Theory and Criticism in Architecture* (Princeton: Princeton Architectural Press, 1993 [1981]): 163–175.
  2. Steve Parnell, "Nairn Mania," *Architectural Review* 235 (May 2014): 118–119. The quote is taken from a BBC documentary produced by Meades that aired in spring 2014, *Bunkers, Brutalism and Bloodymindedness: Concrete Poetry*.
  3. Architect, critic and writer, b. 1923. No relation to the publisher and financier of the same name.
  4. Banham quote: Reyner Banham, "The Atavasm of the Short-Distance Mini-Cyclist," *Living Arts* 3 (1964): 92, also in *Design by Choice: Ideas in Architecture*, ed. Penny Sparke (New York: Rizzoli, 1981): 84–89. Maxwell quote: Maxwell, "Reyner Banham-Historian," in *Sweet Disorder and the Carefully Careless*: 163–175, quote at 165.
  5. The demographic information on the Banham family was provided by Tony Hadland: "1911 England Wales & Scotland Census Transciption"; "1939 National Census"; "England & Wales Marriages, 1837–2008 Transcription." Corinne Julius interview with Mary Banham, Tape 7 (December 2001), British Library Oral Histories, http://sounds.bl.uk/oral-history/Architects-lives/021M-CO-467X0067XX-0007V0. Census records from the period after Peter and Mary met have not yet been released.
  6. Reyner Banham, "How I Learned to Live with the Norwich Union," *New Statesman* 67 (March 6, 1964): 372–373.
  7. Mary Banham: Corinne Julius interview with Mary Banham, Tape 7 (December 2001). John Hewish: Nigel Whiteley, *Reyner Banham: Historian of the Immediate Future* (Cambridge, MA: MIT Press, 2002): 4–5; 413 n2.
  8. Reyner Banham, "Who Is this Pop?," *Motif* 10 (1963): 3–13, reprinted in *Design by Choice*, ed. Penny Sparke (New York: Rizzoli, 1981): 94–96.
  9. Banham, "The Atavasm of the Short-Distance Mini-Cyclist" (1981 version): 87.
  10. Corinne Julius interview with Mary Banham, Tape 15 (February 2002), British Library Oral Histories, http://sounds.bl.uk/oral-history/Architects-lives/021M-CO-467X0067XX-0015V0.
  11. Corinne Julius interview with Mary Banham, Tape 8 (January 2002), British Library Oral Histories, http://sounds.bl.uk/oral-history/Architects-lives/021M-CO-467X0067XX-0008V0.
  12. Whiteley, *Reyner Banham: Historian of the Immediate Future*: 413 n2. Program terminated: Mary Banham, "The 1950s," in Reyner Banham, *A Critic Writes*, ed. Mary Bahnam, et al. (Berkeley: University of California Press, 1996): 1.
  13. Rejected by RAF: Corinne Julius interview with Mary Banham, Tape 7 (December 2001). Invalided out: Whiteley, *Reyner Banham: Historian of the Immediate Future*: 4.
  14. Whiteley, *Reyner Banham: Historian of the Immediate Future*: 7, 9; Penny Sparke, "Introduction," in Reyner Banham, *Design by Choice*, ed. Penny Sparke (New York: Rizzoli, 1981): 8, 17–18.
  15. Banham, "How I Learned to Live with the Norwich Union": 373.
  16. *Ibid.*

17. Mollie Panter-Downes, "Letter from London," *New Yorker* (March 24, 1945): 19–20.
18. Mollie Panter-Downes, "Letter from London," *New Yorker* (September 1, 1945): 23–24.
19. John Lehmann, *The Ample Proposition: Autobiography 3* (London: Eyre & Spotswood, 1966): 30.
20. David Kynaston, *Austerity Britain, 1945–1951* (New York: Walker, 2008): 301.
21. Allan Bullock, *Ernest Bevin, Foreign Secretary* (New York: W. W. Norton, 1983): 49–50.
22. Nick Tiratsoo and Jim Tomlinson, *Industrial Efficency and State Intervention: Labour, 1939–51* (London: Routledge, 1993): 22.
23. Ian Gazeley, "The Leveling of Pay in Britain During the Second World War," *European Review of Economic History* 10 (2006): 175–204.
24. Kynaston, *Austerity Britain, 1945–1951*: 409.
25. In 1938, boys and youth earned 35.5 percent of men; in 1947 this was 33 percent. Gazley, "The Leveling of Pay in Britain During the Second World War," Table 9.
26. Kynaston, *Austerity Britain, 1945–1951*: 406.
27. Kynaston, *Austerity Britain, 1945–1951*: 463.
28. Arthur Marwick, *War and Social Change in the Twentieth Century* (New York: St. Martin's, 1974): 164.
29. Banham, "How I Learned to Live with the Norwich Union": 373.
30. "Witness Seminar: 1949 Devaluation" *Contemporary Record* 4, 3 (Winter 1991): 480–503.
31. J. F. C. Harrison, ""The Man With the Bicycle Wheel [review of J. M. Mogey, *Family and Neighborhood*]," *New Statesman and Nation* 52, 1336 (October 20, 1956): 495–496.
32. *Daily Mirror* (September 30, 1949); Noel Whiteside, "Limits of Americanization," in Becky E. Conekin, Frank Mort and Chris Waters, eds. *Moments of Modernity*, ed. Becky Conekin, et al. (London: Rivers Oram, 1999): 96–113.
33. Amy Sue Bix, *Inventing Ourselves Out of Jobs: America's Debate Over Technological Unemployment, 1929–81* (Baltimore: Johns Hopkins University Press, 2000): 163.
34. Nick Tiratsoo and Jim Tomlinson, "Exporting the Gospel of Productivity: United States Technical Assistance and British Industry 1945–1960," *Business History Review* 71 (Spring 1997): 41–81, quote at 64–65, citing *Iron and Coal Trades Review*, May 9, 1952.
35. Five cities: Nick Tiratsoo, *Reconstruction, Affluence and Labour Politics: Coventry, 1945–60* (London: Routledge, 1990): 108; 160. London statistics, national statistics and survey difficulties: Nicholas Bullock, "Re-assessing the Post-War Housing Achievement: the Impact of War-Damage Repairs Program on the New Housing Programme in London," *Twentieth Century British History* 16, 3 (2005): 256–282; Leslie Ward, ed., *The London County Council Bomb Damage Maps, 1939–1945* (London: Thames and Hudson, 2015): 33. Ward cites statistics of 114,000 A and B buildings, both residential and commercial, in the London Regional Area; 69,000 in the County of London, 2,884 in the square-mile City of London.
36. Kynaston, *Austerity Britain, 1945–1951*: 467; David Kynaston, *Modernity Britain: A Shake of the Dice, 1959–1962* (London: Bloomsbury, 2014): 281–282. Four million homes: Mark Clapson, *Invincible Green Suburbs, Brave New Towns: Social Change and Urban Dispersal in Postwar England* (Manchester: Manchester University Press, 1998): 33.
37. Bullock, "Re-assessing the Post-War Housing Achievement": 278. Bricks: Ward, *The London County Council Bomb Damage Maps, 1939–1945*: 33.
38. Bullock, "Re-assessing the Post-War Housing Achievement": 272; Tiratsoo, *Reconstruction, Affluence and Labour Politics: Coventry, 1945–60*: 108; 160.
39. David Kynaston, *Modernity Britain: Opening the Box, 1957–59* (London: Bloomsbury, 2013): 47–48; Steve Parnell, "Nairn Mania," *Architectural Review* 235, 1407 (May 2014): 118–119. Nairn was debilitated by alcohol addiction. He died of acute liver failure in 1983 at age 53. Nairn was hired as an assistant editor at AR in mid-1954, when Banham was also promoted to assistant editor. Nairn left in early 1964, again about the same time as Banham, to write for *The Observer*. Gillian Darley and David McKie, *Ian Nairn: Words in Place* (Nottingham: Five Leaves, 2013): 21, 47.
40. Mollie Painter-Downes, "Letter From London," *New Yorker* (November 7, 1943): 11.
41. David Kynaston, *Modernity Britain: A Shake of the Dice, 1959–62*: 79.
42. Kirkby: Clapson, *Invincible Green Suburbs, Brave New Towns*: 33. Z-Cars: Kynaston, *Modernity Britain: A Shake of the Dice, 1959–62*: 358. Adam-12: Tim Brooks and Earle Marsh, *The Complete Directory to Prime Time Network TV Shows* (New York Ballantine, 1979): 7. *Adam-12* Ran for five seasons, September 1968 to August 1975.
43. F. T. Burnett and Sheila Scott, "A Survey of Housing Conditions in the Urban Areas of England and Wales," *Sociological Review* 10, 2 (March 1962): 35–79; Tiratsoo, *Reconstruction, Affluence and Labour Politics: Coventry, 1945–60*: 109.
44. James Hinton, *The Mass Observers: A History, 1937–1949* (Oxford: Oxford University Press, 2013): 249.
45. Clapson, *Invincible Green Suburbs, Brave New Towns*: 138–141. Many New Town surveys mention complaints by women about mud, but treated them as an issue of fastitdiousness, completely overlooking the context of the statements in which they are raised, which clearly dealt with mud as a *mobility* problem.
46. Alison Smithson and Peter Smithson, *Ordinariness and Light* (Cambridge: MIT Press, 1970): 20–21, 25.
47. Corinne Julius interview with Mary Banham, Tape 7 (December 2001).
48. Sparke, "Introduction," in *Design by Choice*: 8.
49. Corinne Julius interview with Mary Banham, Tape 7 (December 2001).
50. Rejected by Courtauld the first time: Whiteley, *Reyner Banham: Historian of the Immediate Future*: 7, 9; Banham quotes about Norwich: Banham, "How I Learnt to Live with the Norwich Union," 373.
51. Corinne Julius interview with Mary Banham, Tape 7 (December 2001).
52. Corinne Julius interview with Mary Banham, Tape 7 (December 2001).
53. Mary Banham quote: Alice Twemlow, *Purposes, Poetics and Publics: the Shifting Dynamics of Design Criticism in the U.S. and U.K., 1955–2007* (Ph.D. dissertation, Royal College of Art, 2007): 122. Vidler quote: Anthony Vidler, "Futurist Modernism: Reyner Banham," in *Histories of the Immediate Present: Inventing Architectural Modernism* (Cambridge, MA: MIT Press, 2008): 106–155, quote at 106.

54. Adrian Forty, "Reyner Banham, One Partially Americanized European," 196. Bicycle stolen: Reyner Banham, "Back in the Saddle," *Architects' Journal* 138, 19 (November 1963): 109, 929.

55. Corinne Julius interview with Mary Banham, Tape 10 (January 2002), British Library Oral Histories, http://Sounds.bl.uk/oral-history/Architects-Lives/021M-CO467X0067XX-0010V0.

56. Forty, "Reyner Banham, One Partially Americanized European": 197; Whiteley, *Reyner Banham: Historian of the Immediate Future*: 39–47; Reyner Banham, *Theory and Design in the First Machine Age* (New York: Praeger, 1960); Corinne Julius interview with Mary Banham, Tape 10 (January 2002).

57. Shantel Blakley, "Raise High the Seat Post, Salinger," *A.A. Files* 65 (2012): 122–123.

Chapter 2

1. Martin Pawley, "Building Revisits: Hunstanton School, 1984," *Architects' Journal* (June 20, 1984): 32–36.

2. *Fathers of Pop* (Julian Cooper, director, Arts Council of Britain, producer. 40 min.,1979).

3. Ben Highmore, "Rough Poetry: Patio and Pavilion Revisited," *Oxford Art Journal* (February 2006): 269–290. Hamilton quote: *Fathers of Pop* (1979). Paolozzi's father was expelled in 1939 as an Italian national, and was drowned when his deportation ship was mistakenly torpedoed: Anne Massey, *The Independent Group: Modernism and Mass Culture in Britain, 1945–59* (New York: Manchester University Press, 1995): 34–37. Beatriz Colomina, "Friends of the Future: A Conversation with Peter Smithson," *October* 94 (Fall 2000): 3–30. The "membership" in the IG was always fluid. A fairly inclusive list includes: Lawrence Alloway (critic, writer); Reyner Banham (historian); Frank Cordell (musician): Magda Cordell (artist); Theo Crosby (architectural engineer); Toni del Renzio (illustrator, editor); Richard and Terry Hamilton (artists); Nigel Henderson (photographer): John McHale (artist); Eduardo Paolozzi (artist); Alison and Peter Smithson (architects); James Stirling (architect); William Turnbull (artist); Colin St. John (Sandy) Wilson (architect). See: David Robbins, *The Independent Group: Postwar Britain and the Aesthetics of Plenty* (Cambridge, MA: MIT Press, 1990): *passim*.

4. Nigel Henderson, *Nigel Henderson: Parallel of Life and Art* (London: Thames and Hudson, 2001): 50.

5. Colomina, "Friends of the Future: 10–12.

6. Christoper Woodward, "Drawing the Smithsons: An Artisanal Memoir," In Max Risselada, ed., *Alison & Peter Smithson: A Critical Anthology* (Barcelona: Ediciones Poligrafa, 2011): 258–267; Valerie Grove, "Alison Smithson, Architect," in *The Compleat Woman* (London: Chatto and Windus, 1987): 259–270.

7. Colomina, "Friends of the Future: A Conversation with Peter Smithson": 5, 8, 9, 12. Mary Banham and Magda Cordell quotes: Mark Girouard, *Big Jim: The Life and Work of James Stirling* (London: Chatto and Windus, 1998): 54. Mary Banham repeats the substance of these comments in: Corinne Julius interview with Mary Banham, Tape 9 (January 2002) British Library Oral Histories: http://sounds.bl.uk/oral-history/Architects-Lives/021M-CO467X0067XX-09V0.

8. Colomina, "Friends of the Future: A Conversation with Peter Smithson": 4.

9. Corinne Julius interview with Mary Banham, Tape 9 (January 2002).

10. Robbins, *The Independent Group: Postwar Britain and the Aesthetics of Plenty*: 21–23.

11. Alison and Peter Smithson, "But Today We Collect Ads," *Ark* 18 (November 1956): 26–30.

12. Colomina, "Friends of the Future: A Conversation with Peter Smithson": 4.

13. Massey, *The Independent Group*: 84.

14. Corinne Julius interview with Mary Banham, Tape 9 (January 2002).

15. David Kynaston, *Family Britain, 1951–1957* (New York: Walker, 2009): 49.

16. Robbins, *The Independent Group: Postwar Britain and the Aesthetics of Plenty*: 21, 191. Banham quote: *Fathers of Pop* (1979).

17. Robbins, *The Independent Group: Postwar Britain and the Aesthetics of Plenty*: 197.

18. Reyner Banham, "The New Brutalism," *Architectural Review*," 118 (December 1955): 354–361; Reyner Banham, *The New Brutalism: Ethic or Esthetic?* (London: Architectural Press, 1966). There is continuing controversy over who invented the name "New Brutalism" and what it means: Anthony Vidler, "Troubles in Theory V: The Brutalist Movement(s)," *Architectural Review* 235 (February 2014): 96–101; Irénée Scalbert, "Parallel of Life and Art," *Daidalos* 75 (2000): 75–102. For a chilling reappraisal of the Hunstanton School thirty years on from the people who actually had to live with it, see: Martin Pawley, "Building Revisits: Hunstanton School, 1984," *Architects' Journal* (June 20, 1984): 32–36.

19. Massey, *The Independent Group*: 54–59; Nigel Whiteley *Reyner Banham: Historian of the Immediate Future* (Cambridge, MA: MIT Press, 2002) 123–133; locus classicus: Nigel Whiteley, "Banham and Otherness: Reyner Banham and his Quest for an Architectural Autre," *Architectural History* 33 (1990): 188–221, at 201.

20. The March 1953 lecture was originally titled "Borax, or the Thousand-Horsepower Mink." It was later split into several articles including "Epitaph: Machine Esthetic" (1960, see below).

21. Originally published in Italian: *Civilta della Machine* (1955). Republished in English as: Reyner Banham, "Epitath: Machine Esthetic," *Industrial Design* 7 (March 1960): 45–58 (the title of this article is also variously given as "Industrial Design and Popular Art" and "A Throw-Away Aesthetic.") Banham quote: Reyner Banham, "Design by Choice," *Architectural Review* 130 (July 1961): 43–48.

22. Alice Twemlow, "I Can't Hear You If You Say That, An Ideological Collision at the IDCA, 1970," *Design and Culture* 1, 1 (2009) 23–50. Alice Twemlow, *Purposes, Poetics and Publics: the Shifting Dynamics of Design Criticism in the U.S. and U.K., 1955–2007* (Ph.D. dissertation, Royal College of Art, 2007): 108–114.

23. Deborah Allen, "Cars 55," *Industrial Design* 2 (February 1955): 85–90.

24. Banham compared: Reyner Banham, "Vehicles of Desire," *Art* (September 1, 1955): 3. Banham copied Allen's technique: Tim Benton, "The Art of the Well-Tempered Lecture: Reyner Banham and Le Cour-

bousier," in *The Banham Lectures: Essays in Designing the Future*, ed. Jeremy Aynsley and Harriet Atkinson (Oxford: Berg, 2009): 11–32. *Hers is a Lush Situation* (illustration): Massey, *The Independent Group*: 134–135.

25. Clive Dilnot, "The State of Design History: Part 1: Mapping the Field," *Design Issues*, 1, 1 (Spring 1984): 4–24; Nikolaus Pevsner, *Pioneers of Modern Design* (London: Faber and Faber, 1936).

26. *Fathers of Pop* (1979).

27. Penny Sparke, "Obituary: Peter Reyner Banham (1922–1988)," *Journal of Design History* 1, 2 (1988): 141–142.

28. Corinne Julius interview with Mary Banham, Tape 10 (January 2002).

29. Whiteley, *Reyner Banhan: Historian of the Immediate Future*: 396.

30. Robert Maxwell, "Reyner Banham-Historian," in *Sweet Disorder and the Carefully Careless: Theory and Criticism in Architecture* (Princeton: Princeton Architectural Press, 1993): 163–175 at 166–167.

31. Forty quote: Adrian Forty, "Reyner Banham, One Partially Americanized European," in *Twentieth-Century Architecture and its Histories*, ed. Louise Campbell (Oatley: Society of Architectural Historians of Great Britain, 2000): 195–205, quote at 197. Sparke quote: Penny Sparke, "Introduction," in Reyner Banham, *Design by Choice*, ed. Penny Sparke (New York: Rizzoli, 1981): 8–10, quote at 8. Banham quote: Nigel Whiteley, Olympus and the Marketplace: Reyner Banham and Design Criticism," *Design Issues* 13, 2 (Summer 1997): 24–35, quote at 24.

32. Massey, *The Independent Group*: 130.

33. Hugh Gaitskell, *The Diary of Hugh Gaitskell, 1946–56*, ed. Phillip M. Williams (London: Cape, 1983): 317–318.

34. Anne Massey, "The Independent Group: Towards a Redefinition," *The Burlington Magazine* 129, 1009 (April 1987): 232–242; Massey, *The Independent Group*: 64–66, 94–95. Francis Stonor Saunders, *The Cultural Cold War* (New York: New, 1999): 76, 142. Massey claims that Read objected to the funding for Kloman; Saunders states that Read was an early and willing participant in C.C.F. and Ford Foundation (a principal C.C.F. backer) activities.

35. Banham quote: Massey, *The Independent Group*: 94–95. Mary Banham, "Retrospective Statement" in Robbins, *The Independent Group*: 187–188; Corinne Julius interview with Mary Banham, Tape 9 (January 2002).

36. Careers of IG members: Massey, *The Independent Group*: 95–140; Robbins, *The Independent Group*: passim.

37. Massey, *The Independent Group*: 103–105; Whiteley, "Banham and Otherness": 205–206; Colomina, "Friends of the Future, A Conversation With Peter Smithson": 18.

38. Alison and Peter Smithson, "Caravan—Embryo Appliance House," *Architectural Design* (September 1959): 17–18.

39. Whiteley, "Banham and Otherness": 206.

40. David Kynaston, *Family Britain, 1951–1957*: 64, 666–668.

41. Reyner Banham, "The End of Insolence," *New Statesman* 60 (October 29, 1960): 644–646, quote at 646.

42. Banham, "A Throw-Away Aesthetic" ["Epitath: 'Machine Esthetic'"]: 63.

43. Banham, "Design by Choice": 76.

44. Robbins, *The Independent Group: Postwar Britain and the Aesthetics of Plenty*: 134–161. There is also coverage of the exhibit in Massey: *The Independent Group*, and the documentary *Fathers of Pop* (1979). See especially the floorplan on p. 134 of Robbins. Although nobody admits it, it appears that the relative layout of Team 1, Team 2 and the entry and exit doors anticipated that the Team 1 space—mostly an overhead spaceframe structure by Theo Crosby— would serve as a vestibule between the front door and the Team 2 space, and that many visitors would U-turn and head for the exit door after leaving Team 2. The catalog sales desk lay along this route.

45. *Fathers of Pop* (1979).

46. Corinne Julius interview with Mary Banham, Tape 10 (January 2002).

47. Corinne Julius interview with Mary Banham, Tape 8 (December 2001).

48. Corinne Julius interview with Mary Banham, Tape 11 (February 2002).

49. Corinne Julius interview with Mary Banham, Tape 10 (January 2002).

50. Corinne Julius interview with Mary Banham, Tape 11 (February 2002).

51. Alan Bullock, *Ernest Bevin [Volume 3]: Foreign Secretary, 1945–1951* (New York: W. W. Norton, 1983): 766, 854; Kynaston, *Family Britain*, 162.

52. Kynaston, *Family Britain, 1951–1957*: 416; David Kynaston, *Modernity Britain: Opening the Box, 1957–59* (London: Bloomsbury, 2013): 256.

53. Colin G. Pooley and Jean Turnbull, "Commuting, Transport and Urban Form," *Urban History* 27, 3 (December 2000): 360–372.

54. Nigel Whitely, "Pop Consumerism and the Design Shift," *Design Issues* 2, 2 (Autumn 1985): 31–45. Note that the UK did not convert to a decimal currency until 1970, but for clarity it is used here.

55. Rita Hinton, "Equality with Quality," *Socialist Commentary* 19, 7 (July 1955): 196–208.

56. Banham, "The Atavism of the Short-Distance Mini-Cyclist," in Reyner Banham, *Design by Choice: Ideas in Architecture*, ed. Penny Sparke (New York: Rizzoli, 1981): 84–87, at 85.

57. Whitely, "Pop Consumerism and the Design Shift": 31.

58. Gillian Naylor, "Theory and Design: The Banham Factor," *Journal of Design History* 10, 3 (1997): 241–252, quote at 246.

Chapter 3

1. Reyner Banham, "A Grid on Two Farthings," *New Statesman* (November 1, 1963): 626.

2. Alex Moulton, *From Bristol to Bradford-on-Avon: A Lifetime in Engineering* (Derby: Rolls-Royce Heritage Trust, 2009): 10–11, 13; Alex Moulton, "Innovation: An Address," *Royal Society for the Encouragement of Arts, Manufacturers and Commerce Journal* 128, 5281 (December 1979): 31–44.

3. "Alexander Eric Moulton" [obituary] *New York Times* (December 20, 2012); "Alex Moulton Obituary," *The Guardian* (December 10, 2012); Moulton, "Innovation: An Address": 31–44; Moulton, *From Bristol to Bradford-on-Avon*: 13, 15.

4. Nikolaus Pevsner, *Wiltshire* (Harmondsworth: Penguin, 1963): 121–123; Dan Farrell, *Riding on Rub-*

ber: The Story of Bradford-on-Avon's World Renowned Rubber Industry (manuscript, 2016): n.p.; Moulton, "Innovation: An Address": 31–34. It is probable that the Kingston family retained some ownership and control, hence the name "Kingston Mills."
   5. Farrell, *Riding on Rubber*: n.p.
   6. Farrell, *Riding on Rubber*: n.p.
   7. The Bikeshow from Resonance FM, The Moulton Story, Part 1, Bikeshow_20080929-62kb_M3U. The Bikeshow from Resonance FM, "The Moulton Story, Part 2."
   8. Moulton, "Innovation: An Address": 36; Moulton, *From Bristol to Bradford-on-Avon*: 21–22.
   9. Moulton, *From Bristol to Bradford-on-Avon*: 30–36.
   10. Banham at Bristol Aircraft: Nigel Whiteley, *Reyner Banham: Historian of the Immediate Future* (Cambridge, MA: MIT Press, 2002.): 4; "Introduction: 1950s," in Reyner Banham, *A Critic Writes*, ed. Mary Banham, et al. (Berkeley: University of California Press): 1.
   11. Moulton, *From Bristol to Bradford-on-Avon*: 33; "Address to House of Lords by Lord Brabazon of Tara Regarding Situation of Sir Roy Fedden," *Hansard's Lords Sitting* (17 October 1942). The Wright R3350 used a hybrid setup; the exhaust valve was a traditional poppet valve.
   12. Moulton, "Innovation: An Address": 37.
   13. Volkswagen: Phil Patton, *Bug: The Strange Mutations of the World's Most Famous Automobile* (New York: Simon and Schuster, 2002): 80–91. Fedden: Moulton, *From Bristol to Bradford-on-Avon:* 45–48. A "Boxter" is a horizontally opposed engine; the cylinders, flat, face outward from the crankshaft. Porsche (oil cooled) and Subaru (water cooled) are today the largest makers of boxter engines.
   14. Moulton, *From Bristol to Bradford-on-Avon*: 47.
   15. Farrell, *Riding on Rubber*: n.p.
   16. Gillian Bardsley, *Issigonis: The Official Biography* (Cambridge, UK: Icon, 2005): 117–133. Diarmuid Downs, "Alexander Arnold Constatine Issigonis, 1906–1988," *Biographical Memoirs of Fellows of the Royal Society* 39 (February 1994): 200–211.
   17. Moulton, "Innovation: An Address": 38.
   18. Downs, "Alexander Arnold Constatine Issigonis, 1906–1988": 206; Moulton, "Innovation: An Address": 38; Bardsley, *Issigonis*: 157. Moulton was assistant managing director: Farrell, *Riding on Rubber*: n.p.
   19. Kynaston, *Family Britain, 1951–1957* (New York: Walker, 2009): 270–271.
   20. Bardsley, *Issigonis*: 88–89; 142–145.
   21. In the first two prototypes, the engine pointed left-to-right. After that, it was reversed, for several reasons: it allowed for a better reduction gear set-up, it shortened the exhaust manifolds, and it put the carburetor behind the motor to help prevent frosting. Bardsley, *Issigonis*: 193.
   22. Downs, "Alexander Arnold Constatine Issigonis, 1906–1988": 208–209; Bardsley, *Issigonis*: 213.
   23. Bardsley, *Issigonis*: 230–231.
   24. Farrell, *Riding on Rubber*: n.p.
   25. Farrell, *Riding on Rubber*: n.p.
   26. Bardsley, *Issigonis*: 352.
   27. Bardsley, *Issigonis*: 430–431; Moulton, *From Bristol to Bradford-on-Avon*: 133–135.
   28. Author's interview with Dan Farrell, February 24, 2017.

   29. Moulton, "Innovation: An Address": 40; Tony Hadland, *The Moulton Bicycle, the Story From 1957 to 1981* (Coventry: Author, 1981): 14–15.
   30. Hadland, *The Spaceframe Moultons*: viii; The Bikeshow from Resonance FM, "The Moulton Story, Part 2," Bikeshow_20081005-62kb_M3U. There is a short film of the demonstration on YouTube. The best way of finding it is through the Veteran-Cycle Club of Britain's website.
   31. Author's interview with Tony Hadland and Dan Farrell, February 24, 2017.
   32. The Bikeshow from Resonance FM, "The Moulton Story, Part 1."
   33. The Bikeshow from Resonance FM, "The Moulton Story, Part 1."
   34. Chris Mansell, "The Rallying of Raleigh," *Management Today* (February 1973): 82–93, at 88. Author's interview with Dan Farrell, February 24, 2017.
   35. Moulton, *From Bristol to Bradford-on-Avon*: 154–155.
   36. The Bikeshow from Resonance FM, "The Moulton Story, Part 1."
   37. My thanks to Tony Hadland for his counsel on this confusing technical matter.
   38. Hadland, *The Moulton Bicycle, the Story From 1957 to 1981*: 16; The Bikeshow from Resonance FM, "The Moulton Story, Part 1." The ISO code for these tires was 32–349 thru 37–349. In addition to the Sprite, Dunlop made a blackwall utility tire: Hadland, *The Moulton Bicycle*: 129.
   39. Jack Lauterwasser interview on Tony Hadland's website: https://hadland.woodpress.com/2012/07/01/cycling-history-interviews. Tony Hadland and John Pinkerton, *It's In the Bag: A History in Outline of Portable Cycles in the UK* (Cheltenham: Quorum, 1996): 114–115.
   40. Tony Hadland, *Raleigh: Past and Presence of an Iconic Bicycle Brand* (San Francisco: Cycle, 2012): 207. The joint failure problem was also excaserbated by the different jointing methods Ralieigh used.
   41. Hadland, The Moulton Bicycle:16–17.
   42. "I was quite wrong": Moulton, "Innovation: An Address": 41; "I was horrified": The Bikeshow from Resonance FM, "The Moulton Story, Part 1."
   43. GB-907467-A
   44. Moulton, *From Bristol to Bradford-on-Avon*: 162–163. Hold a bicycle upright by the saddle. Tilt it to the side 45 degrees. If the bicycle is in good maintenance, the front wheel should "flop" to the side. This is one of the factors that makes riding "no hands" possible.
   45. Moulton, *From Bristol to Bradford-on-Avon*: 163, 200.
   46. Moulton, *From Bristol to Bradford-on-Avon*: 164.
   47. Hadland, The Moulton Bicycle, 1957–1981: 18–21.
   48. Christine Murray, "Reinventing the Wheel: An Interview with Alex Moulton," *The Architects' Journal* (September 4, 2008): 95–97.
   49. Moulton, *From Bristol to Bradford-on-Avon*: 164, 166.
   50. "Am I Mad?": The Bikeshow from Resonance FM, "I drifted": "The Moulton Story, Part 1"; Moulton, *From Bristol to Bradford-on-Avon*: 168.
   51. Macnaughtan quote: John Macnaughtan interview, Part 1 (2016), https://www.youtube.com/channel/UC2XzNEKoaCK1INKmSIo6wRA/. "David Duffield

interview" (2008), https://www.youtube.com/channel/UC2XzNEKoaCK1INKmSIo6wRA/.

52. David Duffield interview.

53. David Duffield interview. Moulton later said: Moulton, *From Bristol to Bradford-on-Avon*: 180.

54. Hadland, *The Moulton Bicycle, 1957–1981*: 58–61. Note that this "Speed" model was not the same as the later Speedsix (or Speed Six). The Speedsix was derailleur-equipped, Woodburn's bike had a Sturmey-Archer AC close-range 3-speed hub.

55. David Duffield interview. A photo of the Moulton booth is in Moulton, *From Bristol to Bradford-on-Avon*: 168.

56. David Duffield interview. Short of manpower: Moulton, *From Bristol to Bradford-on-Avon*: 170.

57. The story of Woodburn's record ride is taken from two sources: "John Woodburn interview," https://www.youtube.com/channel/UC2XzNEKoaCK1INKmSIo6wRA/, and Hadland, *The Moulton Bicycle*: 65–66.

58. Email from Tony Hardland to the author, February 16, 2017.

59. John Woodburn interview; Hadland, *The Moulton Bicycle*: 62. The Obree position had the chest resting on the handlebars, torso flat, hands under the chest. The Boardman position had the torso flat or slightly down, arms straight out in front, and for this reason was called the "superman" position. Wind tunnel tests showed the superman position was as much as 17.5 percent more efficient than the Obree position, 67 percent better than riding a standard 1960s track bike down on the drops. Max Glaskin, *Cycling Science: How Rider and Machine Work Together* (Chicago: University of Chicago Press, 2012): 136–137.

60. The decision not to use the Cowl was probably also influenced by two additional reasons: (1) after some equivocation, the Road Records Association decided that it would not recognize a record incorporating such a device, and (2) the use up to that point was sufficient to satisfy the requirements for patenting a bicycle/cowl combination (UK 1,018,962). See: Alex E. Moulton, A. Hadland and Douglas L. Milliken, "Aerodynamic Research Using the Moulton Small-Wheeled Bicycle," *Proceedings of the Institution of Mechanical Engineers*, 220, 3 (May 2006): 189–193.

61. "Vic Nicholson interview," https://www.youtube.com/channel/UC2XzNEKoaCK1INKmSIo6wRA/.

62. John Woodburn interview; Vic Nicholson interview.

63. Author's interview with Dan Farrell, February 24, 2017.

64. The Bikeshow from Resonance FM, "The Moulton Story, Part 2."

65. Vic Nicholson interview; John Woodburn interview.

66. Vic Nicholson interview.

67. Vic Nicholson interview.

68. John Woodburn interview.

69. "Alex Moulton, Pioneer from a Stately Home," *Director* 19, 1 (July 1966): 42–43. Did not catch up until 1964: memo from Moulton Bicycles to British Motor Corporation, November 15, 1965, reproduced as Appendix 3 in Moulton, *From Bristol to Bradford-on-Avon*: 313–314.

70. Moulton, *From Bristol to Bradford-on-Avon*: 171.

71. David Duffield interview.

72. Reyner Banham, "The Atavism of the Short-Distance Mini-Cyclist" in *Design by Choice*, ed. Penny Sparke (New York: Rizzoli, 1981 [1964]): 84–89. Terry Hamilton, wife of IG member and artist Richard Hamilton, was killed in a car accident in 1962. In Stillitoe's story, a reform-school youth who is extraordinarily talented at running is given special privileges so he can prepare for a prestigious national cross-county race, with the promise of more to come with victory. However, he prefers to ostentatiously throw the race (and serve penal duty as retribution) to letting his schoolmaster reap the glory that the victory would bring from his elite colleagues.

73. Banham, "The Atavism of the Short-Distance Mini-Cyclist": 85.

74. Banham, "A Grid on Two Farthings" [1963 version]: 626.

75. Reyner Banham, "Back in the Saddle," *Architects' Journal* 138, 19 (November 6, 1963): 109, 929.

76. Reyner Banham, "A Grid on Two Farthings" [1963 version]: 626.

77. Mark Haworth Booth, "Camera Lucida," *Frieze* 99 (May 2006): 22–23.

78. Somnolent industry: Banham, "Easy Rider": 59; Social conformism: Reyner Banham, "Had I the Wheels of an Angel," *New Society* 18 (August 12, 1971): 463–464. C2-D-E males: Chris Mansell, "The Rallying of Raleigh," *Management Today* (February 1973): 82–93, at 86.

79. Tony Hadland quote: "The Bike Show from Resonance FM: The Moulton Story, Part 1."

80. Mark Abrams, *The Teenage Consumer* (London: London Press Exchange, 1959): 10–12. Plenty of work: David Kynaston, *Family Britain: 1951–1957*: 664.

81. John Barron Mays, "Teen-Age Culture in Contemporary Britain and Europe," *Annals of the American Academy of Political and Social Science* 338 (November 1961): 22–32, at 28–29.

82. Abrams, *The Teenage Consumer*: 12.

83. "Editorial," ARK 36 (Summer 1964): 48.

84. "Letter from Richard Hamilton to Alison and Peter Smithson, January 16, 1957" in *The Collected Words of Richard Hamilton* (London: Bloomsbury, 1982): 28. Ironically, Peter Smithson says he and Alison never received the letter: Beatriz Colomina, "Friends of the Future: A Conversation With Peter Smithson, *October* 94 (Fall 2000): 3–30, at 5.

85. John Davy, "Industry and the Prestigue Cult," *The Observer* (October 18, 1959).

86. Nigel Whiteley, "Pop, Consumerism, and the Design Shift," *Design Issues* 2, 2 (Autumn 1985): 31–45, at 36. Whiteley is quoting Hobsbawn from a speech at the National Union of Teachers, 1960.

87. Hugh Gaitskell, "Understanding the Electorate," *Socialist Commentary* (July 1955): 196–206.

88. Reyner Banham, "Design by Choice," *Architectural Review* 130 (July 1961): 43–48 (see sidebar s.v. "Pop Art"): 47.

89. Lawrence Alloway, "Artists as Consumers," *Image* 3 (1961): 18.

90. F. T. Marinetti (Italian, 1876–1944); A. Sant'Elia (Italian, 1888–1916); S. Giedion (Swiss, 1888–1968).

91. Anthony Vidler, *Histories of the Immediate Present: Inventing Architectural Modernism* (Cambridge, MA: MIT Press, 2008): 121–125; Clive Dilnot, "The State of Design History, Part 1," *Design Issues*, 1, 1 (Spring 1984): 4–23, esp. 8–11.

92. Forty, "Reyner Banham: One Partially Americanized European": 201

93. Corrine Julius interview of Mary Banham, Tape

15 (February 2002), British Library Oral Histories: http://Sounds.bl.uk/oral-history/Architects-Lives/021M-CO467X067XX-0015V0.

94. Eva Diaz, *The Experimenters: Chance and Design at Black Mountain College* (Chicago: University of Chicago Press, 2015): 133.

95. Diaz, *The Experimenters*: 133. Reyner Banham, "The Dymaxicrat," *Arts Magazine* 38 (October 1963): 66–69.

96. Banham, "A Grid on Two Fathings" [1963 version]: 626; Banham, "The Atavism of the Short-Distance Mini-Cyclist": 95, 97. In his popular books, Richard Florida refers to "the Creative Classes," a term I prefer to avoid because I find it at best nebulous, and quite possibly misleading.

97. Man-powered equivalent of a Mini: Banham, "A Grid on Two Fathings" [1963 version]: 626. Twenty minutes off: Banham, "Back in the Saddle": 929. Colin Buchanan, et al., *Traffic in Towns* (London: HMSO, 1963).

98. David Duffield interview.

99. Memo from Moulton Bicycles to British Motor Corporation, November 1965, reproduced as Appendix 3 in Moulton, *From Bristol Bradford-on-Avon*: 313–314.

100. "Policy Memo for B.M.C. Board From Moulton Bicycles Limited, reproduced as Appendix 3 in Moulton, *From Bristol Bradford-on-Avon*.

101. David Gordon Wilson, "Bicycle Manufacturing Defects," *Bicycling* (February 1979): 23–27. Any serial number lower than K64.29 (29th week of 1964) should be considered suspect. Wilson offers a non-destructive test to verify that the front fork is safe. Gently scratch the base of the steerer tube where it enters the fork crown with an awl or other sharp tool. If you see gold, that's braze, and you are okay. Repeat at a minimum of 3 locations around the base of the steerer tube. It appears that insufficient brazing material was used to join the steerer tube and the fork crown on the bad forks.

102. "Interview with Vic Nicholson."

103. Murray, "Reinventing the Wheel: An Interview with Alex Moulton": 95.

104. Hadland, *The Moulton Bicycle*: 43–44.

105. Sheldon Brown's website: http://sheldonbrown.com/mybicycles/.

106. Gregory Houston Bowdon, *History of the Raleigh Cycle* (London: W. H. Allen, 1975): 153.

107. Brazing at Bradford-on-Avon: Hadland, *The Moulton Bicycle*: 35. Moulton production numbers: Tony Hadland, *The Spaceframe Moultons* (Piscataway, NJ: Transaction, 1994): Chapter 1.

108. Duffield figure: David Duffield interview. Moulton figure: "Alex Moulton, Pioneer from a Stately Home": 42. Raleigh: Thomas J. McNichols, "Raleigh Industries," in *Policy Making and Executive Action* (New York: McGraw-Hill, 1967): 485–523, at 502. Company memo: "Policy Memo for B.M.C. Board From Moulton Bicycles Limited, reproduced as Appendix 3 in Moulton, *From Bristol Bradford-on-Avon*. Due to the low figure for 1963 (Kirkby did not come online until July) and the language of the memo, I believe it is likely the memo refers only to Kirkby production. The figures in the memo are:
1963 (April–Dec.): 8,515
1964: 31,184
1965 (Jan.-Nov.): 25,629
1965 (proj.): 30,000

109. Hadland, *The Moulton Bicycle*: 42–43. This would average out to about 478 per week before the buy-out and about 356 per week after the transfer to Raleigh.

110. A fantasy or a joke: Reyner Banham, "Had I the Wheels of an Angel," *New Society* 18 (August 12, 1971): 463–464; media fictions: Reyner Banham, "Alternative Wheels?" *New Society* 46 (December 28, 1971): 712–713.

Chapter 4

1. Paul Rosen, *Framing Production, Technology, Culture and Change in the British Bicycle Industry* (Cambridge, MA: 2002): 102

2. *Who Framed Roger Rabbit* (Robert Zemeckis, director, Steven Spielberg, producer, 103 min. Touchstone Pictures, 1988).

3. Deborah Stone, *Policy Paradox: The Art of Political Decision Making* (New York: Norton, 1997): 132–140.

4. Bradford Snell, "American Ground Transport," Part 4A, U.S. Senate Committee on the Judiciary, *Hearings before the Subcommittee on Antitrust and Monopoly on S.1167, The Industrial Reorganization Act*, 93rd Cong., 2nd Sess. 1974; Johnathan Kwitny, "The Great Transportation Conspiracy," *Harpers Magazine* (February 1981): 14–21. Snell's report also, rather incredibly, accused General Motors of encouraging Nazi officials to use its Opal factories (Opal was a GM subsidiary) to produce bombers and jet engines during World War II—*after* America had entered the war, and claimed both GM and Ford cooperated with the German government in the months prior to Pearl Harbor to convert their factories to military use.

5. Scott Bottles, *Los Angeles and the Automobile* (Berkeley: University of California Press, 1987): *passim*; Sy Adler, "The Transformation of the Pacific Electric Railway: Bradford Snell, Roger Rabbit, and the Politics of Transportation in Los Angeles," *Urban Affairs Quarterly* 27 (September 1991): 51–86; Martha Bianco, "Kennedy, 60 Minutes and Roger Rabbit: Understanding Conspiracy-Theory Explanations of the Decline of Mass Transit," Center for Urban Studies, Portland State University, Discussion Paper 98–11, November, 1998, http:// www.upa.pdx.edu/CUS/publications/docs/DP98–11.pdf. GM did collude with the makers of batteries, spare tires and other suppliers to rig bids submitted to transit agencies for buses they were buying to replace trolleys and interurban cars. However, this was *after* the decision to switch had already been made, largely due to maintenance costs and the problem of traffic blocking streetcars. GM paid a $5,000 civil fine. *American Ground Transport* was part of a long political war that took place over 12 years between Robert and Edward Kennedy, General Motors and Lyndon Johnson. In 1962 Attorney General Robert Kennedy filed a lawsuit accusing GM of price fixing in the sales of buses to municipal transit companies, anticipating a possibly useful campaign issue for his brother in the upcoming 1964 elections. After John Kennedy's 1963 death, now-president Lyndon Johnson ordered the suit dropped because Johnson was wooing GM to build a factory in his home state of Texas. It was dropped, and GM did build a plant in Arlington in 1967. Robert Kennedy and Johnson hated each other passionately. Bradford

Snell was a college roommate of John Kerry, a Kennedy family acquaintance through Kerry's future wife, heiress Theresa Heinz. Edward Kennedy was co-chair of the senate committee that hired Snell. The recommendations in the report were essentially those that had eluded Robert Kennedy in the 1960s. Kennedy's efforts to carve up the Big Three were rendered obsolete by the Arab oil embargos of the 1970s, which bankrupted American automakers and gave foreign makers a 30 percent domestic market share.

6. "But for them": "Alex Moulton: Pioneer from a Stately Home," *Director* 19, 1 (July 1966): 42–43. Between 26K and 39K about 1966: Tony Hadland estimated sales of about 750 per week: *The Spaceframe Moultons* (Picataway, NJ: Transaction, 1994): 8. David Duffied estimated sales of between 500 and 750 per week: "David Duffield interview" (2008), https://www.youtube.com/channel/UC2XzNEKoaCK1INKmSIo6wRA/. Raleigh 1966 production: Thomas J. McNichols, "Raleigh Industries, Ltd.," in *Policy Making and Executive Action* (New York: McGraw-Hill, 2d ed. 1967): 485–524.

7. Alex Moulton, "Innovation: An Address," *Royal Society for the Encouragement of Arts, Manufacturers and Commerce Journal* 128, 5281 (December 1979): 31–44. The official was Jim Harrisson: Tony Hadland, *Raleigh: Past and Presence of an Iconic Bicycle Brand* (San Francisco: Cycle, 2012): 207. See also: Alex Moulton, *From Bristol to Bradford-on-Avon: A Lifetime in Engineering* (Derby: Rolls Royce Heritage Trust, 2009): 180–181.

8. Pinkerton quote: made during Pinkerton's interview of David Duffield, https://www.youtube.com/channel/UC2XzNEKoaCK1INKmSIo6wRA/. Mysterious Sturmey-Archer problems: Tony Hadland, *The Spaceframe Moultons* (Piscataway, NJ: Transaction, 1994): 11; Woolf quote: "The Bike Show from Resonance FM: The Moulton Story, Part 1: bikeshow_20080929_64kb_M3U (2008).

9. "David Duffield interview."

10. This assumption is based on a total output at Bradford-on-Avon of 21,000 from April 1963 to December 1965, with a 30 percent decline after May 1965.

11. Memo From Moulton Bicycles to British Motor Corporation, reproduced as Appendix 3 in Moulton, *From Bristol to Bradford-on-Avon*: 313; Moulton said 25,000: Moulton, *From Bristol Bradford-on-Avon*: 182. The Bradford-on-Avon estimates are consistent with Alex Moulton's statement that The Hall produced between 20,000 and 25,000 bicycles between April 1963 and January or February 1966, when all production except for the deluxe handbuilt "S series" was transferred to Kirkby.

12. "Policy Memo for B.M.C. Board From Moulton Bicycles Limited, in Moulton, *From Bristol Bradford-on-Avon*: 313.

13. Total market: *Bicycles: A Report on the Application by TI Raleigh Industries, Ltd. and TI Raleigh Ltd. of Certain Criteria for Determining Whether to Supply Bicycles to Retail Outlets* (London: Monopolies and Mergers Commission, 1981): 4 [henceforth *MAMC Bicycles Report*]; Raleigh: McNichols, "Raleigh Industries": Tables 17–18.

14. McNichols, "Raleigh Industries": 512–522.

15. Tony Hadland, *The Sturmey Archer Story* (n.p.: The Author, 1987): 142–144. "OEM"—Original Equipment Manufacture. The components a supplier provides to cyclemakers for building up new bicycles, which may or may not be the same equipment it sells retail.

16. Hadland, *The Sturmey Archer Story*: 145.

17. Wolverhampton and Kirkby: Moulton, *From Bristol to Bradford-on-Avon*: 171, 173. Timothy R. Whisler, *The British Motor Industry: 1945–1994: A Study in Decline* (Oxford: Oxford University Press, 1999): 47–49.

18. Wash machines: David Kynaston, *Family Britain 1951–57* (London: Bloomsbury, 2009): 668; David Kynaston, *Modernity Britain, Opening the Box, 1957–59* (London: Bloomsbury, 2013): 185; David Kynaston, *Modernity Britain, A Shake of the Dice, 1959–62* (London: Bloomsbury, 2014): 102–03; Kirkby plant: Kynaston: *Modernity Britain, A Shake of the Dice, 1959–62*: 79–80; "The Cars: Mini Development History, Part 1," Aronline online: http:www.aronline.co.uk/blogs/mini-classic/.

19. Joseph Sykes, "Postwar Distribution of Industry in Great Britain," *Journal of Business of the University of Chicago* 22, 3 (July 1949): 188–199; "Some Effects of the Distribution of Industry Act, 1945," *Manchester School of Economic and Social Studies Journal* 17 (January 1949): 36–48.

20. Kynaston, *Modernity Britain: A Shake of the Dice, 1959–62*: 78–81.

21. Woolf assertion: "The Bike Show from Resonance FM: The Moulton Story, Part 1. Problems from IG welding: Christine Murray, "Reinventing the Wheel: An Interview With Alex Moulton," *The Architects' Journal* (September 4, 2008): 95–97; Tony Hadland, *The Moulton Bicycle: the Story from 1957 to 1981* (Coventry: Author, 2000): 43. Duffield quotes: David Duffield interview. It is possible that the switch from brazing to MIG welding occurred when frame fabrication moved from Wolverhampton to Kirkby.

22. David Duffield interview. The Bradford-on-Avon factory was limited to 10,000 square feet under the 1948 Distribution of Industry Act. Anything larger required local planning authorities to secure a permit from the Board of Trade essentially justifying why the factory was not being built in one of the distressed "Development Areas." See: Joseph Sykes, "Postwar Distribution of Industry in Great Britain," *Journal of Business of the University of Chicago* 22, 3 (July 1949): 188–199. Moulton tubing: Hadland, *The Moulton Bicycle*: 24.

23. Hadland, *The Moulton Bicycle*: 14–23; David Duffield interview. Even without a factory hire-purchase plan, Moultons could be purchased from some shops on time payments: Email from Tony Hadland to the author, February 16, 2016.

24. Roger Lloyd-Jones and M. J. Lewis, *Raleigh and the British Bicycle Industry: An Economic and Business History, 1870–1960* (Aldershot: Ashgate, 2000): 47–60; Tony Hadland, *Raleigh: Past and Presence of an Iconic Bicycle Brand* (San Francisco: Cycle Publishing): 7–23.

25. "Social Survey of Bicycles in War Time Transport," *Board of Trade Journal* (January 12, 1946): 24.

26. Timothy R. Whisler, *At the End of the Road: The Rise and Fall of Austin Healy, MG and Triumph Sports Cars* (Greenwich, CT: Greenwood, 1995): 191–227; Whisler, *The British Motor Industry, 1945–1994*: 13–33.

27. Whisler, *At the End of the Road*: 17–19.

28. *The Economist* (April 20, 1946): 643.

29. Lloyd-Jones and Lewis, *Raleigh and the British Bicycle Industry*: 193 and appendix table E.2.
30. "Tube Investments Limited," *The Economist* (December 15, 1951): 1501.
31. Transcript of author's Interview with Norman Clarke (1998): 28.
32. The Columbia Mfg. Co. was allocated 25,000 bicycles a year, but never made more than 20,000.
33. Transcript of author's Interview with Norman Clarke (1998): 28–29.
34. UK production: Lloyd-Jones and Lewis, *Raleigh and the British Bicycle Industry*: Tables E.2 and E.3; 215–216; Ross Petty: "Peddling Schwinn Bicycles: Marketing Lessons from the Leading Post-WWII U.S. Bicycle Brand," *Quinnipiac University CHARM Symposium Papers 2007*: 162–171; Roger Lloyd-Jones, M.J. Lewis and Mark Easton, "Culture as Metaphor: Company Culture and Business Strategy at Raleigh Industries, 1945–60," *Business History*, 41, 3 (July 1999): 93–127.
35. "Wobble in Bicycles," *Barron's Weekly* (July 25, 1955): 1.
36. Lloyd-Jones and Lewis, *Raleigh and the British Bicycle Industry*: 187–221.
37. Lloyd-Jones and Lewis, *Raleigh and the British Bicycle Industry*: 11, 133, 230–231.
38. Lloyd-Jones, Lewis and Easton, "Culture as Metaphor: Company Culture and Business Strategy at Raleigh Industries": 93–106; "Cycles in Trouble," *The Economist* (September 29, 1956): 1081.
39. Lloyd-Jones and Lewis, *Raleigh and the British Bicycle Industry*: 210, 235. Kynaston, *Family Britain, 1951–1957*: 652–653.
40. "Cycles in Trouble": 1081; "The First Redundancy Strike," *The Economist* (January 18, 1958): 196; "A New, Lower Level?" *The Economist* (May 10, 1958): 524–525.
41. "Raleigh Industries, Ltd.," *The Economist* (December 19, 1955): 1199.
42. Lloyd-Jones, Lewis and Eason, "Culture as Metaphor: Company Culture and Business Strategy at Raleigh Industries": 106; Hadland, *Raleigh: Past and Presence of an Iconic Bicycle Brand*: 110–113.
43. Rosen, *Framing Production*: 108.
44. Lloyd-Jones and Lewis, *Raleigh and the British Bicycle Industry*: Tables E.2 and E.3.
45. Lloyd-Jones and Lewis, *Raleigh and the British Bicycle Industry*: 239; Rosen, *Framing Production*: 104; Chris Mansell, "The Rallying of Raleigh," *Management Today* (February 1973): 83–92.
46. McNichols, "Raleigh Industries": 516.
47. *Motor Cycle and Cycle Trader* (January 2, 1959): 2.
48. Transcript of author's interview with Norman Clarke (1998): 24.
49. Lloyd-Jones and Lewis, *Raleigh and the British Bicycle Industry*: 241.
50. Interview with John Macnaughton, February 24, 2017.
51. "Raleigh Industries Limited," *The Economist* (December 13, 1958): 1025–1026.
52. "One Final Merger," *The Economist* (April 23, 1960): 356. TI paid ⅓ share of TI stock plus 27s 3p (£1.36) per share of Raleigh stock, a cash price of roughly 53s 8d (£2.69).
53. Hadland: *Raleigh, Past and Presence of an Iconic Bicycle Brand*: 156–157.
54. Kynaston, *Modernity Britain: Opening the Box, 1957–59*: 263–264. TI had an aluminum division: "Tube Investments, Limited," *The Economist* (November 22, 1952): 574–575.
55. Michael J. Kolin and Denise M. de la Rosa, *The Custom Bicycle* (Emmaus, PA: Rodale, 1979): 22–26.
56. Lloyd-Jones and Lewis, *Raleigh and the British Bicycle Industry*: 246–249.
57. Lloyd-Jones, Lewis and Easton, "Culture as Metaphor: Culture and Business Strategy at Raleigh Industries, 1945–60": 111; McNichols, "Raleigh Industries": 489; Mansell, "The Rallying of Raleigh": 85.

Chapter 5

1. Chris Mansell, "The Rallying of Raleigh" *Management Today* (February 1973): 83–92, quote at 88.
2. Hadland, *Raleigh: Past and Presence of an Iconic Bicycle Brand* (San Francisco: Cycle, 2012): 155–156.
3. "Contraction," *The Economist* (April 15, 1961): 256.
4. Later story: (2009, 2008): Alex Moulton, *From Bristol to Bradford-on-Avon: A Lifetime in Engineering* (Derby: Rolls Royce Heritage Trust, 2009): 160; The Moulton Story, Part 1," *The Bikeshow from Resonance FM*. Bikeshow_20080929–62kb_M3U. Earlier story (1966): "Alex Moulton: Pioneer from a Stately Home," *Director* 19, 1 (July 1966): 42–43.
5. Roger Lloyd-Jones, M.J. Lewis and Mark Easton, "Culture as Metaphor: Company Culture and Business Strategy at Raleigh Industries, 1945–60," *Business History*, 41, 3 (July 1999): 93–127. For example, a Sturmey-Archer FW rear hub required 21 machine operations using 48 separate machine tools and 46 gages for the axle, and 37 tools and 22 gages for the hub shell—all tools, cutting heads, jigs and gages produced in-house. *Ibid.*, fn. 68.
6. Hadland, *Raleigh: Past and Presence of an Iconic Bicycle Brand*: 207; Thomas J. McNichols, "Raleigh Industries," in *Policy Making and Executive Action* (New York: McGraw Hill, 2d ed. 1967): 485–524.
7. McNichols, "Raleigh Industries": 502–503.
8. Chris Mansell, "The Rallying of Raleigh" *Management Today* (February 1973): 83–92; McNichols, "Raleigh Industries": 502–503; 520–521.
9. Mansell, "The Rallying of Raleigh": 85.
10. Roger Lloyd-Jones and M. J. Lewis, *Raleigh and the British Bicycle Industry: An Economic and Business History, 1870–1960* (Aldershot: Ashgate, 2000): 214–216.
11. Mansell, "The Rallying of Raleigh": 85.
12. Lloyd-Jones and Lewis, *Raleigh and the British Bicycle Industry*: 178, 214–215, 230; McNichols, "Raleigh Industries": 519–520; Monopolies and Mergers Commission, *Bicycles: A Report on the Application by TI-Raleigh Industries Ltd. and TI-Raleigh Ltd. of Certain Criteria for Determining Whether to Supply Bicycles to Retail Outlets* (London: Her Majesty's Stationery Office, December 1981) (hereafter, *MAMC 1981 Bicycles Report*). Raleigh would only sell low-end, mostly kids bikes to discount stores and mail-order houses. *Ibid*.
13. McNichols, "Raleigh Industries": 520–521;
14. Mansell, "The Rallying of Raleigh": 85.
15. Ray Gosling, "Robin Hood Rides Again—A Rebel Scene," *Anarchy* 38 (April 1964): 99–108.
16. McNichols, "Raleigh Industries": 512–515.

17. We started looking: Mansell, "The Rallying of Raleigh": 85. Target audience houswives: email from Tony Hadland to the author, January 31, 2010. Tape recorders and transistor radios: McNichols, "Raleigh Industries": 506–507.
18. David Kynaston, *Modernity Britain: A Shake of the Dice, 1952–1962* (London: Bloomsbury, 2014): 207–212.
19. Letter from Michael Young to Dorothy Goodman Elmhurst and Leonard Elmhurst, October 10, 1957, Dartington Hall Trust Archives, LKE/G/35.
20. McNichols, "Raleigh Industries": 503.
21. Hadland, *Raleigh: Past and Presence of an Iconic Bicycle Brand*: 155–156. Raleigh did introduce U-frame folding bikes starting in the 1970s, but they did not incorporate the Oakley improvements.
22. Hadland, *Raleigh: Past and Presence of an Iconic Bicycle Brand*: 155–60.
23. Hadland, *The Moulton Bicycle: The Story from 1957 to 1981*: 120–121. There is some disagreement over this, with Alex Moulton arguing for a difference of as much as 38 percent. Certainly the RSWs original 2-inch wide, 50 psi tires had a high rolling resistance.
24. "John Macnaughtan interview, Part 2," https://www.youtube.com/channel/UC2XzNEKoaCK1INKmSIo6wRA/.
25. A gimmick: Mansell, "The Rallying of Raleigh": 85.
26. S3B: Tony Hadland, *The Sturmey-Archer Story* (n.p.: Author, 1987): 145. A single, unvarying product: McNichols, "Raleigh Industries": 521.
27. Duffield quote: David Duffield interview. British prices based on the official 1965 exchange rate of $2.40 to the pound. Moulton prices: Hadland, *The Moulton Bicycle*: 98. Raleigh's claimed factory prices for RSW: McNichols, "Raleigh Industries": 507. Hadland's claimed factory prices for RSW: Hadland, *Raleigh: Past and Presence of an Iconic Bicycle Brand*: 209–210. Halford's pricing policies: *MAMC 1981 Bicycles Report*: section 5.41.American prices: *Raleigh in America Retail Catalog for 1967*; *Raleigh in America Retail Catalog for 1968*.
28. Hadland, *The Moulton Bicycle*: 97–100.
29. Mansell, "The Rallying of Raleigh," 88
30. Weeded out: Mansell, "The Rallying of Raleigh," 85. 1962 and 1964 figures: McNichols, "Raleigh Industries": 520; *MAMC 1981 Bicycles Report*: 7 (1981 Raleigh sales 21 percent at Halford's and Curry's).
31. "Policy Memo for B.M.C. Board From Moulton Bicycles Limited, reproduced as Appendix 3 in Alex Moulton, *From Bristol Bradford-on-Avon: A Life in Engineering* (Derby: Rolls Royce Heritage Trust, 2009).
32. Gregory Houston Bowdin, *The History of the Raleigh Cycle* (London: W. H. Allen, 1975): 137–140; "John Macnaughtan interview, Part 1."
33. "John Macnaughtan interview, Part 1."
34. Mansell, "The Rallying of Raleigh," 88. The Twenty first appeared in the Raleigh USA catalog for 1969. Both RSWs appeared in the USA catalog for 1967; the Compact alone in 1968.
35. McNichols, "Raleigh Industries": 506–507.
36. He grabbed everybody: Mansell, "The Rallying of Raleigh," 88. King of product placement: email from Tony Hadland to the author, August 31, 2013; Eleanor Bron, *Life and Other Punctures* (London: Andre Deutsche, 1978).
37. Banham quite: Reyner Banham, "Who Is This Pop?" *Motif* 10 (1963): 3–13.

38. Raleigh attempted to resurrect the RSW 14 in 1975 with something called the "Raleigh Fourteen." According to Tony Hadland, it was "essentially an RSW 14 with a new chunky frame." Email from Tony Hadland to the author, February 16, 2016. Macnaughtan recalls: author's interview with John Macnaughtan, February 24, 2017.
39. Author's interview with John Macnaughton, February 24, 2017.
40. Email from Tony Hadland to the author, January 31, 2010.
41. Reyner Banham, "Had I the Wheels of an Angel," *New Society* 18 (August 12, 1971): 463–464.
42. The problem of Industrial Design: Reyner Banham, "Design by Choice," *Architectural Review* 130 (July 1961): 43–48. The old, standardized and unquestioned: Reyner Banham, "The End of Insolence," *New Statesman* 60 (October 29, 1960) 644–646.
43. Reyner Banham, "Epitaph Machine Esthetic" *Industrial Magazine* 7 (March 1960): 45–58.
44. Seales quote: Mansell, "The Rallying of Raleigh," 88.
45. Nigel Whiteley, "Toward a Throw-Away Culture: Consumerism, Style Obsolescence and Cultural Theory in the 1950s and 1960s," *Oxford Art Journal* 10, 2 (1987): 3–27, esp. 17; David Kynaston, *Modernity Britain*: 337–339.
46. Banham, "Epitaph Machine Esthetic": 50.
47. "The Bike Show from Resonance FM: The Moulton Story, Part 2," Bikeshow_20081005–62kb_M3U.
48. Mansell, "The Rallying of Raleigh," 88.
49. Seales: Mansell, "The Rallying of Raleigh," 86; Moulton, *From Bristol to Bradford-on-Avon*: 181.
50. J. Gordon Lippincott, *Design for Business* (Chicago: P. Theobald, 1947): 2, 14–16.
51. Seales quote: Mansell, "The Rallying of Raleigh," 87. Banham quote: Banham, "Epitaph Machine Esthetic": 51.

Chapter 6

1. Reyner Banham, "Had I the Wheels of an Angel," *New Society* 18 (August 12, 1971): 463–464.
2. Email from Rob van der Plas to the author, August 30, 2013; "Quelques stands remarques a l' I.F.M.A.," *Le Cycle* (November 10, 1956): 21; "Le Salon du Cycle" [Paris Show, 1960], *Le Cycle* (October 1960): 33.
3. Tony Hadland, *Raleigh: Past and Presence of an Iconic Bicycle Brand* (San Francisco: Cycle Publishing, 2012): 206.
4. Hadland, *Raleigh: Past and Presence of an Iconic Bicycle Brand*: 210.
5. Tony Hadland and John Pinkerton, *It's in the Bag: A History in Outline of Portable Cycles in the UK* (Cheltenham: Quorum, 1996): 40–44; email from Tony Hadland to the author, January 31, 2010.
6. Thanks to Tony Hadland for pointing this out.
7. Hadland; *Raleigh: Past and Presence of an Iconic Bicycle Brand*: 218; Monopolies and Mergers Commission, *Bicycles: A Report on the Application by TI-Raleigh Industries Ltd. and TI-Raleigh Ltd. of Certain Criteria for Determining Whether to Supply Bicycles to Retail Outlets* (London: Her Majesty's Stationery Office, December 1981) (hereafter, *MAMC 1981 Bicycles Report*): 8; Interview with John Macnaughtan, February 24, 2017.

8. John H. Goldthorpe, et al., *The Affluent Worker: Part 2, Political Attitudes and Behavior* (Cambridge, UK: Cambridge University Press, 1968): Table 16; *The Affluent Worker: Part 3, The Affluent Worker in the Class Structure* (Cambridge, UK: Cambridge University Press, 1969): 98.

9. John H Goldthorpe, et al., *The Affluent Worker: Part 2, Political Attitudes and Behavior*: Table 16 (car and home ownership); p. 50 and p. 50 n1 (white-collar affiliations); Goldthorpe, et al., *The Affluent Worker: Part 3, The Affluent Worker in the Class Structure*: 91 (white-collar affiliations).

10. David Kynaston, *Modernity Britain: Opening the Box, 1957–59* (London: Bloomsbury, 2013): 232.

11. According to Goldthorpe and his colleagues, the two most significant factors comprising the "new working class consciousness" were: (1) decreased (or at least delayed) fecundity; and, (2) a desire that their children receive the most and best education available, given family and community resource constraints. See: Goldthorpe, et al., *The Affluent Worker: Part 3, The Affluent Worker in the Class Structure*, Table 16, pp. 130–134.

12. Tony Hadland, "The Raleigh Twenty Range: Raleigh's Biggest Seller of the mid-1970s," https://: Hadland.woodpress.com/2012/07/01/articles.

13. *Final Report of the President's Commission on Product Safety* (Washington, 1970); Ross Petty, "The Consumer Product Safety Commission's promulgation of a Bicycle Safety Standard," *Journal of Products Liability* 10 (1987): 25–50.

14. In his book *Raleigh: Past and Presence of an Iconic Brand* (2012, p. 218), Tony Hadland shows a picture of what he claims is a prototype for the R20 Stowaway (folder). Instead of the metal upside-down "L" shaped handle for the folding mechanism on top of the monotube used on the production version, there is a round plastic knob, probably again investigated out of CPSC regulatory concerns.

15. *Raleigh in America Catalog for 1968*; *Raleigh in America Catalog for 1969*; *Raleigh in America Catalog for 1970*.

16. Bicycle Safety Rules: 16 CFR 1512.4(g). *See*: 38 Federal Register (May 10, 1973): 12300–12305; 41 Federal Register (January 28, 1976): 4144–4154. Concerns of folding bike makers regarding hinge handles, *see*: 40 Federal Register (June 16, 1975) at 25482–25488. Eligible for waiver: John Allen, John Dowlin and John Schubert, "Buying a Folding Bicycle," *Bicycling* 22, 2 (March 1981): 50–52, 106.

17. Hadland; *Raleigh: Past and Presence of an Iconic Bicycle* Brand: 297–298; *Raleigh in America Catalog for 1975*; *Raleigh in America Catalog for 1976*; *Raleigh in America Catalog for 1977*.

18. The story of the Stingray is taken from three sources: "The Bike Boom Rises to its Christmas Best," *Business Week* (December 21, 1968): 45–47; Judith Crown and Glenn Coleman, *No Hands: The Rise and Fall of the Schwinn Bicycle Company: An American Institution* (New York: Henry Holt, 1996): 70–81; Jay Pridmore and Jim Hurd, *Schwinn Bicycles* (Osceola: Motorbooks International, 1996): 120–132.

19. Transcript of author's interview of Norman Clarke, April 5, 1998: 17.

20. Frank Jr. and Nancy Wilson marriage: Crown and Coleman, *No Hands: The Rise and Fall of the Schwinn Bicycle Company*: 76–77. The marriage ended amiably after six months. *Ibid.*

21. Reyner Banham, "Had I the Wheels of an Angel," *New Society* 18 (August 12, 1971): 463–464. Graham scholarship: Penny Sparke, "Introduction," in Reyner Banham, *Design by Choice*, ed. Penny Sparke (New York: Rizzoli, 1981): 8, 17–18.

22. "Tube Investments-Raleighing," *The Economist* (February 12, 1963): 79.

23. Ross Petty, "Peddling Schwinn Bicycles: Marketing Lessons from the Leading Post-WWII U.S. Bicycle Brand," *Quinnipiac University CHARM Symposium Papers 2007*: 162–171; Gregory Houston Bowden, *The Story of the Raleigh Bicycle* (London: W. H. Allen, 1975): 169–173.

24. Ignatz Schwinn had two daughters and one son, Frank Sr. While the stock was split equally among them at his death, a trust ensured that only Frank Sr. and his descendents retained operational control of the firm. Frank Jr., was childless. At the time the firm entered bankruptcy in 1992, 14 lineal discendants held stock. Edward R. and Richard C., sons of Frank Jr.'s brother Richard, each held 3.4 percent of the stock, but were president and vice president, and collected salaries of $250,000 and $140,000 respectively. Their aunts, Betty Dombecki and Shirley White, daughters of Frank Sr.'s sister Margaret, each held 16.65 percent, but had no director's votes. Crown and Coleman blame this dysfunctional structure for the ultimate failure of the firm: Crown and Coleman, *No Hands*: 241–244.

25. Transcript of author's interview with Norman Clarke: 11.

26. Alex Moulton, *From Bristol to Bradford-on-Avon: A Lifetime in Engineering* (Derby: Rolls Royce Heritage Trust, 2009): 179. These bicycles were technically sold to an intermediary, Reynolds Trading, Ltd., located at Azusa, probably to circumvent tariff duties.

27. Banham, "Had I the Wheels of an Angel": 463–464.

28. Hadland, *Raleigh: Past and Presence of an Iconic Bicycle Brand*: 221; *Raleigh in America Retail Catalog for 1967*; *Raleigh in America Retail Catalog for 1968*.

29. Oakley's version: Bowden, *The Story of the Raleigh Bicycle*: 170–172. "Raleigh Chopper Designer Alan Oakley Dies from Cancer," BBC.com (May 20, 2012). Stingray Krate series: Pridmore and Hurd, *Schwinn Bicycles*: 110–122. According to Tony Hadland, Oakley knew the Raleigh version of the story was a marketing fable concocted by his employer and found it, and the Chopper itself, distasteful: email from Tony Hadland to the author, February 16, 2016.

30. Ogle Design version of the story: Tom Karen, "Designing the Future: An Industrial Designer's Perspective" (The 17th Reyner Banham Memorial Lecture) in *The Banham Lectures: Essays on Designing the Future*, ed. Jeremy Aynsley and Harriet Atkinson (Oxford: Berg, 2009): 252–263.

31. Banham, "Had I the Wheels of an Angel": 463–464.

32. Hadland, *Raleigh: Past and Presence of an Iconic Bicycle Brand*: 221–227.

33. Tony Hadland, *Raleigh: Past and Presence of an Iconic Name*: 227; David Duffield interview.

34. Crown and Coleman, *No Hands*: 109–110.

35. Crown and Coleman, *No Hands*: 150–151, 270; Pridmore and Hurd, *Schwinn Bicycles*: 128–132; Transcript of author's interview with Norman A. Clarke: 6.

36. Hadland, *Raleigh: Past and Presence of an Iconic Bicycle Brand*: 221; *MAMC 1981 Bicycles Report*: 8; David Duffield interview.

37. Banham quote: Reyner Banham, "A Grid on Two Farthings," in *Design by Choice*, ed. Penny Sparke (New York: Rizzoli, 1981): 119–120. 5,200 units: Tony Hadland was told this by Raleigh production manager Jim Bratby: email from Tony Hadland to the author, February 16, 2016.

38. Peter Knottley, Preview: The Moulton Prototype," *Bicycling* 11, 8 (February, 1970): 24–27; Tony Hadland, "Marketing the Mk3," *The Moultoneer* 70 (2004): 23–27. According to Jim Bratby: Hadland, *Raleigh: Past and Presence of an Iconic Bicycle Brand*: 192.

39. Hadland, *Raleigh: Past and Presence of an Iconic Bicycle Brand*: 217–220.

40. Tony Hadland, "The Doctor's GP," *The Moultoneer* 71 (2004): 42–47.

41. Arthur M. Louis, "How the Customers Thrust Unexpected Prosperity on the Bicycle Industry," *Fortune* 89, 3 (March 1974): 117–124. *MAMC 1981 Bicycles Report*: 8. The UK domestic market: 569K (1969); 739K (1972). Raleigh domestic production: 375K (1970, 1969 not available); 525K (1972). Small wheel sales: Banham, "Had I the Wheels of an Angel": 461.

42. Bowden, *The Story of the Raleigh Cycle*: 173–176; Hadland, *Raleigh: Past and Presence of an Iconic Bicycle Brand*: 294–296.

43. McLarty quote: Hadland, *Raleigh: Past and Presence of an Iconic Bicycle Brand*: 297. Schwinn caps output: Louis, "How the Customers Thrust Unexpected Prosperity on the Bicycle Industry": 118–119. For a while the Le Tour series was built at Schwinn's Greenville, Mississippi, plant before it was shuttered: Crown and Coleman, *No Hands*: 149–150.

44. Hadland, *Raleigh: Past and Presence of an Iconic Bicycle Brand*: 216; Moulton, *From Bristol to Bradford-on-Avon*: 190–192.

45. Tony Hadland, *The Spaceframe Moultons* (Picataway, NJ: Transaction, 1994): *passim*.

46. Who should not be confused with the British-born, California-residing cycling historian and writer of the same name.

47. "Andrew Ritchie and His Brompton," *Spin Asia* (Singapore edition) (May 2010): 9–12; "The Bicycle that Turned into Folding Money," *The Observer* (August 6, 2005).

48. "Andrew Ritchie and his Brompton": 11. To be precise, the name of the church is the London Oratory on Brompton Road.

49. Tony Hadland and John Pinkerton, *It's in the Bag: a History of Portable Cycles in the UK* (Cheltenham: Quorum, 1996): 82–86; made 50 instead of 30: "The Bicycle that Turned into Folding Money."

50. *MAMC 1981 Bicycles Report*: 2.

51. *MAMC 1981 Bicycles Report*: 37.

52. *MAMC 1981 Bicycles Report*: 42.

53. John Macnaughtan interview, Part 2 (2016).

54. "Pedal Power," *The Economist* (July 19, 1980): 70; "Tube Investments-Raleighing," *The Economist* (February 12, 1983): 79.

55. Hadland, *Raleigh: Past and Presence of an Iconic Bicycle Brand:* Chapters 41 and 45.

56. John Macnaughtan interview, Part 2 (2016).

57. "Raleigh Chopper Designer Alan Oakley Dies from Cancer," BBC.com (May 21, 2012).

Chapter 7

1. Frank Dudas, "Flash Gordon and American Auto Design in the 1960s," in *The Banham Lectures: Essays on Designing the Future*, Jeremy Aynsley and Harriet Atkinson, eds. (New York: Berg, 2009 [1991]): 99–110.

2. Reyner Banham, "Unlovable at Any Speed," *Architects' Journal* 144 (December 21, 1966): 1527–1529.

3. In the forty years since Wolfe's comment and *Learning from Las Vegas*, Las Vegas has, to a large degree, evolved into a form of monumentalist Disneylandism, so as to sustain its previously nocturnal effects both day and night. The middle-class money sought by Las Vegas's new generation of multinational corporate owners comes from families with kids who still go to bed at night, even if a little late, not from open-shirted fifty-year-old stag males.

4. Lawrence Alloway, "City Notes," *Architectural Design* 29 (January 1959): 34–35.

5. Reyner Banham, "Encounter with Sunset Boulevard," *The Listener* 80 (August 22, 1968): 235–236; Reyner Banham, *Los Angeles: The Architecture of the Four Ecologies* (New York: Harper & Row, 1976): 237.

6. Nigel Whiteley, *Reyner Banham, Historian of the Immediate Future* (Cambridge, MA: MIT Press, 2002): 239–241. Schlockology: Anthony Vidler, *Histories of the Immediate Present: Inventing Architectural Modernism* (Cambridge, MA: MIT Press, 2008): 142–143. The fashionable sonofabitch: Peter Plagens, "Los Angeles: The Ecology of Evil," *Artforum* (December 1972): 76.

7. Corinne Julius interview of Mary Banham, Tape 9 (January 2002): http://Sounds.bl.uk/oral-history/Architects-Lives/021M-C0467X0067XX-0009V0. Hector Corfiano himself trained at the Beaux-Arts and ran the Bartlett's architectural program from 1946 to 1959. He was intensely unpopular for his habit of critiquing students' work-in-progress with a fat 2B pencil. Experienced students knew to gather up their work and run when they saw him enter the studio.

8. Corinne Julius interview of Mary Banham, Tape 12 (February 2002). Actually, the Architectural Press published two magazines, the monthly *Architectural Review* and the biweekly *Architects' Journal*.

9. James Sloan Allen, *The Romance of Commerce and Culture: Capitalism, Modernism and the Chicago-Aspen Crusade for Cultural Reform* (Boulder: University of Colorado Press, 2002): *passim;* Corinne Julius interview of Mary Banham, Tape 9 (January 2002).

10. Letter from Peter [Reyner] Banham to Mary Banham, June 19, 1970. Reyner Banham Papers, Library and Archives Division, Getty Museum, Los Angeles, CA. Alice Twemlow, "I Can't Talk to You If You Say That: An Ideological Collision at the International Design Conference at Aspen, 1970," *Design and Culture* 1, 1 (2009): 23–50. Twemlow notes that the two sides weren't even speaking a common language. To the traditional designers "environment" meant a "space" "defined place" or "enclosure." To the students and activists it had the now-common meaning of "a natural biotic system."

11. Robert Maxwell, "Reyner Banham—Historian," in *Sweet Disorder and the Carefully Careless: Theory and Criticism in Architecture* (Princeton: Priceton Architectural Press, 1993): 163–183. Peter Hodgkinson, "Drug-In City," *Architectural Design* (November 1969): 586.

12. Twemlow, "I Can't Talk to You If You Say That": 27, 44 n2. Pre-paid registrations were 625. Multiple estimates give total attendance as about 1,000. However, the number of on-site registrations is not known. In a typical year, it was small—about 100—due to limited housing and transport.

13. A land lacking in culture: Twemlow, "I Can't Talk to You If You Say That": 30; "Joint Statement," [Joint Composition of Eleven 1970 Attendees], *IDCA 1970 Special Edition of the Aspen Times* (June 20–24, 1971), Group IV, File 6, Reyner Banham Papers, Division of Archives, Getty Museum, Los Angeles.

14. "Joint Statement" [of the 1970 French attendees]: n.p.

15. Twemlow, "I Can't Talk to You If You Say That": 30.

16. Reyner Banham, "Design by Choice," *Architectural Review* 130 (July 1961): 43–48. Reyner Banham, ed., *The Aspen Papers* (New York: Praeger, 1974).

17. See: Alison Smithson, ed., "CIAM Team 10," *Architectural Design* 30, 5 (May 1960): entire issue; Alison Smithson, ed., "Team 10 Primer 1953–1962," *Architectural Design* 32, 12 (December 1962): entire issue.

18. "Cliff Humphrey," *Conference Proceedings of the International Design Conference at Aspen, 1970*, IDCA Papers, Library and Archives Division, Getty Museum, Los Angeles, CA.

19. "Joint Statement," [of the 1970 French attendees]: n.p.

20. Letter from Peter [Reyner] Banham to Mary Banham, June 19, 1970. Banham Papers, Getty Museum. Banham referred to the contingent as the "Chicago Black Panthers," probably jokingly. They were actually 15 black and Hispanic industrial design students from Chicago under the guidance of black designer Stephen Frazier. Banham's brush-off of the well-respected Frazier and his hard-working, underfinanced contingent was not his finest moment.

21. Letter from Peter [Reyner] Banham to Mary Banham, June 19, 1970. Banham Papers, Getty Museum; Twemlow, "I Can't Talk to You If You Say That": 38. The individual clauses were, with the exception of the last, the widely controversial issues of the day: end the Vietnam War; end the draft, legalize abortion; recognize tribal lands of American Indians; equality for women, blacks, Hispanics, and homosexuals; pledge that designers will not produce material for capitalistic ends.

22. Reyner Banham, ed., *The Aspen Papers* (New York: Praeger, 1974): 222.

23. Twemlow, "I Can't Talk to You If You Say That: 47 n22.

24. Alice Twemlow, "Purposes, Poetics and Publics: the Shifting Dynamics of Design Criticism in the U.S. and UK, 1955, 2007." (Ph.D. dissertation, Royal College of Art, 2013): 206–207, 212."

25. Twemlow, "I Can't Talk to You If You Say That": 38.

26. A community of intellectuals: Whiteley, *Reyner Banham, Historian of the Immediate Future*: 399–400. *Pop Goes the Easel* and *Fathers of Pop*: Anne Massey, *The Independent Group: Modernism and Mass Culture in Britain, 1945–59* (Manchester: Manchester University Press, 1995): 112–113; *Fathers of Pop* (Julian Cooper, director, 47 min., Arts Council of Great Britain), 1979. Massey, Penny Sparke, Gillian Naylor, and even Mary Banham later commented on the title "Fathers of Pop," given that seminal figures in the success of the Independent Group included Magda Cordell, Alison Smithson, Terry Hamilton and ICA director Dorothy Morland: Gillian Naylor, "Theory and Design: The Banham Factor," in *The Banham Lectures: Essays on Designing the Future*, ed. Jeremy Aynsley and Harriet Atkinson (Oxford: Berg, 2009): 47–58; Mary Banham, "Retrospective Statement," in *The Independent Group: Postwar Britain and the Aesthetics of Plenty*, ed. David Robbins (Cambridge, MA: MIT Press, 1990): 187–188.

27. Reyner Banham, "Introduction," in *Design by Choice*, ed. Penny Sparke (New York: Rizzoli: 1981): 3–7.

28. Penny Sparke, "Obituary: Peter Reyner Banham, 1922–1988," *Journal of Design History* 1, 2 (1988): 141–142; Adrian Forty, "Reyner Banham, One Partially Americanized European," in *Twentieth-Century Architecture and its Histories*, ed. Louise Campbell (Oatley: Society of Architectural Historians of Great Britain, 2000): 195–205, at 197; Sparke, "Introduction" in Banham, *Design by Choice*: 17.

29. Whiteley, *Reyner Banham, Historian of the Immediate Future*: 399–400.

30. Sparke, "Introduction," in *Design by Choice*: 17.

31. Maxwell, "Reyner Banham—Historian": 183; "Pedro Ignacio Alonso & Thomas Weaver in Conversation with Tim Street-Porter," *AA Files* 62 (2011): 28–33.

32. Corinne Julius interview of Mary Banham, Tape 18 (May 2002) http://Sounds.bl.uk/oral-history/Architects-Lives/021M-C0467X0067XX-0018V0.

33. Corinne Julius interview of Mary Banham, Tape 17 (February 2002) http://Sounds.bl.uk/oral-history/Architects-Lives/021M-C0467X0067XX-0017V0.

34. Corinne Julius interview of Mary Banham, Tape 17 (February 2002).

35. Reyner Banham, "Alternative Wheels?" *New Society* 46 846/847 (December 28, 1978): 712–713.

36. Driver's license suspension: Pedro Ignacio Alonso and Thomas Weaver, "Deserta," *AA Files* 62 (2011): 20–27. Manufacturing history: Tony Hadland and John Pinkerton, *It's in the Bag! A History in Outline of Portable Bicycles in the UK* (Cheltenham: Quorum, 1996): 67–71.

37. Richard Ballantine, *Richard's Bicycle Book* (New York: Ballantine, 1978): 63–64.

38. Richard Ballantine, "The Bickerton Special," *Bike World* 5, 12 (December 1976): 16–18.

39. "Pedro Ignacio Alonso & Thomas Weaver in Conversation with Tim Street-Porter": 28–29. Corinne Julius interview of Mary Banham, Tape 18 (May 2002).

40. Mary Banham's recollections: Alonso and Weaver, "Deserta": 20–27. Moving handlebars: Ballantine, "The Bickerton Special": 18.

41. Alonso and Weaver, "Deserta": 21.

42. The sales of derailleur-geared bicycles was 610,000 in 1969, 3.5 million in 1975: Frank J. Berto, "The Great American Bicycle Boom" in *Cycle History 10: Proceedings of the 10th International Cycle History Conference, Nijmegan*, ed. Hans Erhard Lessing and Andrew Ritchie (San Francisco: Van der Plas, 2000): 133–148. 40 million bicycles: "U.S. Bicycle Market Statistics," *Schwinn Reporter* (February 1978): n.p. Clarke quote: Norman Clarke, "There Oughta Be a Law" in *Proceedings of the Seminar on Bicycle-Pedestrian Planning and Design, Orlando, Florida, Dec. 12–14, 1973* (New York: ASCE, 1974) 549–551.

43. History of the Schwinn market report: Jeff Mapes, *Pedaling Revolution: How Cyclists Are Changing American Cities* (Oregon State University Press, 2009): 29–31. Clarke quote: Transcript of author's interview with Norman Clarke, April 5, 1998: 16.

44. Frank J. Berto, *The Dancing Chain: History and Development of the Derailleur Bicycle,* 2d ed., (San Francisco: Cycle, 2005): *passim.*

45. Reyner Banham, "Alternative Wheels?": 712–713.

46. Email from Tony Hadland to the author, January 31, 2010.

47. Ray Thomas, "Milton Keynes, City of the Future?" *Built Environment* 9, 3–4(1983): 245–254; Hugh McClintock, "Planning for the Bicycle in Newer and Older Towns and Cities," in *The Bicycle and City Traffic, ed. Hugh McClintock* (London: Belhaven, 1992): 40–61. Nordweststadt and Cumbernauld: Rosemary Wakeman, *Practicing Utopia: An Intellectual History of the New Town Movement* (Chicago: University of Chicago Press, 2016): 270–276.

48. William Oakley, *Winged Wheel: The History of the First Hundred Years of the Cyclists' Touring Club* (Galdalming: CTC, 1977): 27, 41, 91, 165. John Pinkerton, "Who Put the Working Man on a Bicycle," in *Cycle History 8: Proceedings of the Eighth Cycle History Conference, Glasgow,* ed. Nicholas Oddy and Rob van der Plas (San Francisco: Van der Plas, 1998): 101–106.

49. Peter Cox, "A Denial of our Boasted Civilization: Cyclists' Views on Conflicts Over Roads Use in Britain, 1926–35," *Transfers* 2, 3 (Winter 2012): 4–30.

50. Peter Walker, "75 Years After the UK's First Cycle Lane Opened, the Same Debate Rages On," *The Environmental Guardian Online*: http://www.Environmentalguardianonline.com/ (posted December 13, 2009); Oakley, *Winged Wheel*: 190.

51. Oakley, *Winged Wheel*: 178.

52. Manuel Stoffers, "Cycling as Heritage: Representing the History of Cycling in the Netherlands," *Journal of Transport History* 33, 1 (June 2012): 92–114; Ton Welleman, "An Efficient Means of Transport: Experiences with Cycling Transport Policy in the Netherlands" in *Planning for Cycling: Principles, Practices and Solutions for Urban Planners,* ed. Hugh McClintock (Cambridge, UK: Woodhead, 2002): 192–208.

53. Thomas Krag, "Urban Cycling in Copenhagen" in *Planning for Cycling: Principles, Practices and Solutions for Urban Planners,* ed. Hugh McClintock (Cambridge, UK: Woodhead, 2002): 223–236.

54. "London Roads Study" in Oscar Newman, ed., *CIAM 1959 in Otterlo* (London: Alec Tiranti, 1961): 77–79.

55. Nick Tiratsoo, *Reconstruction, Affluence, and Labour Politics: Coventry, 1945–60* (London: Routledge, 1990): 79.

56. Wakeman, *Practicing Utopia: An Intellectual History of the New Town Movement*: 295–296.

57. Tiratsoo, *Reconstruction, Affluence, and Labour Politics: Coventry, 1945–60*: 52.

58. John Gringrod, *Concretopia: A Journey Around the Rebuilding of Postwar Britain* (Brecon: Old Street, 2013): 330–336.

59. Nicholas Bullock, "Building the Socialist Dream or Housing Socialist State? Design Versus Production of Housing in the 1960s," in *Neo-Avant-Garde and Postmodernism: Postwar Architecture in Britain and Beyond,* ed. Mark Crinson and Claire Zimmerman (New Haven: Yale University Press): 321–342.

60. "Robin Hood Gardens, E14," *Architectural Design* (September 1972): 559–572.

61. Robin Bowdler and Peter Smithson: Maxwell Hutchinson, "Rebuilding Britain for Baby Boomers," (Lindsay Leonard, prod., 58 min. BBC Radio 4, 2004), www.bbc.co.uk/programmes/b017187m

62. Gringrod, *Concretopia*: 167–171.

63. Hutchinson, "Rebuilding Britain for Baby Boomers."

64. Hutchinson, "Rebuilding Britain for Baby Boomers."

65. Hutchinson, "Rebuilding Britain for Baby Boomers."

66. Gringrod, *Concretopia*: 362.

67. Hutchinson, "Rebuilding Britain for Baby Boomers."

68. Reyner Banham, "Apropros the Smithsons," *New Statesman* (September 8, 1961): 317–318.

69. Cook quote: Peter Cook, "Regarding the Smithsons," in *Alison & Peter Smithson: A Critical Anthology,* ed. Max Risselada (Barcelona: Ediciones Poligrafa, 2011): 290–307. Avant-guarde: Banham, "Apropros the Smithsons": 317.

70. Smithson quote: Beatriz Colomina, "Friends of the Future: A Conversation with Peter Smithson," *October* 94 (Fall 2000): 3–30.

71. Danny Boyle and Sara Knapton, "Grenfell Tower Victims Poisoned by Cyanide After Insulation Released Highly Toxic Gas," *The Telegraph*, June 22, 2017; David D. Kirkpatrick, Danny Hakim and James Glanz, "An Accident Waiting to Happen: Blame in a Deadly London Fire," *New York Times*, June 25, 2017; Stephen Castle, "Fire Risk in London Flats Highlights a Class Divide," *New York Times*, June 25, 2017; updated through July 12, 2017, via BBC Radio and Television news reports.

Chapter 8

1. Brian Phillips, "The Sea of Crises," *Grantland* (2014), reprinted in, *The Best American Magazine Writing 2015,* ed. Sid Holt (New York: Columbia University Press): 313- 340, at 321.

2. Robert Maxwell, "Reyner Banham—The Man," in *Sweet Disorder and the Carefully Careless* (Princeton: Princeton Architectural Press, 1993): 177–182.

3. Mary Banham interview by Corine Julius, Tape 18 (May 2002): http://Sounds.bl.uk/oral-history/Architects-Lives/021M-C0467X0067XX-0018V0.

4. Mary Banham interview by Corinne Julius, Tape 18 (May 2002).

5. Mary Banham interview by Corinne Julius, Tape 18 (May 2002).

6. Nigel Whiteley, *Reyner Banham: Historian of the Immediate Future* (Cambridge, MA: MIT Press, 2002): 404.

7. Reyner Banham, *A Concrete Atlantis: U.S. Industrial Building and European Modern Architecture, 1900–1925* (Cambridge, MA: MIT Press, 1986).

8. Peter Reyner Banham, *Scenes in American Deserta* (London: Thames and Hudson, 1982): 221; Peter Hall, "Introduction," in *A Critic Writes: Essays by Reyner Banham,* ed. Mary Banham, et al. (Berkeley: University of California Press, 1996): xi–xv.

9. "Pedro Ignacio Alonso and Thomas Weaver in Conversation with Tim Street-Porter," *AA Files* 62 (2011): 28–33.
10. Banham, *Scenes in American Deserta*: 98.
11. "Pedro Ignacio Alonso and Thomas Weaver in Conversation with Tim Street-Porter": 30.
12. "Pedro Ignacio Alonso and Thomas Weaver in Conversation with Tim Street-Porter": 28–29.
13. Banham, *Scenes in American Deserta*: 98–99.
14. "Pedro Ignacio Alonso and Thomas Weaver in Conversation with Tim Street-Porter": 31.
15. Peter Hall, "Introduction": xv; Mary Banham, "Introduction: The 1980s,": 235, both from: *A Critic Writes: Essays by Reyner Banham*, ed. Mary Banham, et al. (Berkeley: University of California Press, 1996). "Reyner Banham, 1922–1988 [obituary]," *AA Files* 16 (Autumn 1987): 33–40. *Note*: this issue carries the cover date "Autumn 1987" although it was published six months after Banham's death.
16. Tony Hadland and John Pinkerton, *It's in the Bag: a History of Portable Cycles in the UK* (Cheltenham: Quorum, 1996): 35.
17. Hadland and Pinkerton, *It's in the Bag*: 82–86; "Andrew Ritchie and his Brompton," *Spin Asia* (Singapore edition) (May 2010): 9–12. Ritchie quote: "The Bicycle that Turned into Folding Money," *The Observer* (August 6, 2005).
18. Tony Hadland, *Raleigh: Past and Presence of an Iconic Bicycle Brand* (San Francisco: Cycle Publishing, 2012): 261; Gregory Houston Bowden, *History of the Raleigh Cycle* (London: W. H. Allen, 1975) 196; Michael J. Kolin and Denise M. de la Rosa, *The Custom Bicycle* (Emmaus, PA: Rodale, 1979): 84–94. Moulton recalled: Alex Moulton, *From Bristol to Bradford-on-Avon: A Lifetime in Engineering* (Derby: Rolls Royce Heritage Trust, 2009): 189.
19. Email from Tony Hadland to the author, February 17, 2016.
20. Will Smale, "Brompton Boss: the Bike-Maker Who Disproved the Doubters," *BBC Business On-line*, June 29, 2009; Jon Excell, "Brompton Managing Director Will Butler-Adams," *The Engineer* (February 28, 2011): n.p; interview with Nick Charlier, Public Relations and Communications Executive, Brompton Bicycles, February 27, 2017.
21. Excell, "Brompton Managing Director Will Butler-Adams": n.p.
22. Hadland and Pinkerton, *It's in the Bag: a History of Portable Cycles in the UK:* 73, 124; Tony Hadland, "Small Wheels for Adult Bicycles" *Cycling Science* (Fall 1997): 3–7; Tony Hadland and Hans-Erhard Lessing: *Bicycle Design* (Cambridge, MA: MIT Press): 447–471. Tikit: Jeffrey Winters, "Origami Cycle," *Mechanical Engineering* 130, 6 (June 2008): 48–49.
23. Nicole Formosa, "Josh Hon Talks Tern One Year In," *Bicycle Retailer and Industry News* 21, 12 (July 15, 2012): 20.
24. Matt Wiebe, "Harry Montague Left Behind Folding Legacy," *Bicycle Retailer and Industry News* 20, 4 (March 15, 2011): 17; Jason Norman, "Folding Bike Makers Report Robust Sales During First Half 2007," *Bicycle Retailer and Industry News* 16, 15 (September 1, 2007): 35.
25. Doug McClellan, "Dahon Forges Ahead with Environmental Plans as it Turns 25," *Bicycle Retailer and Industry News* 17, 5 (April 1, 2008): 47.
26. Sheldon Brown and John S. Allen, "Five New Folding Bicycles," *Bicycling* 25, 5 (May 1984): 16–19; 61; Ben DeLaney, "David Hon: Management With a Human Face," *Bicycle Retailer and Industry News* 14, 6 (April 15, 2005): 15.
27. Author's interview with Alan Issacson, December 9, 2015.
28. "Retailers About Ready to Pack Them in Are Discovering More Consumers Want Them," *Bicycle Retailer and Industry News* (September 2000): 13.
29. DeLaney, "David Hon: Management With a Human Face": 15; "Retailers About Ready to Pack Them in Are Discovering More Consumers Want Them": 13; McClellan, "Dahon Forges Ahead with Environmental Plans as it Turns 25": 47
30. Further complicating matters is the fact that David Hon is a Ph.D. physicist who, prior to starting Hon Bicycles in 1981, was working at Hughes Aerospace on the development of laser-based weapon systems (DeLaney, "David Hon: Management With a Human Face": 15).
31. Nicole Formosa, "Hon Family at Center of Folding Bike Controversy," *Bicycle Retailer and Industry News* 20, 12 (July 15, 2011): 1–2; Nicole Formosa, "Josh Hon Talks Tern One Year In," *Bicycle Retailer and Industry News* 21, 12 (July 15, 2012): 20.
32. Moulton, *From Bristol to Bradford-on-Avon*: 195.
33. Tony Hadland, *The Spaceframe Moultons* (Picataway, NJ: Transaction, 1994); Hadland and Pinkerton, *It's in the Bag!*: 107–116.
34. There are several "20-inch" standards. The two most common are the "American" or "BMX" 406 mm and the "British" 451 mm; the numbers indicate the diameter of the tire's wire bead. (A distant third is the French 550A with a bead diameter of 440 mm.) This makes the BMX rim 16.0 inches at the bead and the British rim 17.75 inches at the bead. In the past, 451 mm tires generally had a narrower, thinner profile, so the overall (tread) diameter of the two was close, even if the rim diameter was quite different at the bead seat. But this is no longer true; 406 mm tires are now available in a wide selection of profiles, treads and pressures; as a result, the 451 mm has become less common. Also, as Tony Hadland notes, 406 mm bead tires now come in a profile so thin that the tread diameter is as small as 18.3 inches, whereas the 369 mm wheel with a thin tubular tire has a tread diameter of 17.5 inches—hard to tell apart, especially if the 369 mm rim is shod with something a little bigger than a racing tubular.
35. Interview with Jack Lauterwasser, https://www.youtube.com/channel/UC2XzNEKoaCK1INKmSIo-6wRA/.
36. "The Bike Show from Resonance FM: The Moulton Story, Part 1," http://bikeshow_20080929_64kb_m3u.
37. Interview with Dan Farrell, February 24, 2017.
38. Interview with Dan Farrell, February 24, 2017, Dan Farrell, "The Moulton Bicycle: Origin and Intent," one-page manuscript, n.l. [Bradford-on-Avon], 2016.
39. Alex Moulton also makes oblique reference to a "leakage" of produced bicycles, which could mean either bicycles shipped without proper billing—or theft: Moulton, *From Bristol to Bradford-on-Avon*: 195.
40. Moulton, *From Bristol to Bradford-on-Avon*: 195.
41. Hadland, *The Spaceframe Moultons*: *passim*;

Christine Murray, "Reinventing the Wheel," *Architects Journal* (September 4, 2009): 95–97.

42. Hadland and Pinkerton, *It's in the Bag!*: 114–115. The license expired in 2016, according to John Macnaughtan.

43. Alex Moulton, Anthony Hadland and Douglas L. Milliken, "Aerodynamic Research Using the Moulton Small-Wheeled Bicycle," *Proceeding of the Institution of Mechanical Engineers*, 220, 3 (2006): 189–193.

44. Interview of John Macnaughtan, Part 2 (2016), https://www.youtube.com/channel/UC2XzNEKoaCK1INKmSIo6wRA/.

45. Doug McClellan, "Folder Sales Unfold as Buyers Opt for More Compact Rides," *Bicycle Retailer and Industry News* 17, 16 (October 1, 2008): 65.

46. "Alex Moulton" (obituary) *The Guardian* (December 10, 2012); interview with Dan Farrell, February 24, 2107.

47. Interview with John Macnaughtan, Part 2 (2016).

48. Interview with John Macnaughtan, Part 2 (2016).

49. "Bicycle Frame," UK Patent GB-2471041-A, to Moulton Developments Ltd. and Alexander Eric Moulton, issued Dec. 15, 2010.

50. Moulton, *From Bristol to Bradford-on-Avon*: 232; Dan Farrell interview, February 24, 2017.

51. Interview with John Macnaughtan, Part 2 (2016); author's interview with John Macnaughtan, February 24, 2017.

52. Author's interview with John Macnaughtan, February 24, 2017.

53. Interview with John Macnaughtan, Part 2 (2016); author's interview with John Macnaughtan, February 24, 2017; author's interview with Dan Farrell, February 24, 2017.

Chapter 9

1. Mary Banham, "Retrospective Statement" in *The Independent Group and the Aesthetics of Plenty*, ed. David Robbins (Cambridge, MA: MI Press, 1990): 187–188.

2. For a recent, technical discussion of the platform concept, especially as applied to the bicycle industry, see Peter Galvin and Andre Morkel, "The Effect of Product Modularity on Industry Structure: the Case of the World Bicycle Industry," *Industry and Innovation* 8, 1 (April 2001): 31–47.

3. Reyner Banham, "Without Mercy or Metaphor," *The Listener* (November 13, 1958): 775–776; Reyner Banham, "Stocktaking," *Architectural Review* 127 (February 1960): 93–100.

4. Nigel Whiteley, *Reyner Banham: Historian of the Immediate Future* (Cambridge, MA: MIT Press, 2002): 173–176; Martin Pawley, "We Shall Not Bulldoze Westminster Abbey: Archigram and the Retreat from Technology," in *The Strange Death of Architectural Crtiticism: Martin Pawley Collected Writings* (London: Black Dog, 2007): 59–69.

5. Reyner Banham, "A Clip-on Architecture," *Design Quarterly* 63 (1965): 3–30, quote at 9; Reyner Banham, "A Clip-on Architecture," *Architectural Design* 35 (November 1965): 524–535; Reyner Banham, "Arts in Society: Zoom Wave Hits Architecture," *New Society* 7, 179 (March 3, 1966): 21–22.

6. Corinne Julius interview of Mary Banham, Tape 10 (January 2002). http://Sounds.bl.uk/oral-history/Architects-Lives/021M-C04670067XX-0010V0.

7. Elizabeth A. T. Smith, *Case Study Houses* (Cologne: Taschen, 2016): 6–7; 23–29. Some published designs were never built. Some were erected later, after significant alteration. One or two were designs commissioned by clients and published only after completion. Case Study #23 was a three-house, three lot group, built as designed. Hence, there were 28 Case Studies and Two Apartment Case Studies. The apartment building was built in Phoenix. About a dozen remain largely unchanged.

8. Whiteley, *Reyner Banham: Historian of the Immediate Future*: 152–153; Reyner Banham, "On Trial 4: CLASP III Met by Clip-Joint?" *Architectural Review* 131 (May 1962): 349–352.

9. Martin Pawley, "Heavy Stuff this Symbolism," *The Guardian* (December 17, 1984); Reyner Banham, "The Writing on the Walls," *Times Literary Supplement* (November 17, 1978): 1337.

10. James E. Starrs, *The Noiseless Tenor: The Bicycle in Literature* (New York: Cornwall, 1982): 127–128. Starrs later became famous for his part in the Olson Investigation. In November 1953 four CIA employees, including Dr. Frank Olson, were, without warning, given large doses of LSD to test their responses. Olson's mental and physical heath subsequently deteriorated. His superiors sought medical care for him in New York, but could find no appropriate specialist with the required security clearance. A week after the drugging, Olson jumped through a closed tenth floor window in his hotel room. Olson's family believed he had been thrown through the window. In 2002 they exhumed Olson's body and Starrs undertook forensic analysis. The autopsy and subsequent analysis proved inconclusive. Albert J. Mauroni, *Chemical and Biological Warfare: A Reference Handbook* (Santa Barbara: ABC-CILO, 2007): s.v. "Operation Artichoke."

11. Patricio del Rio, "Slums Do Stink: Artists, Bricolage, and Our Need for Doses of Real Life," *Art Journal* 67, 1 (Spring 2008): 82–99.

12. Reyner Banham, "Bricologes a la Lanterne," *New Society* 37, 717 (July 1, 1976): 25–26.

13. del Rio, "Slums Do Stink" 85, 99.

14. Banham, "Bricologes a la Lanterne": 25–26.

15. Ross D. Petty, "The Bicycle as a Communications Medium," in *Cycle History 16: Proceedings of the 16th International Cycle History Conference, Davis*, ed. Andrew Ritchie (San Francisco: Van der Plas, 2006): 147–159; Gregory J. Downey, *Telegraph Messenger Boys: Labor, Technology, and Geography, 1850–1950* (New York: Routledge, 2002): 195–197.

16. Benjamin Stewart, "Bicycle Messengers and the Dialectics of Speed" in *Fighting for Time: Shifting Boundaries of Work and Social Life*, ed. Cynthia Fuchs Epstein and Arne L. Kalleberg (New York: Russell Sage Foundation, 2004): 150–190.

17. New York State Department of Labor, "Guidelines for Determining Worker Status in the Message Courier Industry (Albany: Division of Labor Standards, ca. December 2005).

18. Peter Cheney, "Bicycle Couriers in Love with Life on the Mean Streets," *Toronto Star* (March 27, 1993).

19. Stewart, "Bicycle Messengers and the Dialectics of Speed": 153; Rebecca Reiley, *Nerves of Steel* (Kenmore, NY: Spoke and Word, 2000): 15–25; Jef-

fery L. Kidder, "Appropriating the City: Space, Theory and Bike Messengers," *Theoretical Sociology* 38 (2009): 307–328.

20. Elizbieta Drazkiewicz, *On the Bicycle Towards Freedom: Bicycle Messenger's Answer for Identity Crisis* (M.S. Thesis, Lund University, December 2003): 7.

21. Karl Kim Connell, "The Bicycle Messenger," *The North American Review* 265, 4 (December 1980): 34–37.

22. Jack M. Kugelmass, "I'd Rather be a Messenger," *Natural History* 90, 8 (old series) (August, 1981): 66–73.

23. Drazkiewicz, *On the Bicycle Towards Freedom*: 2; Ben Fincham, "Generally Speaking People Are in it for the Cycling and the Beer: Bicycle Couriers, Subculture and Enjoyment," *The Sociological Review* 55, 2 (2007): 189–202.

24. Fincham, "Generally Speaking People Are in it for the Cycling and the Beer": 198.

25. Fincham, "Generally Speaking People Are in it for the Cycling and the Beer": 192; Jeffery L. Kidder, "Style and Action: A Decoding of Bike Messenger Symbols," *Journal of Contemporary Ethnography* 34, 2 (June 2005): 344–367, at 349; Drazkiewicz, *On the Bicycle Towards Freedom*: 9.

26. Jeffrey L. Kidder, *Alleycats, Fixies, and Double Rushes: An Ethnography of New York City Bike Messengers* (B.A. Thesis, University of North Carolina at Greensboro, 1999): 7–8 n7. Benjamin Fincham, *Bicycle Couriers in the New Economy* (Cardiff: Cardiff University Working Paper Series 46, January 2004): 26.

27. Email from Sam Shupe to the author, June 21, 2011.

28. Robert Lipsyte, "Coping: Voices from the Sweatshop of the Streets," *New York Times* (14 May 1995).

29. As Jeffery Kidder explains: "While this is a rather esoteric point, all track bikes are fixed gears, but not all fixed gears are track bikes. A track bike technically refers to frames and components that have been purposefully engineered for velodrome racing. Road bike and mountain bikes can be converted into fixed gears (by altering the ratcheting mechanism connected to the rear hub, or by replacing the hub), but they cannot be made into real track bikes. There are also fixed-gear-specific-frames that are not true road bikes (because they lack fittings for derailleurs), but they do not have track geometry. Fixed [gear bicycles] can also be fitted with brakes, but if a brake is installed on a track bike, it is no longer a real track bike (Jeffrey L. Kidder, *Emotions, Space, and Cultural Analysis: the Case of Bicycle Messengers*, Ph.D. dissertation, University of California at San Diego, 2009, 66 n33).

30. Leighton Klein, "For Cyclists Seeking Simplicity, the Fixed is In," *Boston Globe* (November 9, 2008): 4; Kugelmass, "I'd Rather be a Messenger,": 66–72; Sarah Lyall, "For Manhattan Couriers, Brakeless Bikes Are the Way to Go," *New York Times* (June 14, 1987).

31. Email from Tony Hadland to the author, February 16, 2017. Tom Avena's shop: Dick Teresi, *The Popular Mechanics Book of Bikes and Bicycling* (New York: Hearst, 1975): 115–129. Tariffs: "Wobble in Bicycles: Higher Tariffs Are No Substitute for Enterprise," *Barrons* (July 25, 1955): 1.

32. Peter Nye, *Hearts of Lions: The History of American Bicycle Racing* (New York: W. W. Norton, 1988): 65–130.

33. Henry Miller, "My Best Friend," in *My Bike & Other Friends* (Santa Barbara, CA: Capra, 1978): 105–110.

34. *Twentieth Century Authors*, ed. Stanley Kunitz and Howard Haycraft (New York: H. W. Wilson, 1942): s.v. Miller, Henry.

35. Oscar Wastyn: Teresi, *The Popular Mechanics Book of Bikes and Bicycling*: 117. Seattle: Kidder, *Emotions, Space, and Cultural Analysis: the Case of Bicycle Messengers:* 230.

36. Email from Sam Shupe, Boston University, to the author, June 21, 2011.

37. Jonathan Bodkin, "Fixed Gear Bicycles—A Modern Trend in a Historical Context," in *Cycle History 20: Proceedings of the 20th International Cycle History Conference, Freehold* (No location: John Pinkerton Memorial Publishing Fund, 2010): 168–175.

38. Bodkin, "Fixed Gear Bicycles—A Modern Trend in a Historical Context": 173.

39. According to Ford historian Robert Lacey, the Mustang was put together by the Ford design shop in the summer of 1961 in reaction to Chevrolet's Corvair Monza—unveiled in March 1960—and was introduced in the spring of 1964. Its development costs were about $7 million or $8 million dollars, about a quarter the cost of the Corvair. Robert Lacey, *Ford: The Men and the Machines* (New York: Little, Brown, 1986): 512–515.

40. Banham, "Bricologes a la Lanterne": 25–26; Charles Jencks, "Letter," *New Society* (July 15, 1976): 25; Reyner Banham, "Letter," New Society (July 22, 1976): 141; Charles Jencks, "Letter," *New Society* (July 29, 1976).

41. Reyner Banham, "Two by Jencks: The Tough Life of the Enfant Terrible," *AIA Journal* (December 1980): 50; Whiteley, *Reyner Banham; Historian of the Immediate Future*: 374–375.

42. Reyner Banham, "Alternative Wheels?" *New Society* 46, 846–847 (21/28 December 1978): 712–713.

43. Martin Cadin and Jay Barbree, *Bicycles in War* (New York: Hawthorn, 1974): Chapter 10.

44. Cadin and Barbee, *Bicycles in War*, 143.

45. Grant Petersen, *Just Ride: A Radically Practical Guide to riding Your Bike* (New York: Workman, 2012): 178–181.

46. Leigh F. Wade, "Letter to the Editor," *Bike Report* (May, 1990).

47. Email from Tony Hadland to the author, February 16, 2015.

48. Email from John S. Allen to the author, February 17, 2017; Peterson, *Just Ride*: 179–180.

49. This is one reason Brompton offers an unusual 6-speed combination of 3-speed rear hub and proprietary 2-cog rear derailleur. Asked why they use such a system, Brompton communication executive Nick Charlier answers that that it's the only way they can get the combination of compactness, folding speed, and reliability they need. Nobody's offered to engineer something for them, even for a bike that sells 50,000 units a year? "For Shimano, that's barely a container." Nick Charlier, Greenville, England, March 1, 2017.

50. Tony Hadland and Mike Burrows, *From Bicycle to Superbike* (n.l. [Oxford], Hadland Books, 2016): 85, 161.

51. Reyner Banham, "Had I the Wheels of an Angel," *New Society* 18 (August 12, 1971): 463–64.
52. Banham quote: "Had I the Wheels of an Angel": 463–464. Rix quote: Paul Rosen, *Framing Production: Technology, Culture and Change in the British Bicycle Industry* (Cambridge, MA: MIT Press, 2002): 103.

Chapter 10

1. Reyner Banham, "Introduction," in *Design by Choice*, ed. Penny Sparke (New York: Rizzoli, 1981): 19–22.
2. Tony Hadland and Hans-Erhard Lessing, *Bicycle Design: An Illustrated History* (Cambridge, MA: MIT Press, 2014): Chapter 15; Ben Delaney, "Retailer Hosting Small-Wheel, Folding-Bike Festival," *Bicycle Retailer and Industry News* 12, 8 (May 15, 2003): 16; "Performance Folding Bikes Attract Wide Customer Base," *Bicycle Retailer and Industry News* 13, 15 (September 1, 2004): 62–63.
3. "Andrew Ritchie and his Brompton," *Spin Asia* (Singapore Edition) (May 2010): 10–11; Stephen Krcmar, "With Diverse Offerings, Hip Styling, Folder Sales are on the Rise," *Bicycle Retailer and Industry News* 15, 16 (October 1, 2006): 52; Doug McClellan, "Folder Sales Unfold as Buyers Opt for More Compact Bikes," *Bicycle Retailer and Industry News* 17, 16 (October 1, 2008): 65; Ben DeLaney, "David Hon: Management Has a Human Face": *Bicycle Retailer and Industry News* 14, 6 (April 15, 2005): 15.
4. John Allen, John Dowlin and John Schubert, "Buying a Folding Bicycle," *Bicycling* 22, 2 (March 1981): 50–52, 106.
5. Biographical information on Sheldon Brown and his family has been taken from three sources: two illustrated essays written by Brown, "My Father" and "My Mother" at http://sheldonbrown.com. These were last accessed on March 20, 2015. The third source is "Sheldon C. Brown, Cycle Mechanic and Writer" [obituary], *Times of London* (March 4, 2008).
6. Email from John S. Allen to the author, December 8, 2015.
7. Sheldon Brown, "My 1964 Moulton Deluxe," http://sheldonbrown.com/mybikes.html. http://sheldonbrown.com/mybikes.html. Last accessed March 20, 2015.
8. Brown, "My 1964 Moulton Deluxe."
9. Brown wrote that 1972 was the first year that Twentys were first imported into North America, but it appears 1969 is the correct date.
10. "Sheldon Brown's Raleigh Twenty Site": http://sheldonbrown.com/raleigh-twenty.html.
11. Email from Aaron Gross to John S. Allen to the author, December 4, 2015; email from John S. Allen to the author, December 8, 2015.
12. Email from John S. Allen to the author, December 8, 2015.
13. Email from John S. Allen to the author, October 30, 2010; John S. Allen, "Twenty-Five Years on a Twenty," www.bikeexpert.com/twenty.html (written May 7, 2005).
14. John Allen and John Dowlin, "The Transit Bicycle," *Bicycling* 22, 2 (March 1981): 50–51, 118.
15. Sheldon Brown, "My Raleigh Twenties," http: sheldonbrown.com/Raleigh-twenty.html. A "Q-ship" is an armed fighting ship disguised to look like merchant cargo vessel or tanker, used to trap submarines.
16. Reyner Banham, "A Grid on Two Farthings," *New Statesman* 66 (November 1, 1963): 626–627.
17. "CLASP" stands for Consortium of Local Authorities Special Program [Prefabricated System]. It debuted at the 1960 Milan Tiennale. Eames Case Study houses: Elizabeth A. T. Smith, *Case Study Houses* (Cologne, Taschen, 2016): 6–7, 22–30. Banham quote: Reyner Banham, "A Clip-On Architecture," *Design Quarterly* 63 (1965): 3–30.
18. Service units for houses: Banham, "A Clip-on Architecture": 534–535. Smithson houses: Reyner Banham, "The House of the Future," *Design* (March 1956): 16.
19. Banham quote: Reyner Banham, "A Home is Not a House," *Art in America* 53 (April 1965): 70–79.
20. Anthony Vidler, "Futurist Modernism: Reyner Banham," in *Histories of the Immediate Present: Inventing Architectural Modernism* (Cambridge, MA, 2008): 106–155, quote at 133–134.
21. Reyner Banham, "Not Quite Architecture," On First Looking into Warshawsky," *Architect's Journal* 133 (July 26 1961): 109–111; Reyner Banham, "On Trial 4: CLASP III Met by Clip-Joint?" *Architectural Review* 131 (May 1962): 349–352.
22. Reyner Banham, "Softer Hardware," Ark 44 (Summer 1969): 2–11.
23. David N. Lucsko, *Junkyards, Gearheads and Rust: Salvaging the Automotive Past* (Baltimore: Johns Hopkins University Press, 2016): 46–47.
24. David N. Lucsko, "Junkyard Jamboree: Hunting for Automobile Treasure in the Twentieth Century," *Journal of the Society for Industrial Archeology* 39, 1/2 (2013): 93–111; quote at 103.
25. Lucsko, *Junkyards, Gearheads and Rust:* 61.
26. David N. Lucsko, "Of Clunkers and Camaros: Accelerated Vehicle Retirement Programs and the Automobile Enthusiast, 1990–2009," *Technology and Culture* 55, 2 (April 2014): 390–428. *See also*: Michael Thompson, *Rubbish Theory: The Creation and Destruction of Value* (New York: Oxford University Press, 1979): passim.
27. Luckso, "Of Clunkers and Camaros": the three states were Maine, Illinois and Texas. The "symbiotic vs. parasitic" argument largely rested on assumptions about the price of gas versus the lifetime price of used cars. In the end, the evidence was inconclusive that the clunkers pulled off the road consistently polluted more than the average car. Texas kept it going for decade largely as an economic stimulus measure and a vehicle safety program.
28. Alex Moulton, *From Bristol to Bradford-on-Avon: A Lifetime in Engineering* (Derby: Rolls Royce Heritage Trust, 2009): 27.
29. Interview with Alan Issacson and Eric Hodges, January 11, 2016.
30. Email from John S. Allen to the Author, December 3, 2015.
31. Author's interview with Alan Issacson, December 9th 2015.
32. Allen and Dowlin, "The Transit Bicycle": 118; Peter Cox, "The Role of Human Powered Vehicles in Sustainable Mobility," *Built Environment* 34, 2 (2008): 140–160.
33. Cox, "The Role of Human Powered Vehicles in Sustainable Mobility": 145.
34. Allen and Dowlin, "The Transit Bicycle": 118.
35. Authors interview with Allen Issacson, December 9, 2015.

36. Allen, "Twenty-Five Years on a Twenty."
37. Allen and Dowlin, "The Transit Bicycle": 50–51, 118–119.
38. Email from John S. Allen to the author, October 30, 2010.
39. Allen and Dowlin, "The Transit Bicycle": 50–51, 118–119.
40. Interview with Alan Issacson, December 9th 2015.
41. Interview with Eric Hodges. Interview with Tony Hadland, February 24, 2017. See also: Roderick Watson and Martin Gray, The Penguin Book of the Bicycle (Middlesex: Penguin, 1984 [1978]): 190–194.
42. Tony Hadland and John Pinkerton, *It's in the Bag! A History of Portable Cycles in Outline in the UK* (Cheltenham: Quorum, 1996): 43–46.
43. Hadland; *Raleigh: Past and Presence of an Iconic Bicycle Brand*: 217–219; *Raleigh in America Catalog for 1975; Raleigh in America Catalog for 1976; Raleigh in America Catalog for 1984*; Allen, Dowlin and Schubert, "Buying a Folding Bike": 50–52, 106, 118.
44. *Raleigh in America Retail Catalog for 1976*; Dick Teresi, *Popular Mechanics Book of Bicycles and Bicycling* (New York: Hearst, 1975): 37.
45. Email from John S. Allen to the author, December 3, 2015.
46. Interview with Eric Hodges, December 2, 2015.
47. Interview with Alan Issacson and with Eric Hodges, January 11, 2016.
48. Interview with Dan Farrell, February 24, 2017.
49. Interview with Alan Issacson and with Eric Hodges of Airport Cyclery (pseudonym) January 11, 2016.

Bibliography

Reyner Banham frequently wrote in regular columns with supertitles such as "Arts in Society" in *New Statesman* and "Not Quite Architecture" in *Architects' Journal*. There are, for example, over two hundred articles beginning with "Arts in Society," making alphabetization difficult. Mary Banham, et al., in their definitive bibliography in *A Critic Writes* (1996), did include the supertitles. However, their bibliography was organized chronologically, avoiding the alphabetization problem. Here, the supertitles have been deleted. Readers desiring more bibliographic data are referred to *A Critic Writes* and the supplemental bibliography in Nigel Whiteley's *Reyner Banham: Historian of the Immediate Future* (2002). Those Reyner Banham articles and essays reprinted in the anthology *Design by Choice* (1981) are indicated by "Reprinted in DBC" with the applicable page numbers. Similarly, those articles in *A Critic Writes* (1996) are indicated by "Reprinted in ACW," with page numbers. Those articles that discuss bicycles or bicycling to a significant extent are in **bold**.

Books and Articles

Abrams, Mark. *The Teenage Consumer.* London: London Press Exchange, 1959.
"Address to House of Lords by Lord Brabazon of Tara Regarding the Situation of Sir Roy Fedden." *Hansard's Lords Sitting* (17 October 1942).
Adler, Sy. "The Transformation of the Pacific Electric Railway: Bradford Snell, Roger Rabbit, and the Politics of Transportation in Los Angeles." *Urban Affairs Quarterly* 27 (September 1991): 51–86.
"Alex Moulton Obituary." *The Guardian* (December 10, 2012).
"Alex Moulton—Pioneer from a Stately Home." *Director* 19, 1 (July 1966): 42–43.
"Alexander Eric Moulton." [Obituary.] *New York Times* (December 20, 2012).
Allam, Hannah. "Messenger Chic Is Now a Fashion Statement." *Minneapolis Star-Tribune* (August 6, 1997).
Allen, Deborah. "Cars 55." *Industrial Design* 2 (February 1955): 85–90.
Allen, James Sloan. *The Romance of Commerce and Culture: Capitalism, Modernism and the Chicago-Aspen Crusade for Cultural Reform.* Boulder: University of Colorado Press, 2002.
Allen, John S., and John Dowlin. "The Transit Bicycle." *Bicycling* 22, 2 (March 1981): 50–51, 118.
Allen, John S., John Dowlin and John Schubert. "Buying a Folding Bicycle." *Bicycling* 22, 2 (March 1981): 50–52, 106.
Alloway, Lawrence. "Artists as Consumers." *Image* 3 (1961): 18.
———. "City Notes." *Architectural Design* 29 (January 1959): 34–35.
———. "The Independent Group: Postwar Britain and the Aesthetics of Plenty." In *The Independent Group: Postwar Britain and the Aesthetics of Plenty*, ed. David Robbins. Cambridge, MA: MIT Press, 1990, 49–53.
Alonso, Pedero Ignacio, and Thomas Weaver. "Deserta." *AA Files* 62 (2011): 20–27.
"American Ground Transport." Part 4A, U.S. Senate Committee on the Judiciary, *Hearings before the Subcommittee on Antitrust and Monopoly on S.1167, The Industrial Reorganization Act*, 93rd Cong., 2nd Sess. 1974.
"Andrew Ritchie and his Brompton." *Spin Asia* (Singapore edition) (May 2010): 9–12.
Ballantine, Richard. "The Bickerton Special." *Bike World* 5, 12 (December 1976): 16–18.
———. *Richard's Bicycle Book.* New York: Ballantine, 1978.

Banham, Mary. "The 1950s." In *A Critic Writes*, ed. Mary Banham, et al. Berkeley: University of California Press, 1996, 1.
_____. "The 1980s." In *A Critic Writes*, ed. Mary Banham, et al. Berkeley: University of California Press, 1996, 235.
_____. "Retrospective Statement." In *The Independent Group: Postwar Britain and the Aesthetics of Plenty*, ed. David Robbins. Cambridge, MA: MIT Press, 1990, 187–188.
Banham, Reyner. "Alternative Wheels?" *New Society* 46, 846–847 (21/28 December 1978): 712–713.
_____. "Anti-Technology." *New Society* (May 4, 1967): 645.
_____. "Apropros the Smithsons." *New Statesman* (September 8, 1961): 317–318.
_____. *The Architecture of the Well-Tempered Environment*. Chicago: University of Chicago Press, 1984.
_____. **"The Atavasm of the Short-Distance Mini-Cyclist." *Living Arts* 3 (1964): 91–97. Reprinted in DBC, 84–89.**
_____. **"Back in the Saddle." *Architects' Journal* 138, 19 (November 1963): 928–929.**
_____. "Bricologes a la Lanterne." *New Society* 37, 717 (July 1, 1976): 25–26. Reprinted in ACW, 196–199.
_____. "A Clip-on Architecture." *Design Quarterly* 63 (1965): 3–30.
_____. "A Clip-on Architecture." *Architectural Design* 35 (November 1965): 524–535. Excerpted from the *Design Quarterly* article, with many illustrations omitted.
_____. *A Concrete Atlantis: U.S. Industrial Building and European Modern Architecture, 1900–1925*. Cambridge, MA: MIT Press, 1986.
_____. "Counter-Attack: NY!" *Architects' Journal* (May 4, 1961): 629–630.
_____. *A Critic Writes*, ed. Mary Banham, et al. Berkeley: University of California Press, 1996.
_____. "Design by Choice." *Architectural Review* 130 (July 1961): 43–48. Reprinted in ACW, 67–78.
_____. *Design by Choice*, ed. Penny Sparke. New York: Rizzoli, 1981.
_____. "The Dymaxicrat." *Arts Magazine* 38 (October 1963): 66–69. Reprinted in ACW, 91–95.
_____. **"Easy Rider" (letter). *Design* 181 (August 1964): 59.**
_____. "Encounter with Sunset Boulevard." *The Listener* 80 (August 22, 1968): 235–236.
_____. "The End of Insolence." *New Statesman* 60 (October 29, 1960): 644–646. Reprinted in DBC, 121–122.
_____. "Epitaph Machine Esthetic." *Industrial Design* 7 (March 1960): 45–58. Originally published in Italian as "Industrial Design e arte popolare." In *Civilta della Machine* (1955). *See also* "Industrial Design and Popular Art" and "A Throw Away Aesthetic."
_____. "The Great Gizmo." *Industrial Design* 12 (September 1965): 48–59. Reprinted in DBC, 108–114. Reprinted in ACW, 109–118.
_____. **"A Grid on Two Farthings." *New Statesman* (November 1, 1963): 626.**
_____. **"A Grid on Two Farthings" [with 1979 postscript]. In *Design by Choice*, ed. Penny Sparke. New York: Rizzoli, 1981: 119–120; 205.**
_____. **"Had I the Wheels of an Angel." *New Society* 18 (August 12, 1971): 463–464.**
_____. "A Home is Not a House." *Art in America* 53 (April 1965): 70–79.
_____. "The House of the Future." *Design* (March 1956): 16.
_____. "How I Learned to Live with the Norwich Union." *New Statesman* 67 (March 6, 1964): 372–373. Reprinted in ACW, 100–104.
_____. "Industrial Design and Popular Art." *Industrial Magazine* (March 1960). This is nearly identical to "Epitaph Machine Esthetic" (see above).
_____. "Introduction." In *Design by Choice*, ed. Penny Sparke. New York: Rizzoli: 1981, 3–7.
_____. "Letter." *New Society* (July 22, 1976): 195.
_____. *Los Angeles: The Architecture of the Four Ecologies*. New York: Harper & Row, 1976.
_____. "The New Brutalism." *Architectural Review* 118 (December 1955): 354–361. Reprinted in ACW, 7–15.
_____. *The New Brutalism: Ethic or Esthetic?* London: Architectural Press, 1966.
_____. "On First Looking into Warshawsky." *Architects' Journal* 133 (July 26 1961): 109–111.
_____. "On Trial 4: CLASP III Met by Clip-Joint?" *Architectural Review* 131 (May 1962): 349–352.
_____. **ial Deserta*. Salt Lake City: Gibbs M. Smith, 1982.**
_____. "Softer Hardware." *Ark* 44 (1969): 6–7.
_____. "The Spec Builders: Towards a Pop Architecture." *Architectural Review* 132 (July 1962): 43–46. Reprinted in DBC as "Towards a Pop Architecture," 61–63.
_____. "Stocktaking." *Architectural Review* 127 (February 1960): 93–100. Reprinted in DBC, 48–55. Reprinted in ACW as "1960-Stocktaking": 49–63.
_____. *Theory and Design in the First Machine Age*. New York: Praeger, 1960.
_____. "A Throw Away Aesthetic." *Industrial Design* (March 1960): 61–65. Nearly the same as "Epitaph Machine Esthetic."

_____. "Two by Jencks: The Tough Life of the Enfant Terrible." *AIA Journal* 69 (December 1980): 50.
_____. "Unavoidable Options." *New Society* (September 18, 1969): 446.
_____. "Unlovable at Any Speed." *Architects' Journal* 144 (December 21, 1966): 1527–1529. Reprinted in ACW, 122–123.
_____. "Vehicles of Desire." *Art* (September 1, 1955): 3. Reprinted in ACW, 3–6.
_____. "Who Is this Pop?" *Motif* 10 (1963): 3–13. Reprinted in DBC, 94–96.
_____. "Without Mercy or Metaphor." *The Listener* (November 13, 1958): 775–776.
_____. "The Writing on the Walls," *Times Literary Supplement* (November 17, 1978): 1337.
_____. "Zoom Wave Hits Architecture." *New Society* 7, 179 (March 3, 1966): 21. Reprinted in DBC, 64–65.
_____, ed. *The Aspen Papers.* New York, Praeger, 1974.
Banham, Reyner, Paul Barker, Peter Hall and Cedric Price. "Non Plan: An Experiment in Freedom." *New Society* 13, 338 (March 20, 1969): 435–443.
Bardsley, Gillian. *Issigonis: The Official Biography.* Cambridge, MA: Icon, 2005.
Barker, Paul. "Non-Plan Revisited, or the Real Way Cities Grow." In *The Banham Lectures: Essays on Designing the Future*, ed. Jeremy Aynsley and Harriet Atkinson. New York: Berg, 2009, 179–186.
Benton, Tim. "The Art of the Well-Tempered Lecture: Reyner Banham and Le Courbousier." In *The Banham Lectures: Essays in Designing the Future*, ed. Jeremy Aynsley and Harriet Atkinson. Oxford: Berg, 2009, 11–32.
Berto, Frank J. *The Dancing Chain: History and Development of the Derailleur Bicycle.* San Francisco: Cycle, 2005 [2001].
_____. "The Great American Bicycle Boom." In *Cycle History 10: Proceedings of the 10th International Cycle History Conference, Nijmegan*, ed. Hans Erhard Lessing and Andrew Ritchie. San Francisco: Van der Plas, 2000, 133–148.
"Bicycle Makers Seek Tariff Help." *New York Times* (August 22, 1954).
"The Bicycle that Turned into Folding Money." *The Observer* (August 6, 2005).
Bicycles: A Report on the Application by TI Raleigh Industries, Ltd. and TI Raleigh Ltd. of Certain Criteria for Determining Whether to Supply Bicycles to Retail Outlets. London: Monopolies and Mergers Commission, 1981.
"The Bike Boom Rises to its Christmas Best." *Business Week* (December 21, 1968): 45–47.
Bix, Amy Sue. *Inventing Ourselves Out of Jobs: America's Debate Over Technological Unemployment, 1929–81.* Baltimore: Johns Hopkins University Press, 2000.
Blakley, Shantel. "Raise High the Seat Post, Salinger." *A.A. Files* 65 (2012): 122–123.
Bodkin, Jonathan. "Fixed Gear Bicycles—A Modern Trend in a Historical Context." In *Cycle History 20: Proceedings of the 20th International Cycle History Conference, Freehold.* No location, John Pinkerton Memorial Publishing Fund, 2010, 168–175.
Boekraad, Cees. "The Way Back." In *Alison & Peter Smithson: A Critical Anthology*, ed. Max Risselada. Barcelona: Ediciones Poligrafa, 2011, 268–289.
Booth, Mark Haworth. "Camera Lucida." *Frieze* 99 (May 2006): 22–23.
Bottles, Scott. *Los Angeles and the Automobile.* Berkeley: University of California Press, 1987.
Bowdon, Gregory Houston. *History of the Raleigh Cycle.* London, W. H. Allen, 1975.
Bron, Eleanor. *Life and Other Punctures.* London: Andre Deutsch, 1978.
Brown, Sheldon, and John S. Allen. "Five New Folding Bicycles." *Bicycling* 25, 5 (May 1984): 16–19, 61.
Buchanan, Colin, et al. *Traffic in Towns.* London: Her Majesty's Stationery Office, 1963.
Bullock, Alan. *Ernest Bevin: Foreign Secretary, 1945–1951.* New York: W. W. Norton, 1983.
Bullock, Nicholas. *Building the Post-War World: Modern Architecture and Reconstruction in Britain.* London: Routledge, 2002.
_____. "Building the Socialist Dream or Housing the Socialist State? Design Versus the Production of Housing in the 1960s." In *Neo-Avant-Garde and Postmodernism: Postwar Architecture in Britain and Beyond*, ed. Mark Crinson and Claire Zimmerman. New Haven: Yale University Press, 321–342.
_____. "Re-assessing the Post-War Housing Achievement: the Impact of War-Damage Repairs Program on the New Housing Programme in London." *Twentieth Century British History* 16, 3 (2005): 256–282.
Burnett, F. T., and Sheila Scott. "A Survey of Housing Conditions in the Urban Areas of England and Wales." *Sociological Review* 10, 2 (March 1962): 35–79.
Cadin, Martin, and Jay Barbree. *Bicycles in War.* New York: Hawthorn, 1974.
Cheney, Peter. "Bicycle Couriers in Love with Life on the Mean Streets." *Toronto Star* (March 27, 1993).
Clapson, Mark. *Invincible Green Suburbs, Brave New Towns: Social Change and Urban Dispersal in Post-war England.* Manchester: Manchester University Press, 1998.
Clarke, Norman. "There Oughta Be a Law." In *Proceedings of the Seminar on Bicycle-Pedestrian Planning and Design, Orlando, Florida, Dec. 12–14, 1974.* [sic: s.b. 1973] New York: ASCE, 1974, 549–551.

Colomina, Beatriz. "Friends of the Future: A Conversation with Peter Smithson." *October* 94 (Fall 2000): 3–30.
Connell, Karl Kim. "The Bicycle Messenger." *The North American Review* 265, 4 (December 1980): 34–37.
"Contraction." *The Economist* (April 15, 1961): 256.
Cox, Peter. "A Denial of our Boasted Civilization: Cyclists' Views on Conflicts Over Roads Use in Britain, 1926–35." *Transfers* 2, 3 (Winter 2012): 4–30.
_____. "The Role of Human Powered Vehicles in Sustainable Mobility." *Built Environment* 34, 2 (2008): 140–160.
Crown, Judith, and Glenn Coleman. *No Hands: The Rise and Fall of the Schwinn Bicycle Company: An American Institution.* New York: Henry Holt, 1996.
"Cycles in Trouble." *The Economist* (September 29, 1956): 1081.
Darley, Gillian, and David McKie. *Ian Nairn: Words in Place.* Nottingham: Five Leaves, 2013.
Davy, John. "Industry and the Prestige Cult." *The Observer* (October 18, 1959).
DeLaney, Ben. "David Hon: Management with a Human Face." *Bicycle Retailer and Industry News* 14, 6 (April 15, 2005): 15.
_____. "Retailer Hosting Small-Wheel, Folding-Bike Festival." *Bicycle Retailer and Industry News* 12, 8 (May 15, 2003): 16.
del Rio, Patricio. "Slums Do Stink: Artists, Bricolage, and Our Need for Doses of Real Life." *Art Journal* 67, 1 (Spring 2008): 82–99.
Diaz, Eva. *The Experimenters: Chance and Design at Black Mountain College.* Chicago: University of Chicago Press, 2015.
Dilnot, Clive. "The State of Design History: Part 1: Mapping the Field." *Design Issues* 1, 1 (Spring 1984): 4–24.
Downey, Gregory J. *Telegraph Messenger Boys: Labor, Technology, and Geography, 1850–1950.* New York: Routledge, 2002.
Downs, Diarmuid. "Alexander Arnold Constatine Issigonis, 1906–1988." *Biographical Memoirs of Fellows of the Royal Society* 39 (February 1994): 200–211.
Drazkiewicz, Elizbieta. *On the Bicycle Towards Freedom: Bicycle Messenger's Answer for Identity Crisis.* M.S. Thesis, Lund University, December 2003.
Dudas, Frank. "Flash Gordon and American Auto Design in the 1960s." In *The Banham Lectures: Essays on Designing the Future*, ed. Jeremy Aynsley and Harriet Atkinson. New York: Berg, 2009, 99–110.
"Editorial." ARK 36 (Summer 1964): 48.
Excell, John. "Brompton Managing Director Will Butler-Adams." *The Engineer* (28 February 2011): n.p.
Farrell, Dan. *The Moulton Bicycle: Origin and Intent.* One-page manuscript, no location [Bradford-on-Avon], 2016.
_____. *Riding on Rubber: The Story of Bradford-on-Avon's World Renowned Rubber Industry.* Manuscript (46 pp.), no location [Bradford-on-Avon], no date [2016].
"Feeling the Pinch." *The Economist* 195 (January 21, 1961): 275.
Fincham, Benjamin. *Bicycle Couriers in the New Economy.* Cardiff: Cardiff University Working Paper Series 46, January 2004.
_____. "Generally Speaking People Are in it for the Cycling and the Beer: Bicycle Couriers, Subculture and Enjoyment." *The Sociological Review* 55, 2 (2007): 189–202.
Final Report of the President's Commission on Product Safety. Washington: Government Printing Office, 1970.
"The First Redundancy Strike." *The Economist* (January 18, 1958): 196.
Formosa, Nicole. "Hon Family at Center of Folding Bike Controversy." *Bicycle Retailer and Industry News* 20, 12 (July 15, 2011): 1–2.
_____. "Josh Hon Talks Tern One Year In." *Bicycle Retailer and Industry News* 21, 12 (July 15, 2012): 20.
Forty, Adrian. "Reyner Banham, One Partially Americanized European." In *Twentieth-Century Architecture and its Histories*, ed. Louise Campbell. Oatley: Society of Architectural Historians of Great Britain, 2000, 195–205.
Fuller, R. Buckminster. *Buckminster Fuller: Anthology for the New Millennium*, ed. Thomas Zung. New York: St. Martin's, 2001.
Gaitskell, Hugh. *The Diary of Hugh Gaitskell, 1946–56*, ed. Phillip M. Williams. London: Cape, 1983.
_____. "Understanding the Electorate." *Socialist Commentary* (July 1955): 196–206.
Galvin, Peter, and Andre Morkel. "The Effect of Product Modularity on Industry Structure: the Case of the World Bicycle Industry." *Industry and Innovation* 8, 1 (April 2001): 31–47.
Gazeley, Ian. "The Leveling of Pay in Britain During the Second World War." *European Review of Economic History* 10 (2006): 175–204.
Girouard, Mark. *Big Jim: The Life and Work of James Stirling.* London: Chatto and Windus, 1998.

Glaskin, Max. *Cycling Science: How Rider and Machine Work Together.* Chicago: University of Chicago Press, 2012.
Goldthorpe, John H., et al. *The Affluent Worker: Part 2, Political Attitudes and Behavior.* Cambridge, UK: Cambridge University Press, 1968.
_____. *The Affluent Worker: Part 3, The Affluent Worker in the Class Structure.* Cambridge, UK: Cambridge University Press, 1969.
Gosling, Ray. "Robin Hood Rides Again—A Rebel Scene." *Anarchy* 38 (April 1964): 99–108.
Grindrod, John. *Concretetopia: A Journey Around the Rebuilding of Postwar Britain.* London: Old Street, 2013.
Grove, Valerie. "Alison Smithson, Architect." In *The Compleat Woman.* London: Chatto and Windus, 1987, 259–270.
Hadland, Tony. "The Doctor's GP." *The Moultoneer* 71 (2004): 42–47.
_____. "Marketing the Mk3." *The Moultoneer* 70 (2004): 23–27
_____. *The Moulton Bicycle: The Story from 1957 to 1981.* Coventry: Author, 2000 [1981].
_____. *Raleigh: Past and Presence of an Iconic Bicycle Brand.* San Francisco: Cycle Publishing, 2012.
_____. "Small Wheels for Adult Bicycles." *Cycling Science* (Fall 1997): 3–7.
_____. *The Spaceframe Moultons.* Piscataway, NJ: Transaction, 1994.
_____. *The Sturmey Archer Story.* No location: Author, 1987.
Hadland, Tony, and Hans Erhard Lessing. *Bicycle Design: An Illustrated History.* Cambridge, MA: MIT Press, 2014.
Hadland, Tony, and John Pinkerton. *It's in the Bag! A History in Outline of Portable Bicycles in the UK.* Cheltenham: Quorum, 1996.
Hadland, Tony, and Mike Burrows. *From Bicycle to Superbike.* No location [Oxford]: Hadland, 2016.
Hall, Peter. "Introduction." In *A Critic Writes*, ed. Mary Banham, et al. Berkeley: University of California Press, 1996, xi-xv.
Hamilton, Richard. *The Collected Words of Richard Hamilton.* London: Bloomsbury, 1982.
Harrison, J. F. C. "The Man with the Bicycle Wheel." *New Statesman and Nation* 52, 1336 (October 20, 1956): 495–496.
Henderson, Nigel. *Nigel Henderson: Parallel of Life and Art.* London: Thames and Hudson, 2001.
Highmore, Ben. "Rough Poetry: Patio and Pavilion Revisited." *Oxford Art Journal* (February 2006): 269–290.
Hinton, James. *The Mass Observers: A History, 1937–1949.* Oxford, Oxford University Press, 2013.
Hinton, Rita. "Equality with Quality." *Socialist Commentary* 19, 7 (July 1955): 196–208.
Hodgkinson, Peter. "Drug-In City." *Architectural Design* (November 1969): 586.
"The Humble Bicycle." *Barron's Weekly* (July 19, 1954): 9.
Jencks, Charles. "Letter." *New Society* (July 15, 1976): 25
_____. "Letter." *New Society* (July 29, 1976): 251
"Joint Statement" [of the 1970 French attendees.] *IDCA 1970 Special Edition of the Aspen Times* (June 20–24, 1971).
Karen, Tom. Dr. "Designing the Future: An Industrial Designer's Perspective." In *The Banham Lectures: Essays on Designing the Future*, ed. Jeremy Aynsley and Harriet Atkinson. Oxford: Berg, 2009, 252–263.
Kidder, Jeffery L. *Alleycats, Fixies, and Double Rushes: An Ethnography of New York City Bike Messengers.* B.A. Thesis, University of North Carolina at Greensboro, 1999.
_____. "Appropriating the City: Space, Theory and Bike Messengers." *Theoretical Sociology* 38 (2009): 307–328.
_____. *Emotions, Space, and Cultural Analysis: The Case of Bicycle Messengers.* Ph.D. dissertation, University of California at San Diego, 2009.
_____. "Style and Action: A Decoding of Bike Messenger Symbols." *Journal of Contemporary Ethnography* 34, 2 (June 2005): 344–367.
Klein, Leighton. "For Cyclists Seeking Simplicity, the Fixed Is In." *Boston Globe* (November 9, 2008).
Knottley, Peter. "Preview: The Moulton Prototype." *Bicycling* 11, 8 (February, 1970): 24–27.
Kolin, Michael J., and Denise M. de la Rosa. *The Custom Bicycle.* Emmaus, PA: Rodale, 1979.
Krag, Thomas. "Urban Cycling in Copenhagen." In *Planning for Cycling: Principles, Practices and Solutions for Urban Planners*, ed. Hugh McClintock. Cambridge, UK: Woodhead, 2002, 223–236.
Krcmar, Stephen. "With Diverse Offerings, Hip Styling, Folder Sales are on the Rise." *Bicycle Retailer and Industry News* 15, 16 (October 1, 2006): 52.
Kugelmass, Jack M. "I'd Rather be a Messenger." *Natural History* 90, 8 (first series) (August 1981): 66–73.
Kwitny, Johnathan. "The Great Transportation Conspiracy." *Harpers Magazine* (February 1981): 14–21.
Kynaston, David. *Austerity Britain, 1945–1951.* New York: Walker, 2008.

_____. *Family Britain, 1951–1957.* New York: Walker, 2009.
_____. *Modernity Britain: Opening the Box, 1957–59.* London, Bloomsbury, 2013.
_____. *Modernity Britain: A Shake of the Dice, 1959–1962.* London, Bloomsbury, 2014.
Lacey, Robert. *Ford: The Men and the Machines.* New York: Little, Brown, 1986.
Lehmann, John. *The Ample Proposition: Autobiography 3.* London: Eyre & Spotswood, 1966.
Lippincott, J. Gordon. *Design for Business.* Chicago: P. Theobald, 1947.
Lipsyte, Robert. "Coping: Voices from the Sweatshop of the Streets." *New York Times* (14 May 1995).
Lloyd-Jones, Roger, and M. J. Lewis. *Raleigh and the British Bicycle Industry: An Economic and Business History, 1870–1960.* Aldershot: Ashgate, 2000.
Lloyd-Jones, Roger, M.J. Lewis and Mark Easton. "Culture as Metaphor: Company Culture and Business Strategy at Raleigh Industries, 1945–60." *Business History* 41, 3 (July 1999): 93–127.
"London Roads Study." In *CIAM 1959 in Otterlo,* ed. Oscar Newman. London: Alec Tiranti, 1961, 77–79.
Louis, Arthur M. "How the Customers Thrust Unexpected Prosperity on the Bicycle Industry." *Fortune* 89, 3 (March 1974): 117–124.
Lowndes, G. A. N. *The Silent Revolution.* Oxford: Oxford University Press, 1937.
Lucsko, David N. *Junkyards, Gearheads and Rust: Salvaging the Automotive Past.* Baltimore: Johns Hopkins University Press, 2016.
_____. "Junkyard Jamboree: Hunting for Automobile Treasure in the Twentieth Century." *Journal of the Society for Industrial Archeology* 39, 1/2 (2013): 93–111.
_____. "Of Clunkers and Camaros: Accelerated Vehicle Retirement Programs and the Automobile Enthusiast, 1990–2009." *Technology and Culture* 55, 2 (April 2014): 390–428.
Lyall, Sarah. "For Manhattan Couriers, Brakeless Bikes Are the Way to Go." *New York Times* (June 14, 1987).
Mansell, Chris. "The Rallying of Raleigh." *Management Today* (February 1973): 82–93.
Mapes, Jeff. *Pedaling Revolution: How Cyclists are Changing American Cities.* Oregon State University Press, 2009, 29–31.
Marwick, Arthur. *War and Social Change in the Twentieth Century.* New York: St. Martin's, 1974.
Massey, Anne. *The Independent Group: Modernism and Mass Culture in Britain, 1945–59.* New York: Manchester University Press, 1995.
_____. The Independent Group: Towards a Redefinition." *The Burlington Magazine* 129, 1009 (April 1987): 232–242.
Mauroni, Albert J. *Chemical and Biological Warfare: A Reference Handbook.* Santa Barbara: ABC-CLIO, 2007.
Maxwell, Robert. "Reyner Banham—Historian." In *Sweet Disorder and the Carefully Careless: Theory and Criticism in Architecture.* Princeton: Princeton Architectural Press, 1993, 163–175. Originally published as "The Plentitude of Presence: Reyner Banham." *Architectural Design* 8/9 (1981).
_____. "Reyner Banham—The Man." In *Sweet Disorder and the Carefully Careless: Theory and Criticism in Architecture.* Princeton: Princeton Architectural Press, 1993, 177–183.
Mays, John Barron. "Teen-Age Culture in Contemporary Britain and Europe." *Annals of the American Academy of Political and Social Science* 338 (November 1961): 22–32.
McClellan, Doug. "Dahon Forges Ahead with Environmental Plans as it Turns 25." *Bicycle Retailer and Industry News* 17, 5 (April 1, 2008): 47.
_____. "Folder Sales Unfold as Buyers Opt for More Compact Rides." *Bicycle Retailer and Industry News* 17, 16 (October 1, 2008): 65.
McClintock, Hugh. "Planning for the Bicycle in Newer and Older Towns and Cities." In *The Bicycle and City Traffic,* ed. Hugh McClintock. London: Belhaven, 1992, 40–61.
McKean, John Maule. "The Last of England? Part 2." *Building Design* (August 27, 1976): 26–27.
McNichols, Thomas J. "Raleigh Industries." In *Policy Making and Executive Action.* New York: McGraw-Hill, 3rd ed., 1967, 485–524.
Miller, Henry. "My Best Friend." In *My Bike & Other Friends.* Santa Barbara, CA: Capra, 1978, 105–110.
"Miniaturisation." *The Economist* (November 10, 1962): 601.
"More Power to the Pedal." *The Economist* (April 30, 1955): 399.
Moulton, Alex. "Innovation: An Address." *Journal of the Royal Society for the Encouragement of Arts, Manufacturers and Commerce* 128, 5281 (December 1, 1979): 31–44.
Moulton, Alex, E. A. Hadland and Douglas L. Milliken. "Aerodynamic Research Using the Moulton Small-Wheeled Bicycle." *Proceedings of the Institution of Mechanical Engineers,* 220, 3 (May 2006): 189–193.
Moulton, Alex, J. Grosjean and G. Owen. *The Moulton Formulae and Methods.* London: Professional Engineering, 2005.
Murray, Christine. "Reinventing the Wheel: An Interview with Alex Moulton." *The Architects' Journal* (September 4, 2008): 95–97.

[Nairn, Ian.] "The Euston Murder." *Architectural Review* (April 1962): 234–244. Anonymous, but attributed to Nairn by biographers Gillian Darley and David McKie (see above).
Naylor, Gillian. "Theory and Design: The Banham Factor." *Journal of Design History* 10, 3 (1997): 241–252.
_____. "Theory and Design: The Banham Factor." In *The Banham Lectures: Essays on Designing the Future*, ed. Jeremy Aynsley and Harriet Atkinson. New York: Berg, 2009, 47–58.
"A New, Lower Level?" *The Economist* (May 10, 1958): 524–525.
New York [State] Department of Labor, Division of Labor Standards. "Guidelines for Determining Worker Status in the Message Courier Industry." Albany: State of New York, ca. December 2005.
Norman, Jason. "Folding Bike Makers Report Robust Sales During First Half 2007." *Bicycle Retailer and Industry News* 16, 15 (September 1, 2007): 35.
Nye, Peter. *Hearts of Lions: The History of American Bicycle Racing*. New York: W. W. Norton, 1988.
Oakley, William. *Winged Wheel: The History of the First Hundred Years of the Cyclists' Touring Club*. Galdalming: CTC, 1977.
"One Final Merger." *The Economist* 195 (April 23, 1960): 356.
Padvovan, Richard. "Seven Small Thoughts on the Architects' Writings." *Alison & Peter Smithson: A Critical Anthology*, ed. Max Risselada. Barcelona: Ediciones Poligrafa, 2011, 66–79.
Panter-Downes, Mollie. "Letter from London." *New Yorker* (March 24, 1945): 19–20.
_____. "Letter from London." *New Yorker* (September 1, 1945): 23–24.
Parnell, Steve. "Nairn Mania." *Architectural Review* 235 (May 2014): 118–119.
Partridge, Eric. *A Dictionary of Slang and Unconventional English*. London: Macmillian, 1961 [1937, 1938, 1949, 1951, 1956].
Patton, Phil. *Bug: The Strange Mutations of the World's Most Famous Automobile*. New York: Simon & Schuster, 2002.
Pawley, Martin. "Building Revisits: Hunstanton School, 1984." *Architects' Journal* (June 20, 1984): 32–36.
_____. "Heavy Stuff this Symbolism." *The Guardian* (December 17, 1984).
_____. "The Last of the Piston Engine Men." *Building Design* (October 1, 1971): 6.
_____. "We Shall Not Bulldoze Westminster Abbey: Archigram and the Retreat from Technology." In *The Strange Death of Architectural Crtiticism: Martin Pawley Collected Writings*. London: Black Dog, 2007, 59–69.
"Pedal Power." *The Economist* (July 19, 1980): 70;
"Pedro Ignacio Alonso and Thomas Weaver in Conversation with Tim-Street Porter." *AA Files* 62 (2011): 28–33. *See also* Alonso, Pedro Ignacio.
"Performance Folding Bikes Attract Wide Customer Base." *Bicycle Retailer and Industry News* 13, 15 (September 1, 2004): 62–63.
Petersen, Grant. *Just Ride: A Radically Practical Guide to Riding Your Bike*. New York: Workman, 2012.
Petty, Ross D. "The Bicycle as a Communications Medium." In *Cycle History 16: Proceedings of the 16th International Cycle History Conference, Davis*, ed. Andrew Ritchie. San Francisco: Van der Plas, 2006, 147–159.
_____. "The Consumer Product Safety Commission's Promulgation of a Bicycle Safety Standard." *Journal of Products Liability* 10 (1987): 25–50.
_____. "Peddling Schwinn Bicycles: Marketing Lessons from the Leading Post-WWII US Bicycle Brand." *Quinnipiac University CHARM Symposium Papers* (2007): 162–171.
Pevsnser, Nikolaus. *Pioneers of Modern Design*. London: Faber & Faber, 1936.
_____. *Wiltshire*. Harmondsworth: Penguin, 1963.
Pinkerton, John. "Who Put the Working Man on a Bicycle?" In *Cycle History 8: Proceedings of the Eighth Cycle History Conference, Glasgow*, ed. Nicholas Oddy and Rob van der Plas. San Francisco: Van der Plas, 1998, 101–106.
Plagens, Peter. "Los Angeles: The Ecology of Evil." *Artforum* (December 1972): 76.
Pooley, Colin G., and Jean Turnbull. "Commuting, Transport and Urban Form." *Urban History* 27, 3 (December 2000): 360–372.
Pridmore, Jay, and Jim Hurd. *Schwinn Bicycles*. Osceola: Motorbooks International, 1996.
"Quelques Stands Remarqués à l' I.F.M.A." *Le Cycle* (November 10, 1956): 21.
Raleigh in America Retail Catalog for 1967; *1968*; *1969*; *1970*; *1971*; *1972*; *1973*; *1974*; *1975 1976*; *1977*.
"Raleigh Industries Limited," *The Economist* (December 13, 1958): 1025–1026.
"Raleigh Industries, Ltd." *The Economist* (December 19, 1955): 1199.
"Retailers About Ready to Pack Them in Are Discovering More Consumers Want Them." *Bicycle Retailer and Industry News* (September 2000): 13.
Reiley, Rebecca. *Nerves of Steel*. Kenmore, NY, Spoke and Word Press, 2000.
"Reyner Banham, 1922–1988." *AA Files* 16 (Autumn 1987 [sic]): 33–40.
Risselada, Max. "Introduction." In *Alison & Peter Smithson: A Critical Anthology*, ed. Max Risselada. Barcelona: Ediciones Poligrafa, 2011, 18–34.

Risselada, Max, ed. *Alison & Peter Smithson: A Critical Anthology.* Barcelona: Ediciones Poligrafa, 2001.
Robbins, David, ed. *The Independent Group: Postwar Britain and the Aesthetics of Plenty.* Cambridge, MA: MIT Press, 1990.
Rosen, Paul. *Framing Production, Technology, Culture and Change in the British Bicycle Industry.* Cambridge, MA: MIT Press, 2002.
"Le Salon du Cycle," *Le Cycle* (October 1960): 33.
Saunders, Francis Stonor. *The Cultural Cold War.* New York: New, 1999.
Scalbert, Irénée. "Parallel of Life and Art." *Daldalos* 75 (2000): 75–102.
Schumacker, Thomas L. "Architectural Paradise Postponed." *Harvard Design Magazine* (Fall 1998): 81–83.
"Sheldon C. Brown, Cycle Mechanic and Writer." [Obituary.] *The Times* [of London] (March 4, 2008).
Smith, Elizabeth A. T. *Case Study Houses.* Cologne: Taschen, 2016.
Smithson, Alison, ed. "CIAM Team 10." *Architectural Design* 30, 5 (May 1960): entire issue.
_____, ed. "Team 10 Primer 1953-1962." *Architectural Design* 32, 12 (December 1962): entire issue.
Smithson, Alison, and Peter Smithson. "But Today We Collect Ads." *Ark* 18 (November 1956): 26–30.
_____. "Caravan—Embryo Appliance House." *Architectural Design* (September 1959): 17–18.
_____. *Ordinariness and Light.* Cambridge: MIT Press, 1970.
"Social Survey of Bicycles in War Time Transport." *Board of Trade Journal* (January 12, 1946): 24.
"Some Effects of the Distribution of Industry Act, 1945." *Manchester School of Economic and Social Studies Journal* 17 (January 1949): 36–48.
Sparke, Penny. "From Production to Consumption in Twentieth Century Design." In *The Banham Lectures: Essays on Designing the Future*, ed. Jeremy Aynsley and Harriet Atkinson. New York: Berg, 2009: 127–141.
_____. "Introduction." In Reyner Banham, *Design by Choice*, ed. Penny Sparke. New York: Rizzoli, 1981, 8, 17–18.
_____. "Obituary: Peter Reyner Banham (1922–1988)." *Journal of Design History* 1, 2 (1988): 141–142.
Starrs, James E., ed. *The Noiseless Tenor: The Bicycle in Literature.* New York: Cornwall, 1982, 127–128.
Stewart, Benjamin. "Bicycle Messengers and the Dialectics of Speed." In *Fighting for Time: Shifting Boundaries of Work and Social Life*, ed. Cynthia Fuchs Epstein and Arne L. Kalleberg. New York: Russell Sage Foundation, 2004, 150–190.
Stoffers, Manuel. "Cycling as Heritage: Representing the History of Cycling in the Netherlands." *Journal of Transport History* 33, 1 (June 2012): 92–114.
Stone, Deborah. *Policy Paradox: The Art of Political Decision Making.* New York: Norton, 1997.
Sykes, Joseph. "Postwar Distribution of Industry in Great Britain." *Journal of Business of the University of Chicago* 22, 3 (July 1949): 188–199.
Teresi, Dick. *The Popular Mechanics Book of Bikes and Bicycling.* New York: Hearst, 1975.
Thomas, Ray. "Milton Keynes, City of the Future?" *Built Environment* 9, 3–4 (1983): 245–254.
Thompson, Michael. *Rubbish Theory: The Creation and Destruction of Value.* New York: Oxford University Press, 1979.
Tiratsoo, Nick. *Reconstruction, Affluence and Labour Politics: Coventry, 1945–60.* London: Routledge, 1990.
Tiratsoo, Nick, and Jim Tomlinson. "Exporting the Gospel of Productivity: United States Technical Assistance and British Industry, 1945–1960." *Business History Review* 71 (Spring 1997): 41–81.
_____. *Industrial Efficiency and State Intervention: Labour, 1939–51.* London, Routledge, 1996.
"Tube Investments Limited." *The Economist* (December 16, 1950): 1119.
"Tube Investments Limited." *The Economist* (December 15, 1951): 1501.
"Tube Investments Limited." *The Economist* (November 22, 1952): 574.
"Tube Investments—Raleighing." *The Economist* (February 12, 1983): 79.
Twemlow, Alice. "I Can't Talk to You If You Say That: An Ideological Collision at the International Design Conference at Aspen, 1970." *Design and Culture* 1, 1 (2009): 23–50.
_____. *Purposes, Poetics and Publics: the Shifting Dynamics of Design Criticism in the U.S. and U.K., 1955–2007.* Ph.D. dissertation, Royal College of Art, 2007.
"U.S. Bicycle Market Statistics." *Schwinn Reporter* (February 1978): n.p.
Vidler, Anthony. *Histories of the Immediate Present: Inventing Architectural Modernism.* Cambridge, MA: MIT Press, 2008.
_____. "Troubles in Theory V: The Brutalist Movement(s)." *Architectural Review* 235 (February 2014): 96–101.
Wade, Leigh F. "Letter to the Editor." *Bike Report* (May 1990): 2.
Wakeman, Rosemary. *Practicing Utopia: An Intellectual History of the New Town Movement.* Chicago: University of Chicago Press, 2016.
Ward, Leslie, ed. *The London County Council Bomb Damage Maps, 1939–1945.* London: Thames and Hudson, 2015.

Welleman, Ton. "An Efficient Means of Transport: Experiences with Cycling Transport Policy in the Netherlands." In *Planning for Cycling: Principles, Practices and Solutions for Urban Planners*, ed. Hugh McClintock. Cambridge, UK: Woodhead, 2002, 192–208.
Whisler, Timothy R. *At the End of the Road: The Rise and Fall of Austin Healy, MG and Triumph Sports Cars*. Greenwich, CT: Greenwood, 1995.
____. *The British Motor Industry: A Case Study in Industrial Decline*. New York: Oxford University Press, 1999.
Whiteley, Nigel. "Banham and Otherness: Reyner Banham (1922–1988) and his Quest for an Architectural Autre." *Architectural History* 33 (1990): 188–221.
____. "Olympus and the Marketplace: Reyner Banham and Design Criticism," *Design Issues* 13, 2 (Summer 1997): 24–35
____. "Pop, Consumerism, and the Design Shift." *Design Issues* 2, 2 (Autumn 1985): 31–45.
____. *Reyner Banham: Historian of the Immediate Future*. Cambridge, MA: MIT Press, 2002.
____. "Toward a Throw-Away Culture: Consumerism, Style Obsolescence and Cultural Theory in the 1950s and 1960s." *Oxford Art Journal* 10, 2 (1987): 3–27.
Whiteside, Noel. "Limits of Americanization." In Becky E. Conekin, Frank Mort and Chris Waters, eds., *Moments of Modernity*. London: Rivers Oram, 1999: 96–113.
Wiebe, Matt. "Harry Montague Left Behind Folding Legacy." *Bicycle Retailer and Industry News* 20, 4 (March 15, 2011): 17.
Wilks, Neil. "Bright Future Unfolds." [Brompton] *Professional Engineering* 17, 5 (March 10, 2005): 32.
Wilson, David Gordon. "Bicycle Manufacturing Defects." *Bicycling* (February 1979): 23–27.
Winters, Jeffrey. "Origami Cycle." [Bike Friday Tikit] *Mechanical Engineering* 130, 6 (June 2008): 48–49.
"Witness Seminar: 1949 Devaluation." *Contemporary Record* 4, 3 (Winter 1991): 480–503.
Witte, Griffe. "A Rough Ride for Schwinn Bicycle." *Washington Post* (December 3, 2004).
"Wobble in Bicycles." *Barron's Weekly* (July 25, 1955): 1.
Woodward, Christopher. "Drawing the Smithsons: An Artisanal Memoir." In *Alison & Peter Smithson: A Critical Anthology*, ed. Max Risselada. Barcelona, Ediciones Poligrafa, 2011, 258–267.

Audiovisual Media

Fathers of Pop. Documentary film. Julian Cooper, director. 47 min., Arts Council of Great Britain, 1979.
"Pop Goes the Easel." *Monitor*. Documentary television show. Ken Russell, director, 49 min., BBC Television, March 25, 1962.
Reyner Banham Loves Los Angeles. Documentary television show. Julian Cooper, director. 67 min, BBC Television, 1972.
Who Framed Roger Rabbit Cinematic film. Robert Zemeckis, director, Steven Spielberg, producer. 103 min. Touchstone Pictures, 1988.

Internet Resources

Single-person achival interviews accessed online are listed under "interviews."

Allen, John S. "Twenty-Five Years on a Twenty." (Text and illustrations.) www.bikeexpert.com/twenty.html. May 7, 2005.
Bianco, Martha. "Kennedy, 60 Minutes and Roger Rabbit: Understanding Conspiracy-Theory Explanations of the Decline of Mass Transit." (Text.) http://www.upa.pdx.edu/CUS/publications/docs/DP98-11.pdf, 1988.
Brown, Sheldon. "My Father." (Text and illustrations.) www.http://sheldonbrown.com, August 6, 1996.
____. "My Mother." (Text and illustrations.) http://sheldonbrown.com, November 10, 1997.
____. "My 1964 Moulton Deluxe." (Text and illustrations.) http://sheldonbrown.com/mybikes.html, no date.
____. "My Raleigh Twenties." (Text and illustrations.) http://sheldonbrown.com/Raleigh-twenty.html, no date.
____. "Sheldon Brown's Raleigh Twenty Website." (Text and illustrations.) http://sheldonbrown.com/raleigh-twenty.html, no date.
"The Cars: Mini Development History, Part 1." (Text and illustrations.) Aronline, http:www.aronline.co.uk/blogs/mini-classic/, no date.
Hadland, Tony. "The Raleigh Twenty Range: Raleigh's Biggest Seller of the Mid–1970s." (Text and illustrations.) http:// hadland.woodpress.com/2012/07/01/articles, 2012.
"The Moulton Story, Part 1." *The Bikeshow from Resonance FM*. (Audio podcast.) (56 min.), Bikeshow_

20080929–62kb_M3U. (Alex Mouton, Tony Hadland, Michael Woolf, others). Originally broadcast September 2008.
"The Moulton Story, Part 2." *The Bikeshow from Resonance FM.* (Audio podcast.) (56 min.), Bikeshow_ 20081005–62kb_M3U. (Alex Mouton, Tony Hadland, Michael Woolf, others). Originally broadcast September 2008.
Penner, Barbara. "The Man Who Wrote Too Well." (Text and illustrations.) *Places Journal* (September 2015), https://placesjournal.org/article/future-archive-the-man-who-wrote-too-well/.
"Raleigh Chopper Designer Alan Oakley Dies from Cancer." (Text.) BBC, http://www.BBC.com (posted May 20, 2012).
Rebuilding Britain for Baby Boomers." (Radio broadcast.) (Maxwell Hutchinson writer, reporter; Lindsay Leonard, producer. 58 min. BBC Radio 4), www.bbc.co.uk/programmes/b017187m. (Bill Berrett, Roger Bowdler, Andrew Derbyshire, Peter Smithson, others.) Originally broadcast 2004.
Smale, Will. "Brompton Boss: The Bike-Maker Who Disproved th Doubters." (Text.) *BBC Business Online* (posted June 29, 2009).
Walker, Peter. "75 Years After the UK's First Cycle Lane Opened, the Same Debate Rages On." (Text.) *The Environmental Guardian Online*, http://www.Environmentalguardianonline.com/ (posted December 13, 2009).

Interviews

Banham, Mary. Interviewed by Corinne Julius, British Library Sounds Project, Tape 7 (December 2001), http://sounds.bl.uk/oral-history/Architects-Lives/021M-CO467X0067XX-007V0; Tape 8 (December 2001), http://sounds.bl.uk/oral-history/Architects-Lives/021M-CO467X0067XX-008V0; Tape 9 (January 2002), http://sounds.bl.uk/oral-history/Architects-Lives/021M-CO467X0067XX-009V0; Tape 10 (January 2002), http://sounds.bl.uk/oral-history/Architects-Lives/021M-CO467X0067XX-010V0; Tape 11 (January 2002), http://sounds.bl.uk/oral-history/Architects-Lives/021M-CO467X 0067XX-011V0; Tape 12 (February 2002), http://sounds.bl.uk/oral-history/Architects-Lives/021M-CO467X0067XX-012V0; Tape 15 (February 2002), http://sounds.bl.uk/oral-history/Architects-Lives/ 021M-CO467X0067XX-015V0; Tape 17 (May 2002), http://sounds.bl.uk/oral-history/Architects-Lives/021M-CO467X0067XX-017V0; Tape 18 (May 2002), http://sounds.bl.uk/oral-history/Architects-Lives/021M-CO467X0067XX-018V0.
Clarke, Norman. Interviewed by the author, Cape Cod, Massachusetts, April 5, 1998.
Duffield, David. Interviewed by John Pinkerton. https://www.youtube.com/channel/UC2XzNEKoaCK1I NKmSIo6wRA/.
Farrell, Dan. Interviewed by the author, Bradford-on-Avon, February 24, 2017.
Hodges, Eric. Interviewed by the author, Fort Lauderdale, Florida. December 2, 2015.
Issacson, Allen. Interviewed by the author, Hollywood, Florida, December 9, 2015.
Issacson, Alan, and Eric Hodges. Interviewed by the author, Fort Lauderdale, Florida, January 11, 2016.
Lauterwasser, Jack. Interviewed by Tony Hadland, 1997. https://www.youtube.com/channel/UC2Xz NEKoaCK1INKmSIo6wRA/.
Macnaughtan, John. Interviewed by the author, Bradford-on-Avon, February 24, 2017.
Nicholson, Vic. Interviewed by Tony Hadland. https://www.youtube.com/channel/UC2XzNEKoaCK1I NKmSIo6wRA/.
Woodburn, John. Interviewed by Tony Hadland. https://www.youtube.com/channel/UC2XzNEKoaCK1I NKmSIo6wRA/.

Index

Abby Mill *see* Kingston Mill
Alcoa 88, 137–139
Allen, Deborah 34–35
Allen, John S. 12, 170–171; on Raleigh Twenty 176–178, 182, 184, 185, 186
Alloway, Lawrence 1, 35, 37, 40, 67, 121–122, 126
Alvis Motors 49–51
AM series *see* Moulton bicycle (AM series)
An American Deserta (PRB, 1986) 141–144
Angois, Paul 80
Archigram 7, 41, 132, 137, 159
Architects' Journal 1, 35, 41, 64, 69, 133, 179
Architectural Design 122, 123, 159
Architectural Press 9, 41, 122
Architectural Review 1, 10, 22, 27, 33, 158, 122, 178
Architecture, modern: criticized by Jencks 160–162; loss of faith in 134–139; postwar controversy over 22–25
Arconic *see* Alcoa
Arnold, Schwinn and Co. *see* Schwinn Bicycle Co
Art News and Reviews 27
Aspen Design Conference 122–126, 169
Aspen Institute 123
The Aspen Papers (PRB, 1974) 124, 125
"Atavism of the Short Distance Mini-Cyclist" (PRB, 1963) 3, 9, 63
Aubert, Jean 124
Austin Motors 50–51, 52, 77; *see also* British Motor Corp.; Mini automobile
Avila, Tony 167

Baglian, Riccardo 120
Ballentine, Richard 118, 129
Banham, Ben (PRB's son) 40, 127, 130
Banham, Charles (PRB's grandfather) 15

Banham, Debbie (PRB's daughter) 40
Banham, Mary (PRB's wife) 5, 7, 16, 30, 31–32, 37, 121, 122, 123, 127, 129, 130, 140, 144, 158, 159; attends art school in London 25; birth of children 40; cancer, recovery, 40–41; grows up in Dagenham 25; Independent Group and Smithsons 31–32, 37; life in California 67–68, 140–142; move to United States 127–128; moves to Norwich, meets Banham 25–26; on PRB's writing style 35
Banham, Percy (PRB's father) 15
Banham, Peter Reyner (PRB) 98, 31, 34, 158; admitted to Courtard Institute 26; on American culture 36, 37, 42, 43, 66–67; on bicycle industry 65, 172; on bicycles—chopper 112, generally 44, 63, high-tech bicycles 130–131, 169–170, 177, Moulton F-frame 28, 63, 177, Moulton Mk. III 6, RSW 16, 99–100; on "borax" car styling 34–35; childhood, schooling in Norwich 10, 15, 16; death of 144; debates Jencks over postmodernism 168–169; discord at 1970 Aspen conference 123–126; early writings and lectures 26, 27; featured in "Fathers of Pop" 126; first writes about Moulton bicycle 63; friendship with Bucky Fuller 67–68; hired at Architectural Press (AP) 27; on industrial design 67, 99; on "hot-rod" culture 160, 179–180; leaves AP, teaches at Bartlett School 122; on Los Angeles 121–122; meets, marries Mary Mullett, 25–26; moves to California 67, 140–142; moves to USA (Buffalo) 36, 127, 128; on New Brutalism 34; on Pop Culture 10, 16,

36, 39, 42; participates in Independent Group 29–40; on "plug-in/clip-on" technology 7–8, 158–159, 179–180; on postmodernism 160–162; proficiency and skill as writer 2, 35–36, 126; on radical politics 6; return to Norwich, Maddermarket Theatre 17, 25; the Smithsons 137; urban executive radical cyclists 6, 68–69; visits L.A. for first time 121–122; work at Bristol Aeroplane 9, 16–17; work at Norwich remand house 17; writes about American deserts 142–144
Banham, Violet (PRB's mother) 15
Barre, François 124
Bartlett School of Architecture 122, 141
Baudrillard, Jean 124
Bendix Corp. 77, 108, 182
Berenson, Bernard 34
Berrett, Bill 136
Bevin, Ernest 20
Bickerton, Harry 128, 130
Bickerton Portable bicycle 128–130, 142–144, 148, 173
Bicycle bombs 169–170, 186
Bicycle industry (UK): decline and collapse 119–120; forced export policy (1950s) 82–84; stuffy and conservative 65, 172; wages in 19
Bicycle industry (USA) 83, 84, 110; *see also* Schwinn Stingray; Bike boom (1970s)
Bicycle messengers 162–168
Bicycle shops (UK) 79
Bicycling magazine (USA) 114, 175, 184, 186
Bike boom (1970s) (USA) 110, 115–116; effect on bicycle technology 131; statistics of 130–131; Raleigh reacts to 145–146
Bike Friday (bicycle) 146–147, 173, 177, 185

219

Bike World magazine (USA) 129, 175
Blake, Peter 126
Blakel, Shantel 28
Blunt, Anthony 26
BMX bicycles 74, 112–113
Boardman, Chris 61, 170
Bodkin, Jonathin 168
Boshier, Derek 126
Boty, Pauline 126
Boulstridge, Jim 86
Bowden, Frank 80
Bowden, Harold 80, 88
Bowdler, Robin 135–139
Bratby, Jim 114
Bricolage 159–169, 180; applied to bicycle technology 166–168; as contrasted to bicycle "hot-rodding" 180; defined 161–162
Bridgestone bicycles 154, 157, 170
Bristol Aeroplane 16, 17, 47
British Aluminum Co. (BACO) 88, 94
British Cycle Corp. *see* Tube Investments
British Cycling Federation 118
British Leyland *see* British Motor Corp
British Motor Corp. 53, 63, 71, 75, 77, 96; Mini 50–51; *see also* Austin Motors; Fisher and Ludlow; Morris Motors
Brompton bicycle 12 79, 145, 147, 148, 173, 178, 185; development of 117; production of 146; rejected by Raleigh 118; *see also* Ritchie, Andrew
Bron, Eleanor 5
Brooke, Charles 44
Brooks saddles 89, 92, 119–120, 168
Brown, Denise Scott 121
Brown, Sheldon 12, 70, 174, 175, 176–177
BSA Cycles 84–85
Buffalo, NY 127–128, 141–142; *see also A Concrete Atlantis*
Bugatti automobile 99–100
Buick automobile 34–35, 100, 101
Burrows, Mike 171–172

Cage, John 68
Campagnolo components 8, 176
Carbon fiber bicycles 170
Cardiff-to-London road record 61–63
Carlton bicycles 145, 146
Carney, Frank 122
Carteret Street 9, 13, 144
Case Study Homes 159–160, 178
Catting, Jack 97
Celotex Corp. 137–139
Centaurus aeroengine (Bristol) 47–48
Centenary Bldg. 48, 52; *see also* Kingston Mill

Central Park 5, 98
Chalk, Warren 159
Chapman, John M. 167
Charlier, Nick 12, 146
Chopper (Raleigh) 4, 98, 101, 111–112, 114, 115; *see also* Raleigh
Christie, Julie 5
Clarke, Norman 83, 107, 109–110, 131
CLASP buildings 159–160, 178
"Clip-on" technology 7, 123, 158–160, 178–180; *see also* plug-in technology; hot-rod technology
Colon and Schein 158
Columbia Mfg. Co. 83–84, 109–110; *see also* Clarke, Norman
A Concrete Atlantis (PRB, 1986) 141
Congress for Cultural Freedom 37
Connell, Karl Kim 164
Consumer Product Safety Commission (USA) 105, 106
Cook, Peter 41, 132, 137, 159
Cordell, Magda 30, 39–40
Corfiato, Hector 122
Courtauld Institute of Art 1, 10, 25, 26, 27
The Cowl 61–62, 154
Cox, Peter 14, 132
Cresswell, Melvyn 119, 145
Cripps, Stafford 20
Crompton, Denis 132, 159
Crowther, Tom 62
Crystal, Jimmy 48
Curry's 79, 86, 90–91, 96–97; *see also* Halfords; High Street shops
Cycling magazine (UK) 46
Cyclists' Touring Club (UK) 118, 132–133

Dahon bicycles 148–151, 173; *see also* Hon bicycles; Hon, David
Daily Express 18
Daily Mail Ideal Home Exhibit (1956) 37–38, 178; *see also* "House of the Future"
Daily Mirror 20
Davy, John 66
Dawes Cycles Ltd. (UK) 71, 103, 172
Dawes Kingpin bicycle 8, 103, 185–186
Day, Roy 59
de Bure, Gilles 124
de Guingand, Francis 88
del Renzio, Tony 31, 37
del Rio, Patricio 161
Denby, Elizabeth 22–23
Department stores *see* Curry's; Halfords; High Street shops
Derby Investment Holdings 119
Derbyshire, Andrew 136–139
Derrida, Jacques 161

Deserts (landscapes) 141–144
Design magazine 1, 118
Design Quarterly 159
Design Studies 1
Development Area Program 77
Dilnot, Clive 35
Docker, Leonard and Norah 85
Doyle, Michael 125
Drazkiewicz, Elizbieta 164–169
Duchamp, Marcel 161
Duffield, David 58, 59, 60, 61, 63, 69, 75, 79, 87–88, 95, 98, 99, 112–113
Dunlop Co. 55, 89, 100, 152
Dymaxion Deployment Units 67, 160, 179; *see also* Fuller, Buckminster; Wichita House

Eames, Charles and Ray 159–160
Earl's Court cycle show 1, 3, 60, 63, 69, 86, 169
Earl's Court motor show 18, 41, 85
Economics, postwar (UK): class structure 19, 41, 103–104; exports 21, 81–83; industrial productivity 20–21; inflation 99; unemployment 19, 41; wages 19, 41, 65, 81–84, 99, 104; *see also* consumption; housing; rationing; youth; women
The Economist 83, 85, 87, 89
Ellis, William 80
Emanuel, Martin 14
Encounter magazine 37
Entenza, John 159–160
Environment Monthly 123

F-frame *see* Moulton bicycle (F-frame)
Farrell, Dan 11–12, 53, 55, 152, 156, 157
Fathers of Pop (1979 documentary) 126
Fedden, Roy 9, 47–48
Fedden automobile 48
Fell, Harriet 13, 77
Fichtel and Sachs components 131, 183
Fincham, Ben 164–167
Finden-Crofts, Alan 119
Fisher and Ludlow ("Kirkby factory") 3, 63, 69–71, 77–78, 96–97; Moulton production figures 69–71; rumors of sabotage 75–76; quality control problems 75, 76–79
Flexitor springs 9, 48
Folder forum 173
Folding Society 173, 186
Forty, Adrian 36, 67
Friss, Evan 14
Fritz, Al 107, 113; *see also* Schwinn Stingray
Fry, Maxwell 133
Fulbright, William 169

Index

Fuller, R. Buckminster ("Bucky") 67, 123, 130, 160, 178, 179
Furness, Zack 14
Futurists 28, 67

Gaitskell, Hugh 36–37, 66–67
Gazelle (Dutch bicycle co.) 115, 120
Gazelle (UK bicycle co.) 80
General Motors 50, 74
George Spencer, Moulton and Co. 9, 45–46, 48, 49, 51–52
Giant Mfg. Co. (bicyclemaker) 116, 172
Giap, Vo Nguyen 169
Goodyear, Charles 44–45, 48
Gosling, Ray 92
Gottesman, Edward 119
Gramophone Co. (EMI) 84
Green, David 159
Green and Green (architects) 27
Greene, Downes 44
Grenfell Towers fire (2017) 137–139
Grenville, David 51–52
Gringrod, John 136
Gropius, Walter 28, 67, 129, 141
Gross, Aaron 176
Guggenheim, Peggy 29–30

Hadland, Tony 11, 13, 55, 60, 78, 97, 99, 103–104, 114, 131, 144, 152, 154, 157, 167, 170–171, 185; on Raleigh RSW 16, 99; on Raleigh Twenty 103–104, 170
Halfords 79, 90–91
The Hall (Bradford-on-Avon) 10, 12, 45–46, 52, 71, 89, 152, 156–157
Hall, John 45
Hall, Peter 2, 41
Hamilton, Richard 5, 29–30, 37, 39–40
Hammer, Kath 73
Hanappe, Odille 124
Hancock, Thomas 45
Hanstock, Fred 145
Harriman, George 63
Harris, Aaron 175
Harrisson, Jim 75, 90
Harry Wilson Agency *see* Wilson, Harry
Heathcote, Edwin 136–139
Henderson, Nigel 29–30, 32
Hercules Cycles Ltd. 84, 89, 105, 108
Herne Hill Velodrome (UK) 72
Herron, Ron 159
Hertz-Rent-a-Bike 110
Hewish, John 16
High Street shops 3, 79, 90–91; *see also* Curry's; Halfords; RSW 16
Hinton, Rita 42
Hochschule at Ulm 126
Hodge, Ivy 134
Hodges, Eric 182, 185, 186–188

Hodgkinson, Peter 123
Hogg, Quintin 5, 71–72, 98
Holt Road factory ("Bradford-on-Avon factory"): built (1962–63) 63, 69–71, 157; plant sold 152; production figures (1963–65) 69–71
Hon, David 148, 149–150
Hon, Henry 148, 151
Hon, Jason 147, 150–151
Hon bicycles (Dahon bicycles) 148–151, 173; *see also* Dahon bicycles; Shen, Florence
Hot-rod technology (autos) 160, 179–182
"House of the Future" (1956) 37–39, 178
Houseman, William 123
Housing 21–24, 92–93, 104, 134–139
Huffman Mfg. Co. (USA bicyclemaker) 84, 107, 108; imports Moultons 110, 175
Humber bicycles 80, 84
Humphrey, Cliff 124
Hunstanton School 29, 30–31, 33
Hutchinson, Maxwell 136–139
Hydragas suspension 53
Hydrolastic suspension 49–51, 52, 53, 55, 90, 152

Independent Group 5–6, 7, 10, 29–40, 67, 121, 126
Industrial Design magazine 34
Institute for Contemporary Arts 5–6, 25, 29–40, 67, 126; and CIA funding 37
International Design Conference at Aspen *see* Aspen Design Conference
International Human Powered Vehicle Assn. (IHPVA) 154
Issacson, Alan 149, 150, 182, 185, 186–188
Issigonis, Alec 9, 49–51, 53

Jencks, Charles 160–169
Johnson, Phillip 122
Jollant-Braunstein, Francoise 124–125

Karen, Tom 111
Kelso, CA 142
Kidder, Jeffery 164–169
King, Emily 3, 65
Kingpin (Dawes) 8, 103, 185–186
Kingston Mill 45–46, 48, 52
Kirkby (the city) 23–24, 77–78
Kirkby (the factory) *see* Fisher and Ludlow
Kirsten, Lincoln 37
Kloman, Anthony 37
Knottley, Peter 114, 157
Korean War 5, 42
Krag, Thomas 133
Kugelmass, Jack 164, 166

Lannoy, Richard 31, 32
Lauterwasser, Jack 56, 152
Le Corbusier 22, 67, 129, 136, 141
Lehmann, John 17–18
Levi-Strauss, Claude 160–162, 168
Ling, Arthur 133
Living Arts magazine 1, 64
Llewelyn-Davies, Richard 122, 141
London Roads Study (Smithsons) 133–134
Longhurst, James 14
Lord, Leonard 63
Los Angeles, CA 74, 121–122, 132
Los Vegas, NV 121
Lowenthal, Helen 26
Luckso, David 180–182, 187
Lyall, Sarah 166
Lypsyte, Robert 166

Macnaughtan, John 11, 20, 58, 94, 97, 98, 104, 119–120, 155–157
"Man, Machine and Motion" (1955 exhibit) 33
Marinetti, F.T. 28, 67
Marlborough College 3, 10, 46
Marwick, Alan 19
Mass Observation 19, 24
Massey, Anne 36
Maxwell, Robert 1, 7, 15, 36, 126
Mays, John Barron 23, 66, 78
McCullough, Robert 14
McGettian, Michael 173
McGurn, Jim 171
McHale, John 10, 33, 37, 39–40, 67
McLarty, Archibold 89, 115–116
McNamara, Robert 168
Meades, Jonathan 15
Messengers, bicycle *see* bicycle messengers
Metal Inert Gas (MIG) welding 70, 78
Milan Triennale 123, 124, 125, 159
Mill and Factory magazine 20–21
Miller, Henry 167
Milliken Research Associates (Bill and Doug) 154
Mini automobile (Austin/Morris/BMC) 9, 49–51, 77, 169
Mogey, J.M. 20
Monopolies and Mergers Commission (UK) 118–119
Montague bicycle (Harry Montague) 148
Montgomery Ward 91, 108
Moore, Marion 107
Moreland, Dorothy 29, 31, 37
Morris, William (Viscount Nuffield) 49, 50
Morris Motors 50–51, 52
Moulton, Alexander Eric: attends Cambridge 9, 47, 181;

Index

attends Marlborough College 46–48; belief that RSW introduced to kill off his firm 6–7, 74–79; birth, childhood, immediate family 44; builds Holt Road factory 63–64; death of 154–155; develops F-Frame Moulton bicycle 53–57; develops Flexitor 49; develops M-Dev 154–156; Moulton and AM series 151–153; parting of Moulton and Raleigh 96–97, 116; personality of 61–62, 152; as poor production manager 69–70; Raleigh takeover, introduction of MK.III 113–114; reasons for developing bicycle 53–54; sells second company to Pashley 154; starts at George Spencer, Moulton 48–49; starts production of F-Frame 57–59; on technological "jump change" 10–11, 47–48; work at Bristol Aeroplane and Fedden auto 9, 16–17, 47–48; works with Issigonis on Mini auto 49–53
Moulton, Beryl (AEM's mother) 44
Moulton, Dione (AEM's sister) 44, 46
Moulton, John (AEM's brother) 44, 46
Moulton, John Coney (AEM's father) 44
Moulton, Steven 44–45
Moulton bicycle (AM series, 1983) 144; ATB version 56, 153–154; developed 7, 116–117, 173; double pylon 152; and Pashley Bicycles Ltd. 7, 153; price 7, 117, 151
Moulton bicycle (F-frame, 1962): design of 100; development of 53–59, 62; exported to USA 110; introduction at Earle's Court show 60, 63; Mini-Moulton 56, 114; ownership demographics 93; production of 1, 58, 63, 69, 71, 75–79; quality control problems 69–70, 75–79; Stowaway model 58, 110, 144, 174, 175; suspension details 54, 56–57; volume produced 69–71, 75–76
Moulton bicycle (M-Dev, 2012) 155–156
Moulton bicycle (Raleigh-Moulton Mk. III, 1970) 6, 98, 113–114, 116, 132, 146, 157, 175
Moulton bicycle (Y-frame, 1979) 116, 157
Moulton Developments, Ltd. 53, 154
Moulton Preservation Society 70, 75, 100, 188
Museum of Modern Art (MoMA) 31

Nairn, Ian 22–23
Naylor, Gillian 9–10, 43
New Brutalism (style) 34, 41, 130, 134–139
The New Brutalism (PRB, 1966) 41
New Society 35, 41, 130, 159, 168
New Statesman 1, 2, 35, 63, 126
New towns 23–25, 132
New Yorker magazine 17, 23
Newlands, Andy 176
Nicholson, Vic 60–63, 69, 100
Norton, James 166
Norwich (UK) 2–3, 10, 15–16, 25–27
"Not Quite Architecture" column *see Architects' Journal*
Noyes, Eliot 123, 124, 125

Oakley, Alan 54, 90, 92, 95, 98, 101, 102, 114, 158; and Chopper 111–112; death of 120; and RSW 92–94
Oddy, Nicholas 14
O'Donovan, Daniel, and Gerald 145; *see also* Raleigh/Carlton bicycles
Ogle Design 111
Okawa, Russ 113
Oldenziel, Ruth 14
Orwell, George 28

Paddock site *see* Kingston Mill
Paepcke, Walter, and Elizabeth 123; *see also* Aspen Design Conference
Panter-Downes, Mollie 17, 23
Paolozzi, Eduardo 29–34
"Parallel of Life and Art" (1953 exhibit) 33–34
Pashley Bicycles Ltd. 7, 117, 153–155, 156, 157
Peck, Chris 133
Penguin New Writing 17–18
Penrose, Roland 31
Petersen, Grant 170–172
Peugeot (Cycles Peugeot) 72, 131, 170
Pevsner, Nikolas 1, 10, 17, 26, 35, 67
Phillips (components) 58–59, 84, 87, 119, 185
Phillips, Brian 140
Phillips, Peter 126
Pickles, Sam 48
Pinkerton, John 13, 75, 154
Plowden, Edwin 20
"Plug-in" technology *see* technology, in bicycles
Pollack, Jackson 33
Pop culture 5, 10, 16, 17, 32, 35, 43, 66, 67, 159; American origins of 42–43; defined by Banham 39, 67; defined by Richard Hamilton 5, 66; embodied in RSW 98–101; product of rising affluence 5, 39, 42

Pop Goes the Easel (1963 TV documentary) 126
Pope-Hennessy, John 65
Post-War Resettlement of the Motor Industry (1945) 82
Postmodernism 159; *see also* bricolage; Jencks, Charles

Quant, Mary 30, 97

Raleigh (Raleigh Cycle Co. Ltd.; Raleigh Industries, Ltd.; TI-Raleigh): and American bike boom, 1969–73 115–116; bicycle shops 79, 90–91; Brompton bicycle rejection 117–118; Carlton bicycles 114–116; Chopper 4, 98, 101, 114, 112, 115; cited for anti-competitive practices 118–119; and Curry's/Halfords 3, 79, 90–91; decline, end of production 74, 119–120; early history (pre–TI, 1960) 80–87; production statistics 3, 74, 75, 76, 86, 115; Raleigh-Moulton Mk. III bicycle 6, 98, 113–114, 116, 132, 146, 157, 175; RSW 11 bicycle 11, 96, 98, 106; RSW 14 bicycle 96, 98, 100, 114; RSW 16 bicycle 4, 8, 73, 79, 90–91, 95, 96, 97, 98, 99, 100, 106, 110, 158, 175; TI–Raleigh merger 59, 87–88; Twenty bicycle 8, 12, 96, 98, 101, 103–104, 114–115 146, 172, 174, 176–180 182–185 Rampar label (Raleigh) 106
Randall, Reg 63
Randel, Gene 107
Raphel, Adam 122
Rationing, postwar 3, 5, 17–19, 31–32, 41–42
Rauschenberg, Robert 161
Read, Herbert 31, 37
Recumbent bicycles 53
Reyner Banham Loves L.A. (1970 TV documentary) 121–122
Reynobond *see* Alcoa
Reynolds Aluminum (USA) 88
Reynolds 531 tubing 88, 145, 146, 151, 153
Reynolds Tube Co. (UK) 88
Richards, J.M. 22, 27
Richards, Keith 143
Ritchie, Andrew 79, 117–118, 144, 146; *see also* Brompton bicycle
Rix, Yvonne 98, 172
Roberts, Leslie 88, 90
Robin Hood (Raleigh trademark) 80, 108
Roger Rabbit myth 73–74
Rogers, Ernesto 133
Ronan Point Towers 134–135
Rowlinson, Steve 129
RSW 16 bicycle 4, 8, 98, 158, 175; compared with Moulton

95–96; competition not other bicycles 92, 99, 105; development of 93–95; export to USA 110; intended to kill off Moulton 73, 79, 100; introduction of 96; price of 95; sales volume 97; targeted at High Street shops 79, 90–91; targeted at women 79, 92, 98; withdrawal of 98, 100–101, 106
Rubirosa, Danny and Tony 166
Rudge-Witworth Cycles, Ltd. 84

Saarinen, Eero 159
Salinger, J.D. 5
Salisbury, Harrison 5
Santa Cruz, CA 129, 140, 142
Sant'Elia, Antonio 28, 67
Scenes in American Deserta (PRB, 1982) 141–143
Scholtz, Hans 146–147
Schwinn, Edward 113
Schwinn, Frank, Jr. 107
Schwinn, Frank, Sr. 107, 109
Schwinn, Richard 113
Schwinn Bicycle Co. (Arnold, Schwinn and Co.) 90, 109, 106–108, 113, 116, 131, 182, 186–187
Scott Sports Group 113
Seales, Peter (Raleigh) 6, 88, 90, 91, 92, 94, 95, 97, 98, 99, 100
Selle Royal 120
Sellers, Peter 5
Shen, Florence 150–151
Shimano (components) 87, 94, 110, 119, 176
"Shoppers" *see* Kingpin (Dawes); Twenty (Raleigh)
Shupe, Sam 165, 167–168
Sillitoe, Alan 92, 136
Silurian Lake, CA 143–144
Silver, Nathan 161
Simpson, Tommy 72
Sleeve valve engine 47–48
Smith, Charles 88
Smithson, Alison and Peter 25, 133–134, 178; death of 137; and Independent Group 29–40; and Robin Hood Gardens flats 134–139
Snow, C.P. 28
Socialist Commentary 42
Solo Polo saddles 106–108, 182
Sparke, Penny 10, 35, 126, 127
Spencer, Jack 46
Spielberg, Steven 73
Starrs, James 161
State University of New York–Buffalo *see* SUNY-Buffalo

Stedeford, Ivan (TI) 83, 84, 85, 88
Stingray (Schwinn) 4, 106–108, 110, 113, 131, 168, 186–188
Stoffers, Manuel 14, 133
Stratford-on-Avon 7, 44, 46, 154, 156
Street-Porter, Tim 127, 129, 141, 142, 143–144
Sturmey-Archer (components) 12, 75, 76, 80, 84, 86, 90, 94–96, 106, 111–112 115, 119, 120, 131, 176, 178; controversy over "transfer price" to Raleigh 87, 109–110; equips Moulton 90; equips RSW 16, 94–96; sold to SunRace 119–120
Sun (Raleigh brand) 104, 118
SunRace (components) 119–120
SUNY-Buffalo 127–128, 140

Tallon, Roger, and Nicole 124
Taylor, D. Courtney 88
Technology 66; in bicycles 130, 131, 169–172
Thayer Capital Partners 116
Theory and Design in the First Machine Age (PRB, 1960) 1, 28, 35, 67
"This Is Tomorrow" (1956) 1, 30, 33, 39–40
TI *see* Tube Investments
TI-Raleigh *see* Raleigh
TI-South Africa 97
Tikit bicycles *see* Bike Friday
Tiratsoo, Nick 20, 24
Toby the Cat 61, 62, 155
Townley, Jay 131
Transportation mode split, postwar UK 42
Triumph (Raleigh brand) 104, 108, 118
Tube Investments 58, 83, 84, 85, 86, 87, 88; merger with Raleigh 87–88; *see also* Phillips
Turbojet engines 11, 47–48
Twenty (Raleigh) 12, 96, 98, 101, 105, 106 146, 182; introduction of 103–104; labeled "Shopper" 105; Moulton designs replacement for (1971) 114–115; platform for "hot-rodding" 8, 172, 174, 176–180, 182–188; relation to RSW series 104; sales of 105, 115, 185–186

UC–Santa Cruz 2, 140–141
Unité d'Habitation 22
Unocal 181–182

van der Ryn, Sim 123
VeloSolex moped 102
Venturi, Robert 121
Vereker, Julian 145
Victoria and Albert Museum 3, 26, 65, 127
Vidler, Anthony 27, 179
Vietnam War 169
Voelcker, Joel 39–40
Vogue models 6, 71

Wachsmann, Konrad 67
Wade, Leigh 170
Washing machines 77, 99
Wastyn, Oscar 167
Weaver, Thomas 143
Webb, Mike 159
Webster, Harry 53
Welleman, Ton 133
Welwyn Garden City 129, 132
Western Auto 91, 108, 109
Westland Helicopter 120
Wheeler, Sydney 63
Whitechapel Gallery 37, 39–40
Whiteley, Nigel 5, 36, 38, 66
Whizzers (gas motors) 188
Who Framed Roger Rabbit? (1988 movie) 73–74
Wichita House 67, 170
Williams, Adrian 120
Wilson, Colin St. John ("Sandy") 33, 41
Wilson, George (Raleigh) 80, 85, 87, 88, 89–90, 91
Wilson, Harry (Agency) 106–107
Wilson, Nancy 107
Winn, Dan 176
Wolfe, Tom 36, 121
Women: employment and wages, postwar (UK) 41, 92–93, 104; as target market for RSW 16 and Twenty 3, 92, 104
Woodburn, John 60–63, 64, 100, 152, 155
Woodhead, R.M. 80
Woodhead and Angois 80
Woodward, Christopher 30
Woolfe, Michael 75, 76, 79, 100; *see also* Moulton Preservation Society
Workers Education Association 26
Wright, Joe (Dunlop) 89–90
Wymondham (UK) 26

Youth, income of (UK) 42, 65–66

Z Cars (TV series) 23
Zzyzx, CA 142

www.ingramcontent.com/pod-product-compliance
Lightning Source LLC
Chambersburg PA
CBHW081554300426
44116CB00015B/2880